Trends in Mathematics

Trends in Mathematics is a series devoted to the publication of volumes arising from conferences and lecture series focusing on a particular topic from any area of mathematics. Its aim is to make current developments available to the community as rapidly as possible without compromise to quality and to archive these for reference.

Proposals for volumes can be submitted using the Online Book Project Submission Form at our website www.birkhauser-science.com.

Material submitted for publication must be screened and prepared as follows:

All contributions should undergo a reviewing process similar to that carried out by journals and be checked for correct use of language which, as a rule, is English. Articles without proofs, or which do not contain any significantly new results, should be rejected. High quality survey papers, however, are welcome.

We expect the organizers to deliver manuscripts in a form that is essentially ready for direct reproduction. Any version of T_EX is acceptable, but the entire collection of files must be in one particular dialect of T_EX and unified according to simple instructions available from Birkhäuser.

Furthermore, in order to guarantee the timely appearance of the proceedings it is essential that the final version of the entire material be submitted no later than one year after the conference.

For further volumes:
http://www.springer.com/series/4961

Geometric Methods in Physics

XXXI Workshop, Białowieża, Poland,
June 24–30, 2012

Piotr Kielanowski
S. Twareque Ali
Alexander Odesskii
Anatol Odzijewicz
Martin Schlichenmaier
Theodore Voronov
Editors

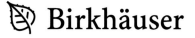 Birkhäuser

Editors

Piotr Kielanowski
Departamento de Física
CINVESTAV
Mexico City
Mexico

S. Twareque Ali
Department of Mathematics and Statistics
Concordia University
Montreal
Canada

Alexander Odesskii
Department of Mathematics - MC J415
Brock University
St. Catharines
Canada

Anatol Odzijewicz
Institute of Mathematics
University of Bialystok
Bialystok
Poland

Martin Schlichenmaier
Mathematics Research Unit, FSTC
University of Luxembourg
Luxembourg-Kirchberg
Luxembourg

Theodore Voronov
School of Mathematics
University of Manchester
Manchester
United Kingdom

ISBN 978-3-0348-0781-4   ISBN 978-3-0348-0645-9 (eBook)
DOI 10.1007/978-3-0348-0645-9
Springer Basel Heidelberg New York Dordrecht London

Printed on acid-free paper

Springer Basel is part of Springer Science+Business Media (www.birkhauser-science.com)

Contents

Preface

This volume has its origin in the XXXI Workshop on Geometric Methods in Physics (XXXI WGMP) held in Białowieza, Poland, during the period June 24–30, 2012. The Workshop is an annual conference on mathematical physics and mathematics organized by the Department of Mathematical Physics of the University of Białystok, Poland.

The scientific program of our workshops is focused on such subjects as quantization, integrable systems, coherent states, non-commutative geometry, Poisson and symplectic geometry and infinite-dimensional Lie groups and Lie algebras. In this year's workshop we also held a special session devoted to discussing the achievements and legacy of the late Boris Vasil'evich Fedosov, the outstanding Russian mathematician, who passed away in 2011. Let us note that B.V. Fedosov in 2003 participated and was a plenary speaker in our workshop.

The workshop was followed for the first time by a week long *School on Geometry and Physics*. The aim of the school was to present to students and young research workers, in an accessible way, some of the most important research topics in mathematical physics. It consisted of 8 courses, each of a duration of 3 hours.

The workshops are traditionally held in Białowieza, a small village which is located in eastern Poland at the edge of the "Białowieza Forest", the only remnant of an ancient European primeval forest. The place is unique and of world renown for its unspoiled environment. Such beautiful surroundings help create a special atmosphere of collaboration and mutual understanding during formal and informal activities.

The organizers of the XXXI WGMP gratefully acknowledge the financial support from the following sources:

- The University of Białystok.
- European Science Foundation (ESF) Research Networking Programme *Interaction of Low-Dimensional Topology and Geometry with Mathematical Physics (ITGP)*.

The XXXI WGMP was included as a Satellite Conference of the 6*th European Congress of Mathematics* held in Kraków, Poland, July 2–7, 2012.

Last but not the least, we would also like to thank the students and young researchers from the Department of Mathematics of the University of Białystok for their enthusiastic help in the daily running of the workshop.

December 2012   The Editors

Photo by Dr. Tomasz Goliński

Boris Vasil'evich Fedosov
Photo by Dr. Leonid L. Lazutkin

Geometric Methods in Physics. XXXI Workshop 2012
Trends in Mathematics, ix–x
© 2013 Springer Basel

A Word About Boris Vasil'evich Fedosov

Theodore Voronov

Abstract. This is an edited version of the introductory speech at the session dedicated to the memory of Boris Fedosov.

Mathematics Subject Classification (2010). Primary 01A70; Secondary 19E99, 19K56, 46L80, 58J20, 53D55.

Keywords. Index theorem, deformation quantization, symplectic manifold, symbol, trace.

The outstanding Russian mathematician Boris Vasil'evich Fedosov departed this life in October last year. It is an immense loss for world mathematics and particularly mathematical physics. Fedosov contribution was great. I will not speak here about the details of his life and achievements; let me draw your attention to the obituary published by the *Russian Mathematical Surveys* [1], to which there is a link from the conference webpage and which copy was posted during our Workshop. There you will find biographic facts and a sketch of his work.

I will tell you just one thing. Fedosov is most universally known for what we call now the "Fedosov quantization". This is a geometric construction of deformation quantization for arbitrary symplectic manifolds; but it is not only that. Fedosov's quantization includes an index theorem and a construction of a so-called asymptotic operator representation for quantum observables. It is worth noting that Fedosov's quantization originated from his earlier work on the index theorem for elliptic operators. Fedosov developed his own approach to the Atiyah–Singer index theorem. His aim was to obtain a direct proof for the index theorem, so that one can get an expression for the index in terms of analytic data for an operator. As a step in this direction, he introduced a non-commutative algebra of formal symbols, whose algebraic properties mimicked the algebraic properties of pseudo-differential operators. The notion of index on this algebra was introduced purely algebraically, in terms of traces. Exactly from here Fedosov came to his quantization construction, after he had realized that instead of the cotangent bundle one can develop a similar construction on an arbitrary symplectic manifold.

Everybody who met Boris Fedosov was fascinated by his great mathematical power, which literally emanated from him. Particularly impressing was a complete lack in his manner of any pretense and "actor's tricks", so to say, which unfortunately are sometimes used by some famous mathematicians. He had always tried to make complicated things simple, not the other way round. In his talks, when lecturing about most advanced subjects such as, for example, K-theory and the index theorem, or his quantization theory, Fedosov was doing that in a wonderful manner of a lecturer of the first year mathematical analysis of good old days. Only the shining eyes and a smile at the end of an argument could tell the audience that Fedosov had just managed to prove on the blackboard some particularly powerful statement.

Fedosov was a plenary speaker here at the Białowieża Workshop in 2003, nine years ago. In this volume we include some photos of him and his wife Elena Stanislavovna, in particular, during the campfire, like the one we have here every year.

Boris Vasil′evich was an absolutely remarkable person. For those who had a privilege to have known him personally, it is hard to believe that we will no longer meet with him in this life; in particular, that he will not come to Białowieża again.

As soon as the members of the organizing committee learned about Fedosov's departure, a decision was made to devote a special session of Białowieża 2012 to his memory. Talks that were given during the Workshop are related to areas that were dear to Fedosov, such as quantization and symplectic geometry. I would like to thank Pierre Bieliavsky, Michel Cahen, Simone Gutt, Mikhail Vasiliev, Akira Yoshioka, and others who gave talks during our special session, for coming to this conference and thus contributing to the memory of Boris Fedosov[1]. Thank you.

References

[1] M.S. Agranovich, L.A. Aĭzenberg, G.L. Alfimov, M.I. Vishik, Th.Th. Voronov, A.V. Karabegov, M.V. Karasev, A.A. Komech, V.P. Maslov, V.V. Pukhnachev, D.E. Tamarkin, N.N. Tarkhanov, B.L. Tsygan, and M.A. Shubin. Boris Vasil′evich Fedosov (obituary). *Uspiekhi Mat. Nauk* **67**(1(403)):169–176, 2012. Russian. (Translated: *Russian Math. Surveys* **67**:1 167–174, 2012.)

[2] P. Bieliavsky, A. de Goursac, and F. Spinnler. Non-formal deformation quantization and star-exponential of the Poincaré Group.

[3] M. Cahen and S. Gutt. Spin^c, Mp^c and symplectic Dirac operators.

[4] M. Iida, C. Tsukamoto, and A. Yoshioka. Star products and certain star product functions.

Theodore Voronov
School of Mathematics
University of Manchester
Manchester, M13 9PL, UK
e-mail: theodore.voronov@manchester.ac.uk

---

[1] See [2, 3] and [4] in this volume.

Part I

Quantization, the Scientific Legacy of B.V. Fedosov

B.V. Fedosov in his office at the University of Warwick

B.V. Fedosov in Białowieża

Geometric Methods in Physics. XXXI Workshop 2012
Trends in Mathematics, 3–12
© 2013 Springer Basel

Non-formal Deformation Quantization and Star-exponential of the Poincaré Group

Pierre Bieliavsky, Axel de Goursac and Florian Spinnler

To the memory of Boris Fedosov

Abstract. We recall the construction of non-formal deformation quantization of the Poincaré group $ISO(1,1)$ on its coadjoint orbit and exhibit the associated non-formal star-exponentials.

Mathematics Subject Classification (2010). 46L65; 46E10; 42B20.

Keywords. Deformation quantization, exponential.

1. Introduction

Quite generally, a theory – either physical or mathematical – often consists in the kinematical data of an algebra **A** together with some dynamical data encoded by some specific, say "action", functional $\mathfrak{S}$ on **A**. In most cases, the algebra is associative because it represents the "observables" in the theory, i.e., operators whose spectrality is associated with measurements. Also, the understanding of the theory passes through the determination of critical points of the action functional $\mathfrak{S}$, or sometimes equivalently, by the determination of first integrals: elements of **A** that will be preserved by the dynamics. Disposing of a sufficient numbers of such first integrals yields a satisfactory understanding of the system. In particular, the consideration of symmetries appears as a necessity – rather than a simplifying hypothesis.

Now, assume that has enough of such symmetries in order to entirely determine the dynamics. In that case, the system $(\mathbf{A}, \mathfrak{S})$ could be called "integrable" or even, quite abusively, "free". Of course, once a "free" system is understood, one wants to pass to a perturbation or "singularization" of it, for instance by implementing some type of "interactions". This is rather clear within the physical context. Within the mathematical context, such a singularization could for example correspond to implementing a foliation. However, once perturbed, the

4   P. Bieliavsky, A. de Goursac and F. Spinnler

problem remains the same: determining symmetries. One may naively hope that the symmetries of the unperturbed "free" system would remain symmetries of its perturbation. It is not the case: perturbing generally implies symmetry breaking.

One idea, due essentially to Drinfel'd, is to define the perturbation process through the data of the symmetries themselves [1] in the framework of deformation quantization [2, 3]. This allows, even in the case of a symmetry breaking, to control the "perturbed symmetries". More specifically, let G be a Lie group with Lie algebra $\mathfrak{g}$ whose enveloping (Hopf) algebra is denoted by $\mathcal{U}(\mathfrak{g})$. Consider the category of $\mathcal{U}(\mathfrak{g})$-module algebras, i.e., associative algebras that admit an (infinitesimal) action of G. A (formal) **Drinfel'd twist** based on $\mathcal{U}(\mathfrak{g})$ is an element F of the space of formal power series $\mathcal{U}(\mathfrak{g}) \otimes \mathcal{U}(\mathfrak{g})[[\nu]]$ of the form $F = I \otimes I + \cdots$ satisfying a specific cocycle property (see, e.g., [4]) that ensures that for every $\mathcal{U}(\mathfrak{g})$-module algebra $(\mathbf{A}, \mu_{\mathbf{A}})$ the formula $\mu_{\mathbf{A}}^F := \mu_{\mathbf{A}} \circ F$ defines an associative algebra structure $\mathbf{A}_F$ on the space $\mathbf{A}[[\nu]]$. Possessing a Drinfel'd twist then allows to deform the above-mentioned category. However, as expected, the deformed objects are no longer $\mathcal{U}(\mathfrak{g})$-module algebras. But, the data of the twist allows to define a Hopf deformation of the enveloping algebra: keeping the multiplication unchanged, one deforms the co-product Δ of $\mathcal{U}(\mathfrak{g})$ by conjugating under F. This yields a new co-multiplication Δ_F that together with the undeformed multiplication underly a structure of Hopf algebra $\mathcal{U}(\mathfrak{g})_F$ on $\mathcal{U}(\mathfrak{g})[[\nu]]$. The latter so-called *non-standard quantum group* $\mathcal{U}(\mathfrak{g})_F$ now acts on every deformed algebra $\mathbf{A}_F$.

At the **non-formal** level, the notion of Drinfel'd twist based on $\mathcal{U}(\mathfrak{g})$ corresponds to the one of **universal deformation formula** for the actions α of G on associative algebras $\mathbf{A}$ of a specified topological type such as Fréchet- or C*-algebras. This – roughly – consists in the data of a two-point kernel $K_\theta \in C^\infty(G \times G)$ ($\theta \in \mathbb{R}_0$) satisfying specific properties that guarantee a meaning to integral expressions of the form $a \star_\theta^{\mathbf{A}} b := \int_{G \times G} K_\theta(x,y)\mu_{\mathbf{A}}(\alpha_x a \otimes \alpha_y b)\, dxdy$ with $a, b \in \mathbf{A}$. Once well defined, one also requires associativity of the product $\star_\theta^{\mathbf{A}}$ as well as the semi-classical limit condition: $\lim_{\theta \to 0} \star_\theta^{\mathbf{A}} = \mu_{\mathbf{A}}$ in some precise topological context. This has been performed for abelian Lie groups in [5] and for abelian supergroups in [6, 7].

In [8], such universal deformation formulae have been constructed for every Piatetskii–Shapiro normal J-group $\mathbb{B}$. For example, the class of normal J-groups (strictly) contains all Iwasawa factors of Hermitean type non-compact simple Lie groups. A universal deformation formula in particular yields a left-invariant associative function algebra on the group $\mathbb{B}$. It is therefore natural to ask for a comparison with the usual group convolution algebra. Following ideas mainly due to Fronsdal in the context of the $\star$-representation program (representation theory of Lie group in the framework of formal $\star$-products), one may expect that (a non-formal version of) the notion of **star-exponential** [2, 9] plays a crucial role in this comparison. Moreover, such a star-exponential can give access to the spectrum of operators [10, 11] determining possible measurements of a system.

In this paper, we recall the construction of the non-formal deformation quantization (see [12]) and exhibit its star-exponential for the basic case of the Poincaré group $ISO(1,1)$. Such a low-dimensional case illustrates the general method developed for normal J-group $\mathbb{B}$ (which are Kählerian), in [8] for star-products and in [13] for star-exponentials. However, the Poincaré group is solvable but of course not Kählerian, so this paper shows also that the method introduced in [8, 13] can be extended to some solvable but non-Kählerian Lie groups.

2. Geometry of the Poincaré group

We recall here some features concerning the geometry of the Poincaré group $G = ISO(1,1) = SO(1,1) \ltimes \mathbb{R}^2$ and of its coadjoint orbits. First, it is diffeomorphic to $\mathbb{R}^3$, so let us choose a global coordinate system $\{(a, \ell, m)\}$ of it. Its group law can be read as

$$(a, \ell, m) \cdot (a', \ell', m') = (a + a', e^{-2a'}\ell + \ell', e^{2a'}m + m').$$

Its neutral element is $(0, 0, 0)$ and the inverse is given by:

$$(a, \ell, m)^{-1} = (-a, -e^{2a}\ell, -e^{-2a}m).$$

By writing $(a, \ell, m) = \exp(aH)\exp(\ell E)\exp(mF)$, we can determine its Lie algebra $\mathfrak{g}$:

$$[H, E] = 2E, \qquad [H, F] = -2F, \qquad [E, F] = 0.$$

Let us have a look to the coadjoint orbit of G. After a short calculation, one can find that

$$\mathrm{Ad}_g^*(\alpha H^* + \beta E^* + \gamma F^*) = (\alpha + 2\beta\ell - 2\gamma m)H^* + \beta e^{-2a}E^* + \gamma e^{2a}F^*$$

if $g = (a, \ell, m) \in G$ and $\{H^*, E^*, F^*\}$ is the basis of $\mathfrak{g}^*$ dual of $\{H, E, F\}$. A generic orbit of G is therefore a hyperbolic cylinder. We will study in particular the orbits associated to the forms $k(E^* - F^*)$ with $k \in \mathbb{R}_+^*$, which will be denoted by $\mathbb{M}_k$ or simply by $\mathbb{M}$. The Poincaré quotient $\mathbb{M}_k$ is globally diffeomorphic to $\mathbb{R}^2$, so we choose the following coordinate system:

$$(a, \ell) := \mathrm{Ad}_{(a,\ell,0)}^* k(E^* - F^*) = k(2\ell H^* + e^{-2a}E^* - e^{2a}F^*). \tag{1}$$

$\mathbb{M}$ is a G-homogeneous space for the coadjoint action:

$$(a, \ell, m) \cdot (a', \ell') := \mathrm{Ad}_{(a,\ell,m)}^* k(2\ell' H^* + e^{-2a'}E^* - e^{2a'}F^*)$$
$$= (a + a', \ell' + e^{-2a'}\ell + e^{2a'}m). \tag{2}$$

Remark 1. Note that the affine group $\mathbb{S}$ (connected component of the identity of "ax+b") is the subgroup of G generated by H and E, i.e., by simply considering the two first coordinates (a, ℓ) of G. Actually the identification (1) yields a diffeomorphism between $\mathbb{S}$ and $\mathbb{M}$ which is $\mathbb{S}$-equivariant with respect to the left action of $\mathbb{S}$ and its action on $\mathbb{M}$ by (2) as a subgroup of G. This identification $\mathbb{S} \simeq \mathbb{M}$ is useful to construct star-products.

The fundamental fields of the action (2), defined by

$$X^*_{(a,\ell)}f = \frac{\mathrm{d}}{\mathrm{d}t}\Big|_{t=0} f(\exp(-tX)\cdot(a,\ell)),$$

for $X \in \mathfrak{g}$, $(a,\ell) \in \mathbb{M}$, $f \in C^\infty(\mathbb{M})$, are given by

$$H^*_{(a,\ell)} = -\partial_a, \qquad E^*_{(a,\ell)} = -e^{-2a}\partial_\ell, \qquad F^*_{(a,\ell)} = -e^{2a}\partial_\ell.$$

This permits to compute the **Kostant–Kirillov–Souriau symplectic form** of $\mathbb{M}$, $\omega_\varphi(X^*_\varphi, Y^*_\varphi) := \langle \varphi, [X,Y]\rangle$, for $\varphi \in \mathbb{M} \subset \mathfrak{g}^*$, and $X, Y \in \mathfrak{g}$. One finds

$$\omega_{(a,\ell)} = 2k\,\mathrm{d}a \wedge \mathrm{d}\ell. \tag{3}$$

For different values of $k \in \mathbb{R}^*_+$, the $(\mathbb{M}_k, \omega)$ are symplectomorphic, so we set $k = 1$ in the following. Since the action of G on its coadjoint orbit $\mathbb{M}$ is strongly hamiltonian, there exists a Lie algebra homomorphism $\lambda : \mathfrak{g} \to C^\infty(\mathbb{M})$ (for the Poisson bracket on $\mathbb{M}$ associated to ω), called the **moment map** and given by

$$\lambda_H = 2\ell, \qquad \lambda_E = e^{-2a}, \qquad \lambda_F = -e^{2a}. \tag{4}$$

Proposition 1. *The exponential of the group G is given by*

$$e^{tX} = \left(\alpha t, \frac{\beta}{\alpha}e^{-\alpha t}\sinh(\alpha t), \frac{\gamma}{\alpha}e^{\alpha t}\sinh(\alpha t) \right)$$

for $X = \alpha H + \beta E + \gamma F$.

Proof. It is a direct calculation using the semigroup property $e^{(t+s)X} = e^{sX}e^{tX}$ and by differentiating with respect to s. $\qquad\square$

For $g = (a, \ell, m) \in G$, we can obtain straightforwardly the logarithm by inversing the above equation:

$$\log(a, \ell, m) = aH + \frac{ae^a\ell}{\sinh(a)}E + \frac{ae^{-a}m}{\sinh(a)}F$$

and the **BCH expression**:

$$\begin{aligned}
\mathrm{BCH}(X_1, X_2) = \log(e^{X_1}e^{X_2}) = {} & (\alpha_1 + \alpha_2)H \\
& + \frac{(\alpha_1 + \alpha_2)}{\sinh(\alpha_1 + \alpha_2)}\left(\frac{\beta_1}{\alpha_1}e^{-\alpha_2}\sinh(\alpha_1) + \frac{\beta_2}{\alpha_2}e^{\alpha_1}\sinh(\alpha_2) \right)E \\
& + \frac{(\alpha_1 + \alpha_2)}{\sinh(\alpha_1 + \alpha_2)}\left(\frac{\gamma_1}{\alpha_1}e^{\alpha_2}\sinh(\alpha_1) + \frac{\gamma_2}{\alpha_2}e^{-\alpha_1}\sinh(\alpha_2) \right)F
\end{aligned} \tag{5}$$

for $X_i = \alpha_i H + \beta_i E + \gamma_i F \in \mathfrak{g}$.

3. Deformation quantization

3.1. Star-products

Due to the identification $\mathbb{M} \simeq \mathbb{R}^2$, we can endow the space of Schwartz functions $\mathcal{S}(\mathbb{R}^2)$ with the Moyal product associated to the constant KKS symplectic form (3):

$$(f \star_\theta^0 h)(a, \ell) = \frac{4}{(\pi\theta)^2} \int \mathrm{d}a_i \mathrm{d}\ell_i \, f(a_1 + a, \ell_1 + \ell) h(a_2 + a, \ell_2 + \ell) e^{-\frac{4i}{\theta}(a_1 \ell_2 - a_2 \ell_1)} \quad (6)$$

for $f, h \in \mathcal{S}(\mathbb{R}^2)$. It turns out that this associative star-product is **covariant** for the moment map (4), formally in the deformation parameter θ:

$$\forall X, Y \in \mathfrak{g} \quad : \quad [\lambda_X, \lambda_Y]_{\star_\theta^0} = -i\theta\lambda_{[X,Y]}. \quad (7)$$

Nonetheless, it is not G-**invariant**, one does not have:

$$\forall g \in G \quad : \quad g^*(f \star_\theta^0 h) = (g^*f) \star_\theta^0 (g^*h) \quad (8)$$

in general, where g^* means the pullback of the action (2) of G on $\mathbb{M}$: $g^*f := f(g\cdot)$. In the following, we exhibit intertwining operators T_θ (see [14]) in order to construct invariant star-products on $\mathbb{M}$, i.e., satisfying (8).

We consider $\mathcal{P}_\theta$ an invertible multiplier on $\mathbb{R}$: $\mathcal{P}_\theta \in \mathcal{O}_M^\times(\mathbb{R}) = \{f \in C^\infty(\mathbb{R})$, $\forall h \in \mathcal{S}(\mathbb{R})$ $f.h \in \mathcal{S}(\mathbb{R})$ and $f^{-1}.h \in \mathcal{S}(\mathbb{R})\}$, and ϕ_θ defined by:

$$\phi_\theta(a, \ell) = \left(a, \frac{2}{\theta} \sinh\left(\frac{\theta\ell}{2}\right)\right), \qquad \phi_\theta^{-1}(a, \ell) = \left(a, \frac{2}{\theta} \mathrm{arcsinh}\left(\frac{\theta\ell}{2}\right)\right).$$

We define the operator $T_\theta = \mathcal{P}_\theta(0)\mathcal{F}^{-1} \circ (\phi_\theta^{-1})^* \circ \mathcal{P}_\theta^{-1} \circ \mathcal{F}$, from $\mathcal{S}(\mathbb{R}^2)$ to $\mathcal{S}'(\mathbb{R}^2)$, where $\mathcal{P}_\theta^{-1}$ acts by multiplication by $\mathcal{P}_\theta(\ell)^{-1}$ and the partial Fourier transformation is given by:

$$\mathcal{F}f(a, \xi) = \hat{f}(a, \xi) := \int \mathrm{d}\ell \, e^{-i\xi\ell} f(a, \ell). \quad (9)$$

The normalization is chosen so that $T_\theta 1 = 1$. On its image, T_θ is invertible. The explicit expressions are:

$$T_\theta f(a, \ell) = \frac{\mathcal{P}_\theta(0)}{2\pi} \int \mathrm{d}t \mathrm{d}\xi \, \cosh\left(\frac{\theta t}{2}\right) \mathcal{P}_\theta(t)^{-1} e^{\frac{2i}{\theta}\sinh\left(\frac{\theta t}{2}\right)\ell - i\xi t} f(a, \xi)$$

$$T_\theta^{-1} f(a, \ell) = \frac{1}{2\pi\mathcal{P}_\theta(0)} \int \mathrm{d}t \mathrm{d}\xi \, \mathcal{P}_\theta(t) e^{-\frac{2i}{\theta}\sinh\left(\frac{\theta t}{2}\right)\xi + it\ell} f(a, \xi).$$

In [14], it has been shown that this intertwining operator yields an associative product on $T_\theta(\mathcal{S}(\mathbb{R}^2))$: $f \star_{\theta,\mathcal{P}} h := T_\theta((T_\theta^{-1}f) \star_\theta^0 (T_\theta^{-1}h))$ which is G-invariant. Its explicit expression is: $\forall f, h \in T_\theta(\mathcal{S}(\mathbb{R}^2))$,

$$(f \star_{\theta,\mathcal{P}} h)(a, \ell) = \frac{4}{(\pi\theta)^2} \int \mathrm{d}a_i \mathrm{d}\ell_i \cosh(2(a_1 - a_2)) \frac{\mathcal{P}_\theta\left(\frac{4}{\theta}(a_1 - a)\right) \mathcal{P}_\theta\left(\frac{4}{\theta}(a - a_2)\right)}{\mathcal{P}_\theta\left(\frac{4}{\theta}(a_1 - a_2)\right) \mathcal{P}_\theta(0)}$$

$$\times e^{\frac{2i}{\theta}(\sinh(2(a_1-a_2))\ell + \sinh(2(a_2-a))\ell_1 + \sinh(2(a-a_1))\ell_2)}$$

$$\times f(a_1, \ell_1)h(a_2, \ell_2). \quad (10)$$

3.2. Schwartz space

In [8], a Schwartz space adapted to $\mathbb{M}$ has been introduced, which is different from the usual one $\mathcal{S}(\mathbb{R}^2)$ in the global chart $\{(a, \ell)\}$ (1).

Definition 1. The **Schwartz space** of $\mathbb{M}$ is defined as

$$\mathcal{S}(\mathbb{M}) = \Bigg\{ f \in C^\infty(\mathbb{M}) \quad \forall \alpha = (k, p, q, n) \in \mathbb{N}^4,$$

$$\|f\|_\alpha := \sup_{(a,\ell)} \left| \frac{\sinh(2a)^k}{\cosh(2a)^p} \ell^q \partial_a^p \partial_\ell^n f(a, \ell) \right| < \infty \Bigg\}.$$

The space $\mathcal{S}(\mathbb{M})$ corresponds to the usual Schwartz space in the coordinates (r, ℓ) with $r = \sinh(2a)$. It is stable by the action of G:

$$\forall f \in \mathcal{S}(\mathbb{M}), \ \forall g \in G \quad : \quad g^* f \in \mathcal{S}(\mathbb{M})$$

due to the formulation of the action of G in the coordinates (r, ℓ):

$$(r, \ell, m)(r', \ell')$$
$$= \left(r\sqrt{1 + r'^2} + r'\sqrt{1 + r^2}, \ell' + \left(\sqrt{1 + r'^2} - r' \right)\ell + \left(\sqrt{1 + r'^2} + r' \right)m \right).$$

Moreover, $\mathcal{S}(\mathbb{M})$ is a Fréchet nuclear space endowed with the seminorms $(\|f\|_\alpha)$. For $f, h \in \mathcal{S}(\mathbb{M})$, the product $f \star_{\theta, \mathcal{P}} h$ is well defined by (10). However, it is not possible to show that it belongs to $\mathcal{S}(\mathbb{M})$ unless we consider this expression as an oscillatory integral. Let us define this concept. For $F \in \mathcal{S}(\mathbb{M}^2)$, one can show using integrations by parts that:

$$\int \mathrm{d}a_i \mathrm{d}\ell_i \ e^{\frac{2i}{\theta}(\sinh(2a_2)\ell_1 - \sinh(2a_1)\ell_2)} F(a_1, a_2, \ell_1, \ell_2)$$

$$= \int \mathrm{d}a_i \mathrm{d}\ell_i \ e^{\frac{2i}{\theta}(\sinh(2a_2)\ell_1 - \sinh(2a_1)\ell_2)} \left(\frac{1 - \frac{\theta^2}{4}\partial_{\ell_2}^2}{1 + \sinh^2(2a_1)} \right)^{k_1} \left(\frac{1 - \frac{\theta^2}{4}\partial_{\ell_1}^2}{1 + \sinh^2(2a_2)} \right)^{k_2}$$

$$\times \left(\frac{1 - \frac{\theta^2}{16\cosh^2(2a_2)}\partial_{a_2}^2}{1 + \ell_1^2} \right)^{p_1} \left(\frac{1 - \frac{\theta^2}{16\cosh^2(2a_1)}\partial_{a_1}^2}{1 + \ell_2^2} \right)^{p_2} F(a_1, a_2, \ell_1, \ell_2)$$

$$= \int \mathrm{d}a_i \mathrm{d}\ell_i \ e^{\frac{2i}{\theta}(\sinh(2a_2)\ell_1 - \sinh(2a_1)\ell_2)} \frac{1}{(1 + \sinh^2(2a_1))^{k_1}}$$

$$\times \frac{DF(a_1, a_2, \ell_1, \ell_2)}{(1 + \sinh^2(2a_2))^{k_2}(1 + \ell_1^2)^{p_1}(1 + \ell_2^2)^{p_2}} \tag{11}$$

for any $k_i, p_i \in \mathbb{N}$, and where D is a linear combination of products of bounded functions (with all derivatives bounded) in (a_i, ℓ_i) with powers of ∂_{ℓ_i} and $\frac{1}{\cosh(2a_i)}\partial_{a_i}$. The first expression of (11) is not defined for non-integrable functions F bounded by polynomials in $r_i := \sinh(2a_i)$ and ℓ_i. However, the last expression of (11) is well defined for k_i, p_i sufficiently large. Therefore it gives a sense to the first expression, now understood as an **oscillatory integral**, i.e., as being equal to the last expression. This definition of oscillatory integral [8, 15] is unique, in particular unambiguous in the powers k_i, p_i because of the density of $\mathcal{S}(\mathbb{M})$ in polynomial functions in (r, ℓ) of a given degree. Note that this corresponds to the usual oscillatory integral [16] in the coordinates (r, ℓ).

The first part of the next theorem shows that this concept of oscillatory integral is necessary [8] for $\mathcal{S}(\mathbb{M})$ to obtain an associative algebra, while the other parts have been treated in [12].

Theorem 2. *Let $\mathcal{P} : \mathbb{R} \to C^\infty(\mathbb{R})$ be a smooth map such that $\mathcal{P}_0 \equiv 1$, and $\mathcal{P}_\theta(a)$ as well as its inverse are bounded by $C \sinh(2a)^k$, $k \in \mathbb{N}$, $C > 0$.*

- *Then, expression (10), understood as an oscillatory integral, yields a G-invariant non-formal deformation quantization. In particular, $(\mathcal{S}(\mathbb{M}), \star_{\theta, \mathcal{P}})$ is a Fréchet algebra.*
- *For $f, h \in \mathcal{S}(\mathbb{M})$, the map $\theta \mapsto f \star_{\theta, \mathcal{P}} h$ is smooth and admits a G-invariant formal star-product as asymptotic expansion in $\theta = 0$.*
- *Every G-invariant formal star-product on $\mathbb{M}$ can be obtained as an expansion of a $\star_{\theta, \mathcal{P}}$, for a certain $\mathcal{P}$.*
- *For $\mathcal{P}_\theta(a) = \mathcal{P}_\theta(0)\sqrt{\cosh(a\theta/2)}$, one has the tracial identity: $\int f \star_{\theta, \mathcal{P}} h = \int f \cdot h$.*

We denote $\star_\theta$ the product $\star_{\theta, \mathcal{P}}$ with $\mathcal{P}_\theta(a) = \sqrt{\cosh(a\theta/2)}$, therefore satisfying the tracial identity.

3.3. Schwartz multipliers

Let us consider the topological dual $\mathcal{S}'(\mathbb{M})$ of $\mathcal{S}(\mathbb{M})$. In the coordinates (r, ℓ), it corresponds to tempered distributions. By denoting $\langle -, - \rangle$ the duality bracket between $\mathcal{S}'(\mathbb{M})$ and $\mathcal{S}(\mathbb{M})$, one can extend the product $\star_\theta$ (with tracial identity) as $\forall T \in \mathcal{S}'(\mathbb{M})$, $\forall f, h \in \mathcal{S}(\mathbb{M})$,

$$\langle T \star_\theta f, h \rangle := \langle T, f \star_\theta h \rangle \quad \text{and} \quad \langle f \star_\theta T, h \rangle := \langle T, h \star_\theta f \rangle,$$

which is compatible with the case $T \in \mathcal{S}(\mathbb{M})$. Then, we define [13]:

$$\mathcal{M}_{\star_\theta}(\mathbb{M}) := \big\{ T \in \mathcal{S}'(\mathbb{M}), f \mapsto T \star_\theta f,$$
$$f \mapsto f \star_\theta T \text{ are continuous from } \mathcal{S}(\mathbb{M}) \text{ into itself} \big\}$$

and the product can be extended to $\mathcal{M}_{\star_\theta}(\mathbb{M})$ by:

$$\forall S, T \in \mathcal{M}_{\star_\theta}(\mathbb{M}), \ \forall f \in \mathcal{S}(\mathbb{M}) \quad : \quad \langle S \star_\theta T, f \rangle := \langle S, T \star_\theta f \rangle = \langle T, f \star_\theta S \rangle.$$

We can equip $\mathcal{M}_{\star_\theta}(\mathbb{M})$ with the topology associated to the seminorms:

$$\|T\|_{B, \alpha, L} = \sup_{f \in B} \|T \star_\theta f\|_\alpha \quad \text{and} \quad \|T\|_{B, \alpha, R} = \sup_{f \in B} \|f \star_\theta T\|_\alpha$$

where B is a bounded subset of $\mathcal{S}(\mathbb{M})$, $\alpha \in \mathbb{N}^4$ and $\|f\|_\alpha$ is the Schwartz seminorm introduced in Definition 1. Note that B can be described as a set satisfying $\forall \alpha$, $\sup_{f \in B} \|f\|_\alpha$ exists.

Proposition 3. $(\mathcal{M}_{\star_\theta}(\mathbb{M}), \star_\theta)$ *is an associative Hausdorff locally convex complete and nuclear algebra, with separately continuous product, called the* **multiplier algebra**.

4. Construction of the star-exponential

4.1. Formal construction

Let us follow the method developed in [13]. We want first to find a solution to the following equation

$$\partial_t f_t(a, \ell) = \frac{i}{\theta}(\lambda_X \star_\theta^0 f_t)(a, \ell) \tag{12}$$

for $X = \alpha H + \beta E + \gamma F \in \mathfrak{g}$, with initial condition $\lim_{t \to 0} f_t(a, \ell) = 1$. To remove the integral of this equation, we apply the partial Fourier transformation (9) to obtain

$$\mathcal{F}(\lambda_H \star_\theta^0 f) = \left(2i\partial_\xi + \frac{i\theta}{2}\partial_a\right)\hat{f}, \qquad \mathcal{F}\left(\lambda_E \star_\theta^0 f\right) = e^{-2a - \frac{\theta\xi}{2}}\hat{f},$$

$$\mathcal{F}\left(\lambda_F \star_\theta^0 f\right) = -e^{2a + \frac{\theta\xi}{2}}\hat{f}$$

so that the equation (12) can be reformulated as

$$\partial_t \hat{f}_t(a, \xi) = \frac{i}{\theta}\left[2i\alpha\partial_\xi + \frac{i\theta\alpha}{2}\partial_a + \beta e^{-2a - \frac{\theta\xi}{2}} - \gamma e^{2a + \frac{\theta\xi}{2}}\right]\hat{f}_t(a, \xi).$$

The existence of a solution of this equation which satisfies the BCH property directly relies on the covariance (7) of the Moyal product. We have the explicit following result.

Proposition 4. *For $X = \alpha H + \beta E + \gamma F \in \mathfrak{g}$, the expression*

$$E_{\star_\theta^0}(t\lambda_X)(a, \ell) = e^{\frac{i}{\theta}\left(2\ell\alpha t + \frac{1}{\alpha}\sinh(\alpha t)(\beta e^{-2a} - \gamma e^{2a})\right)}$$

is a solution of the equation (12) with initial condition $\lim_{t \to 0} f_t(a, \ell) = 1$. Moreover, it satisfies the BCH property: $\forall X, Y \in \mathfrak{g}$,

$$E_{\star_\theta^0}(\lambda_{BCH(X,Y)}) = E_{\star_\theta^0}(\lambda_X) \star_\theta^0 E_{\star_\theta^0}(\lambda_Y). \tag{13}$$

Proof. By performing the following change of variables $b = -\frac{a}{\alpha} - \frac{\theta\xi}{4\alpha}$ and $c = -\frac{a}{\alpha} + \frac{\theta\xi}{4\alpha}$, we reformulate the equation as

$$\partial_t \hat{f}_t = \partial_b \hat{f}_t + \frac{i\beta}{\theta}e^{2\alpha b} - \frac{i\gamma}{\theta}e^{-2\alpha b},$$

whose solution is given by

$$\hat{f}_t(b, c) = \exp\left(-\int_0^b \left(\frac{i\beta}{\theta}e^{2\alpha s} - \frac{i\gamma}{\theta}e^{-2\alpha s}\right)ds\right)h(b + t, c),$$

where h is an arbitrary function. By assuming the initial condition, we obtain the expression of $E_{\star_\theta^0}(tX)(a, \ell)$. The BCH property is given by direct computations from the expression of the product (6) and from (5). □

Finally, we push this solution by T_θ:

$$E_{\star_{\theta,\mathcal{P}}}(tX)(a, \ell) := T_\theta E_{\star_\theta^0}(tT_\theta^{-1}\lambda_X)(a, \ell)$$

$$= \frac{\mathcal{P}_\theta(0)\cosh(\alpha t)}{\mathcal{P}_\theta\left(\frac{2\alpha t}{\theta}\right)} e^{\frac{i}{\theta}\sinh(\alpha t)\left(2\ell + \frac{\beta}{\alpha}e^{-2a} - \frac{\gamma}{\alpha}e^{2a}\right) + \frac{2\mathcal{P}_\theta'(0)}{\theta \mathcal{P}_\theta(0)}\alpha t}.$$

It also satisfies the BCH property (13).

4.2. Multiplier property

For the star-product with tracial property, we want to define the star-exponential at the non-formal level. We can use the oscillatory integral in the star-product to show [13]:

Theorem 5. *For any $X \in \mathfrak{g}$, the function*

$$E_{\star_\theta}(tX)(a, \ell) = \sqrt{\cosh(\alpha t)}e^{\frac{i}{\theta}\sinh(\alpha t)\left(2\ell + \frac{\beta}{\alpha}e^{-2a} - \frac{\gamma}{\alpha}e^{2a}\right)}$$

lies in the multiplier algebra $\mathcal{M}_{\star_\theta}(\mathbb{M})$.

As it belongs to a specific "functional space", the function $E_{\star_\theta}$ is called the **non-formal star-exponential** of the group G for the star-product $\star_\theta$. The BCH property

$$E_{\star_\theta}(\text{BCH}(X, Y)) = E_{\star_\theta}(X) \star_\theta E_{\star_\theta}(Y)$$

now makes sense in the topological space $\mathcal{M}_{\star_\theta}(\mathbb{M})$. This functional framework is useful for applications of the star-exponential discussed in the introduction.

References

[1] V.G. Drinfeld, *Quasi-Hopf algebras*. Leningrad Math. J. **1** (1989) 1419–1457.

[2] F. Bayen, M. Flato, C. Fronsdal, A. Lichnerowicz, and D. Sternheimer, *Deformation theory and quantization*. Ann. Phys. **11** (1978) 61–151.

[3] B. Fedosov, *A simple geometrical construction of deformation quantization*. J. Differential Geom. **40** (1994) 213–238.

[4] A. Giaquinto and J. Zhang, *Bialgebra actions, twists, and universal deformation formulas*. J. Pure Appl. Algebra **128** (1998) 133.

[5] M.A. Rieffel, *Deformation Quantization of Heisenberg Manifolds*. Commun. Math. Phys. **122** (1989) 531–562.

[6] P. Bieliavsky, A. de Goursac, and G. Tuynman, *Deformation quantization for Heisenberg supergroup*. J. Funct. Anal. **263** (2012) 549–603.

[7] A. de Goursac, *On the Hopf algebra setting of the flat superspace's deformation*. arXiv:1105.2420 [math.QA].

[8] P. Bieliavsky and V. Gayral, *Deformation Quantization for Actions of Kählerian Lie Groups Part I: Fréchet Algebras*. arXiv:1109.3419 [math.OA].

[9] F. Bayen and J.M. Maillard, *Star exponentials of the elements of the inhomogeneous symplectic Lie algebra.* Lett. Math. Phys. **6** (1982) 491–497.

[10] M. Cahen and S. Gutt, *Discrete spectrum of the hydrogen atom: an illustration of deformation theory methods and problems.* J. Geom. Phys. **1** (1984) 65–83.

[11] M. Cahen, M. Flato, S. Gutt, and D. Sterheimer, *Do different deformations lead to the same spectrum?* J. Geom. Phys. **2** (1985) 35–49.

[12] P. Bieliavsky, *Non-formal deformation quantizations of solvable Ricci-type symplectic symmetric spaces.* J. Phys. Conf. Ser. **103** (2008) 012001.

[13] P. Bieliavsky, A. de Goursac, and F. Spinnler, *Non-formal star-exponential of Kählerian Lie groups.* In progress.

[14] P. Bieliavsky, *Strict Quantization of Solvable Symmetric Spaces.* J. Sympl. Geom. **1** (2002) 269–320.

[15] P. Bieliavsky and M. Massar, *Oscillatory integral formulae for left-invariant star products on a class of Lie groups.* Lett. Math. Phys. **58** (2001) 115–128.

[16] L. Hörmander, *The Weyl calculus of pseudo-differential operators.* Comm. Pure Appl. Math. **32** (1979) 359.

Pierre Bieliavsky, Axel de Goursac and Florian Spinnler
IRMP, Université Catholique de Louvain
Chemin du Cyclotron, 2
B-1348 Louvain-la-Neuve, Belgium
e-mail: `Pierre.Bieliavsky@uclouvain.be`
        `Axelmg@melix.net`
        `Florian.Spinnler@uclouvain.be`

Geometric Methods in Physics. XXXI Workshop 2012
Trends in Mathematics, 13–28
© 2013 Springer Basel

Spin^c, Mp^c and Symplectic Dirac Operators

Michel Cahen and Simone Gutt

*We are happy to dedicate our talks and this summary to the memory of
Boris Fedosov; we chose a subject which is close to operators and quantization,
two fields in which Boris Fedosov brought essential new contributions.*

Abstract. We advertise the use of the group Mp^c (a circle extension of the symplectic group) instead of the metaplectic group (a double cover of the symplectic group). The essential reason is that Mp^c-structures exist on any symplectic manifold. They first appeared in the framework of geometric quantization [4, 10]. In a joint work with John Rawnsley [1], we used them to extend the definition of symplectic spinors and symplectic Dirac operators which were first introduced by Kostant [9] and K. Habermann [6] in the presence of a metaplectic structure. We recall here this construction, stressing the analogies with the group Spin^c in Riemannian geometry. Dirac operators are defined as a contraction of Clifford multiplication and covariant derivatives acting on spinor fields; in Riemannian geometry, the contraction is defined using the Riemannian structure. In symplectic geometry one contracts using the symplectic structure or using a Riemannian structure defined by the choice of a positive compatible almost complex structure. We suggest here more general contractions yielding new Dirac operators.

Mathematics Subject Classification (2010). 53D05, 58J60, 81S10.
Keywords. Symplectic spinors, Dirac operators, Mp^c structures.

1. The group Spin^c

• On an oriented Riemannian manifold of dimension m, (M, g), the tangent space at any point $x \in M$, $T_x M$, is modeled on a Euclidean vector space $(V, \tilde{g})$. The automorphism group of this model is the special orthogonal group

$$SO(V, \tilde{g}) := \{ A \in Gl(V) \,|\, \tilde{g}(Au, Av) = \tilde{g}(u, v) \det A = 1 \}$$

and there exists a natural principal bundle $\mathcal{B}(M, g) \xrightarrow{p} M$, with structure group $SO(V, \tilde{g})$, which is the bundle of oriented orthonormal frames of the tangent bun-

This work has benefited from an ARC Grant from the communauté française de Belgique.

dle; the fiber above $x \in M$, $p^{-1}(x)$, consists of all linear isomorphisms of Euclidean spaces preserving the orientation

$$f : (V, \tilde{g}) \to (T_x M, g_x)$$

with $SO(V, \tilde{g})$ acting on the right on $\mathcal{B}(M, g)$ by composition

$$f \cdot A := f_\circ A \quad \forall f \in \mathcal{B}(M, g), A \in SO(V, \tilde{g}).$$

The tangent bundle TM is associated to the bundle of oriented orthonormal frames for the standard representation st of the structure group $SO(V, \tilde{g})$ on V (i.e., $st(A)v = Av$),

$$TM = \mathcal{B}(M, g) \times_{(SO(V, \tilde{g}), st)} V$$

with

$$f(v) \simeq [(f, v)] = [(f_\circ A, A^{-1} v)].$$

• The **Clifford Algebra** $Cl(V, \tilde{g})$ is the associative unital algebra generated by V such that $u \cdot v + v \cdot u = -2\tilde{g}(u, v)1$ for all $u, v \in V$. To simplify notations, we shall assume here to be in the even-dimensional situation $m = 2n$. Then the complexification of the Clifford algebra is identified with the space of complex linear endomorphisms of the exterior algebra built on a maximal isotropic subspace $W \subset V^{\mathbb{C}} = W \oplus \overline{W}$

$$Cl(V, \tilde{g})^{\mathbb{C}} \simeq \operatorname{End}(\Lambda W).$$

Indeed, one associates to an element $w \in W$ the endomorphism

$$cl(w)\alpha = \sqrt{2}\, w \wedge \alpha \qquad \forall \alpha \in \Lambda W$$

and to the conjugate element $\overline{w} \in \overline{W}$ the endomorphism

$$cl(\overline{w})\alpha = -\sqrt{2}\, i(\overline{w}_{\tilde{g}})\alpha \qquad \forall \alpha \in \Lambda W, \quad \text{with} \quad v_{\tilde{g}}(v') = \tilde{g}(v, v').$$

• The **Spinor space** S is a complex vector space with a Hermitian scalar product $\langle \cdot, \cdot \rangle$, carrying an irreducible representation cl of $Cl(V, \tilde{g})$, so that each element of V acts in a skewhermitian way

$$\langle cl(v)\alpha, \beta \rangle + \langle \alpha, cl(v)\beta \rangle = 0.$$

Here $S = \Lambda W$, and the Hermitian scalar product is the natural extension of

$$\langle w, w' \rangle = \tilde{g}(w, \overline{w'}) \qquad w, w' \in W.$$

• On a Riemannian manifold (M, g), one defines – when possible – a spinor bundle $\mathcal{S}(M, g)$ and a fiberwise Clifford multiplication Cl of the tangent bundle TM acting on the spinor bundle $\mathcal{S}(M, g)$ by gluing the above construction. For this, one needs:

– a principal bundle $\mathcal{B} \xrightarrow{p_{\mathcal{B}}} M$ with structure group G;
– a group homomorphism $\sigma : G \to SO(V, \tilde{g})$ so that the tangent space is associated to $\mathcal{B}$ for the representation $st_\circ \sigma$ of G on V:

$$TM = \mathcal{B} \times_{(G, \sigma)} V;$$

it is equivalent to ask for the existence of a map

$$\Phi : \mathcal{B} \to \mathcal{B}(M, g)$$

which is fiber-preserving (i.e., $p_B = p_\circ \Phi$) and (G, σ)-equivariant, i.e.,

$$\Phi(\tilde{f} \cdot \tilde{A}) = \Phi(\tilde{f}) \cdot \sigma(\tilde{A}) \qquad \forall \tilde{f} \in \mathcal{B}, \tilde{A} \in G;$$

– a unitary representation r of G on the spinor space S to define **the spinor bundle**

$$\mathcal{S}(M, g) = \mathcal{B} \times_{(G,r)} S.$$

– The Clifford multiplication is well defined via

$$Cl([(\tilde{f}, v)]) \left([(\tilde{f}, s)] \right) := [(\tilde{f}, cl(v)s)] \quad \forall \tilde{f} \in \mathcal{B}, v \in V, s \in S$$

if and only if $cl(\sigma(\tilde{A})v) \left(r(\tilde{A})s \right) = r(\tilde{A})(cl(v)s)$, i.e., iff

$$cl(\sigma(\tilde{A})v) = r(\tilde{A})_\circ cl(v)_\circ r(\tilde{A})^{-1} \quad \forall \tilde{A} \in G, v \in V. \tag{1.1}$$

• There is no representation r of $SO(V, \tilde{g})$ on S satisfying (1.1). Indeed, any representation r of G on S is given by a homomorphism n of G into $Cl(V, \tilde{g})^{\mathbb{C}} \simeq$ $\mathrm{End}(\Lambda W) = \mathrm{End}(S)$ and condition (1.1) is equivalent to

$$\sigma(\tilde{A})v = n(\tilde{A}) \cdot v \cdot n(\tilde{A})^{-1}.$$

The differential of n yields a homomorphism n_* of the Lie algebra $\mathfrak{g}$ of G into $Cl(V, \tilde{g})^{\mathbb{C}}$ endowed with the bracket

$$[\alpha, \beta]_{Cl} = \alpha \cdot \beta - \beta \cdot \alpha$$

and condition (1.1) yields

$$[n_*(B), v]_{Cl} = \sigma_* Bv.$$

The Lie algebra $\mathfrak{so}(V, \tilde{g})$ naturally embeds in $Cl(V, \tilde{g})$:

$$\nu(v_{\tilde{g}} \otimes w - \underline{w}_{\tilde{g}} \otimes v) = \frac{1}{4}(v \cdot w - w \cdot v) \quad \text{with} \quad \underline{w}_{\tilde{g}}(v) = \tilde{g}(w, v) \tag{1.2}$$

and this satisfies:

$$[\nu(B), v]_{Cl} = Bv \quad [\nu(B), \nu(B')]_{Cl} = \nu([B, B'])$$

but this does not lift to a homomorphism of $SO(V, \tilde{g})$ into $Cl(V, \tilde{g})^{\mathbb{C}}$ since

$$\exp \left(2\pi \nu(e_{1_{\tilde{g}}} \otimes e_2 - \underline{e_2}_{\tilde{g}} \otimes e_1) \right) = -\mathrm{Id}$$

for e_1, e_2 two orthonormal vectors.

• One way to proceed is to define **the group** Spin as the connected subgroup of $Gl(S)$ with Lie algebra $\nu(\mathfrak{so}(V, \tilde{g}))$; it is a double cover of $SO(V, \tilde{g})$. A pair consisting of a principal Spin bundle $\mathcal{B} \xrightarrow{p_B} M$ and a map $\Phi : \mathcal{B} \to \mathcal{B}(M, g)$ which is fiber-preserving and (Spin, σ)-equivariant is called a **Spin structure** on the manifold M. Such a structure only exists if a cohomology class (the second Stiefel Whitney class) vanishes.

• Another way to proceed is to stress the importance of the fundamental equation (1.1) and to consider all unitary transformations of S for which this equation is satisfied. More precisely, for any $A \in SO(V, \tilde{g})$, the maps $cl(v)$ and $cl(Av)$ extend to two representations of $Cl(V, \tilde{g})$ on the spinor space S which are equivalent and one defines **the group** Spin^c as the set of unitary intertwiners between those representations:

$$\text{Spin}^c = \{(U, A) \in \mathcal{U}(S) \times SO(V, \tilde{g}) \,|\, U cl(v) U^{-1} = cl(Av)\, \forall v \in V\} \qquad (1.3)$$

with multiplication defined componentwise, the natural unitary representation r on S defined by $r(U, A) = U$, and the obvious homomorphism

$$\sigma : \text{Spin}^c \to SO(V, \tilde{g}) : (U, A) \to A \qquad (1.4)$$

with kernel $U(1)$. This defines a short exact sequence:

$$1 \to U(1) \xrightarrow{i} \text{Spin}^c \xrightarrow{\sigma} SO(V, \tilde{g}) \to 1 \qquad (1.5)$$

which does not split.

At the level of Lie algebras, the sequence splits and

$$\mathfrak{spin}^c = \nu(\mathfrak{so}(V, \tilde{g})) \oplus \mathfrak{u}(1). \qquad (1.6)$$

Observe that $[cl(\nu(\sigma_* X)), cl(v)] = cl(\sigma_*(X)v) = [r_* X, cl(v)]$ for all X in $\mathfrak{spin}^c$ so that $r_* X - cl(\nu(\sigma_* X)) = c(X)\text{Id}$ where c is a unitary infinitesimal character of $\mathfrak{spin}^c$.

This presentation of the group Spin^c as the set of intertwiners can be generalized for any signature in any dimension. To relate this to the Spin group (which historically came first), let us observe that

$$\text{Spin}^c = (\text{Spin} \times U(1)) / {\pm\{\text{Id}\}}$$

(and this is the usual presentation of the group Spin^c). The character

$$\eta : \text{Spin}^c \to U(1) : [(A, \lambda)] \mapsto \lambda^2 \qquad \forall A \in \text{Spin}, \lambda \in U(1) \qquad (1.7)$$

is the squaring map on the central $U(1)$ and has for kernel Spin. We have

$$r_* X = cl(\nu(\sigma_* X)) + \tfrac{1}{2}\eta_*(X)\text{Id} \qquad \forall X \in \mathfrak{spin}^c. \qquad (1.8)$$

• Observe that the short exact sequence 1.5 splits over the unitary group; indeed, given a complex structure j on V which is an isometry for the metric $\tilde{g}$, the unitary group $U(V, \tilde{g}, j)$ – which is identified to the subgroup of j-linear endomorphisms in $SO(V, \tilde{g})$ – injects into Spin^c in the following way. Let us choose $W \subset V^{\mathbb{C}}$ to be the $+i$ eigenspace for j in $V^{\mathbb{C}}$ and realize Spin^c in the group of unitary endomorphisms of $S = \Lambda W$ as above; an element $U \in U(V, \tilde{g}, j)$ acting on $V^{\mathbb{C}}$ stabilizes W and induces a unitary endomorphism denoted $\Lambda(U)$ of ΛW; one has

$$\left.\begin{array}{l} \Lambda(U)cl(w)\Lambda(U)^{-1} = cl(Uw) \\ \Lambda(U)cl(\overline{w})\Lambda(U)^{-1} = cl(U\overline{w}) \end{array}\right\} \forall w \in W$$

so that $\Lambda(U)$ is in Spin^c and we have the injection:

$$\Lambda : U(V, \tilde{g}, j) \to \text{Spin}^c : U \mapsto \Lambda(U). \qquad (1.9)$$

At the level of Lie algebras, we have:

$$\Lambda_*(X) = cl(\nu(X)) + \tfrac{1}{2}\text{Trace}_j(X)\text{Id} \qquad \forall X \in \mathfrak{u}(V, \tilde{g}, j). \qquad (1.10)$$

Let $SpU^c(V, \tilde{g}, j)$ be the inverse image of $U(V, \tilde{g}, j)$ under σ and define the character

$$\lambda : SpU^c(V, \tilde{g}, j) \to U(1) : \Lambda(U)\tilde{\lambda} \mapsto \tilde{\lambda} \qquad \forall U \in U(V, \tilde{g}, j), \tilde{\lambda} \in U(1). \qquad (1.11)$$

The isomorphism

$$\sigma \times \lambda : SpU^c(V, \tilde{g}, j) \to U(V, \tilde{g}, j) \times U(1) \qquad (1.12)$$

has inverse $\Lambda \times i$. The complex determinant defines another character $\det_{j} \circ \sigma$ on $SpU^c(V, \tilde{g}, j)$ and the three characters are related by

$$\eta = \lambda^2 \det{}_{j} \circ \sigma. \qquad (1.13)$$

• A pair consisting of a principal Spin^c bundle $\mathcal{B} \xrightarrow{p_{\mathcal{B}}} M$ and a fiber-preserving and (Spin^c, σ)-equivariant map $\Phi : \mathcal{B} \to \mathcal{B}(M, g)$ is called a Spin^c **structure** on the manifold M. Such a structure does not always exist.

If it exists, it is not unique. Indeed given a Spin^c structure $(\mathcal{B}, \Phi)$ and a $U(1)$ principal bundle $L^1 \xrightarrow{p_1} M$, one builds a new Spin^c structure $(\mathcal{B}, \Phi)$ by

$$\mathcal{B}' = \left(\mathcal{B} \times_M L^1 \right) \times_{(\text{Spin}^c \times U(1))} \text{Spin}^c$$

for the homomorphism of Spin$^c \times U(1)$ into Spin^c given by the natural injection i of $U(1)$ and multiplication, and by

$$\Phi' : \mathcal{B}' \to \mathcal{B}(M, g) : [((\tilde{f}, s), (U, A))] \mapsto \Phi(\tilde{f}) \cdot A.$$

Since 1.5 splits over $U(V, \tilde{g}, j)$, a Riemannian manifold (M, g) which admits an almost complex structure J so that $g(JX, JY) = g(X, Y)$, always admits Spin^c-structures. In particular Kähler manifolds always admit Spin^c-structures. Denoting by $\mathcal{B}(M, g, J) \subset \mathcal{B}(M, g)$ the $U(V, \tilde{g}, j)$ principal bundle of complex orthonormal frames of TM then

$$\mathcal{B} := \mathcal{B}(M, g, J) \times_{U(V, \tilde{g}, j)} \text{Spin}^c$$

and

$$\Phi : \mathcal{B}(M, g, J) \times_{U(V, \tilde{g}, j)} \text{Spin}^c \to \mathcal{B}(M, g) : [(f, (U, A)] \mapsto f \cdot A$$

define a Spin^c structure on (M, g).

2. The symplectic Clifford algebra and the group Mp^c

• On a symplectic manifold (M, ω) of dimension $m = 2n$, the tangent space at any point is modeled on a symplectic vector space (V, Ω).
The isomorphism group of this model is the symplectic group

$$Sp(V, \Omega) := \{ A \in Gl(V) \mid \Omega(Au, Av) = \Omega(u, v) \, \forall u, v \in V \}$$

and there exists a natural principal bundle $\mathcal{B}(M, \omega) \xrightarrow{p} M$, with structure group $Sp(V, \Omega)$, which is the bundle of symplectic frames of the tangent bundle; the fiber above $x \in M$, consists of all linear isomorphisms of symplectic spaces

$$f : (V, \Omega) \to (T_x M, \omega_x).$$

with $Sp(V,\Omega)$ acting on the right on $\mathcal{B}(M,\omega)$ by composition. The tangent bundle TM is associated to the bundle of symplectic frames for the standard representation st of the structure group $Sp(V,\Omega)$ on V:

$$TM = \mathcal{B}(M,\omega) \times_{(Sp(V,\Omega),st)} V.$$

• The **symplectic Clifford Algebra** $Cl(V,\Omega)$ is the associative unital algebra generated by V such that

$$u \cdot v - v \cdot u = \frac{i}{\hbar}\Omega(u,v)1 \qquad \forall u, v \in V, \quad \hbar = \frac{h}{2\pi} \; h \in \mathbb{R}.$$

The Heisenberg group is $H(V,\Omega) = V \times \mathbb{R}$ with multiplication defined by

$$(v_1, t_1)(v_2, t_2) = (v_1 + v_2, t_1 + t_2 - \tfrac{1}{2}\Omega(v_1, v_2));$$

its Lie algebra is the space $\mathfrak{h}(V,\Omega) = V \oplus \mathbb{R}$ with brackets $[(v,\alpha),(w,\beta)] = (0, -\Omega(v,w))$, so that a representation of $Cl(V,\Omega)$ is a representation of $\mathfrak{h}(V,\Omega)$ with central character equal to $-\frac{i}{\hbar}$.

• Let $\mathcal{H}, \rho$ be an irreducible unitary representation of the Heisenberg group $H(V,\Omega)$ on a complex separable Hilbert space $\mathcal{H}$ with central character $\rho(0,t) = e^{-\frac{i}{\hbar}t}\mathrm{Id}$; by Stone–von Neumann's theorem, such a representation is unique up to unitary equivalence. The space $\mathcal{H}^\infty$ of smooth vectors of this representation (or its dual $\mathcal{H}^{-\infty}$) is the **symplectic spinor space** S. It carries a Hermitian scalar product and the symplectic Clifford multiplication scl of V on S is given by the skewhermitian operators

$$scl(v) := \rho_* (v, 0).$$

• To glue the above construction on a symplectic manifold and build a spinor bundle endowed with a symplectic Clifford multiplication by elements of the tangent bundle, one uses, as before:

  – a principal bundle $\mathcal{B} \xrightarrow{p_\mathcal{B}} M$ with structure group G;
  – a group homomorphism $\sigma : G \to Sp(V,\Omega)$;
  – a fiber-preserving (G,σ)-equivariant map $\Phi : \mathcal{B} \to \mathcal{B}(M,\omega)$, i.e., $p_\mathcal{B} = p_\circ \Phi$ and $\Phi(\tilde{f} \cdot \tilde{A}) = \Phi(\tilde{f}) \cdot \sigma(\tilde{A})$ for all $\tilde{f} \in \mathcal{B}, \tilde{A} \in G$;
  – a unitary representation r of G on $\mathcal{H}$ stabilizing the spinor space S and satisfying the fundamental equation

$$\rho(\sigma(\tilde{A})v, 0) = r(\tilde{A})_\circ \rho(v,0)_\circ r(\tilde{A})^{-1} \quad \forall \tilde{A} \in G, v \in V \tag{2.1}$$

which implies on S

$$scl(\sigma(\tilde{A})v) = r(\tilde{A})_\circ scl(v)_\circ r(\tilde{A})^{-1} \quad \forall \tilde{A} \in G, v \in V.$$

The **symplectic spinor bundle** is then $\mathcal{S} = \mathcal{B} \times_{(G,r)} S$; and the **symplectic Clifford multiplication** $sCl : TM \times_M \mathcal{S} \to \mathcal{S}$ is given by

$$sCl([(\tilde{f}, v)]) \left([(\tilde{f}, s)]\right) := [(\tilde{f}, scl(v)s)] \quad \forall \tilde{f} \in \mathcal{B}, v \in V, s \in S.$$

• There is an embedding of the symplectic Lie algebra into the symplectic Clifford algebra given by

$$\nu(\underline{v}_\Omega \otimes w + \underline{w}_\Omega \otimes v) = \frac{-i\hbar}{2}(v \cdot w + w \cdot v) \quad \text{with} \quad \underline{w}_\Omega(v) = \Omega(w,v) \qquad (2.2)$$

and one has

$$[\nu(B), v]_{sCl} = Bv \quad [\nu(B), \nu(B')]_{sCl} = \nu([B,B'])$$

but there is no lift to a representation of the symplectic group on $\mathcal{H}$ satisfying equation (2.1).

On the other hand, the symplectic group $Sp(V, \Omega)$ acts as automorphisms of the Heisenberg group via

$$A \cdot (v,t) := (Av, t).$$

Given an element $A \in Sp(V, \Omega)$, the two representations $\rho(v,t)$ and $\rho(Av,t)$ of the Heisenberg group on $\mathcal{H}$ have the same central character, so they are equivalent. The fundamental equation (2.1) states that $r(\tilde{A})$ is an intertwiner of these two representations. **The group** Mp^c is defined as the set of all such unitary intertwiners:

$$\text{Mp}^c = \{(U, A) \in \mathcal{U}(\mathcal{H}) \times Sp(V, \Omega) \,|\, U\rho(v,0)U^{-1} = \rho(Av, 0)\,\forall v\} \qquad (2.3)$$

with multiplication defined componentwise, the natural unitary representation r on $\mathcal{H}$ defined by $r(U, A) = U$, and the obvious homomorphism

$$\sigma : \text{Mp}^c \to Sp(V, \Omega) : (U, A) \mapsto A \qquad (2.4)$$

which has kernel $U(1)$ by irreducibility of ρ. The short exact sequence

$$1 \to U(1) \overset{i}{\longrightarrow} \text{Mp}^c \overset{\sigma}{\longrightarrow} Sp(V, \Omega) \to 1 \qquad (2.5)$$

does not split. At the level of Lie algebras, the sequence splits and

$$\mathfrak{mp}^c = \nu(\mathfrak{sp}(V, \Omega)) \oplus \mathfrak{u}(1). \qquad (2.6)$$

Observe that $[scl(\nu(\sigma_* X)), scl(v)] = scl(\sigma_*(X)v) = [r_* X, scl(v)]$ for all X in $\mathfrak{mp}^c$ so that $r_* X - scl(\nu(\sigma_* X)) = c(X)\text{Id}$ where c is a unitary infinitesimal character of $\mathfrak{mp}^c$.

2.1. Explicit description of the group Mp^c

To get the nice Fock description of the irreducible unitary representation of the Heisenberg group with prescribed central character, one chooses a positive compatible complex structure on (V, Ω).

A **compatible complex structure** j is a (real) linear map of V which is symplectic, $\Omega(jv, jw) = \Omega(v, w)$, and satisfies $j^2 = -I_V$. Given such a j, the map $(v, w) \mapsto \Omega(v, jw)$ is a non-degenerate symmetric bilinear form and j is **positive** if this form is positive definite.

The choice of j gives V the structure of a complex vector space, with $(x + iy)v = xv + yj(v)$ and when j is positive, one has a Hermitian structure on V defined by

$$\langle v, w \rangle_j = \Omega(v, jw) - i\Omega(v, w), \qquad |v|_j^2 = \langle v, v \rangle_j.$$

The set of positive compatible j's is the contractible homogeneous space

$$Sp(V, \Omega)/U(V, \Omega, j).$$

Having chosen a positive compatible j, Fock's description of the Hilbert space carrying an irreducible unitary representation ρ_j of the Heisenberg group with central character $\rho_j(0, t) = e^{-\frac{i}{\hbar}t}\mathrm{Id}$ is denoted $\mathcal{H}(V, \Omega, j)$; it is the space of holomorphic functions $f(z)$ on (V, Ω, j) which are L^2 in the sense of the norm $\|f\|_j$ given by

$$\|f\|_j^2 = h^{-n} \int_V |f(z)|^2 e^{-\frac{|z|_j^2}{2\hbar}} \, dz.$$

The unitary and irreducible action of the Heisenberg group $H(V, \Omega)$ on $\mathcal{H}(V, \Omega, j)$ is given by

$$(\rho_j(v, t)f)(z) = e^{-it/\hbar + \langle z, v \rangle_j/2\hbar - |v|_j^2/4\hbar} f(z - v).$$

The space $\mathcal{H}(V, \Omega, j)$ has the nice property to possess a family of **coherent states** e_v parametrized by V:

$$(e_v)(z) = e^{\frac{1}{2\hbar}\langle z, v \rangle_j} \quad \text{such that} \quad f(z) = (f, e_z)_j$$

so that any unitary operator U on $\mathcal{H}(V, \Omega, j)$ is entirely characterized by its Berezin kernel

$$(Ue_v, e_w)_j$$

which is a holomorphic function in w and an antiholomorphic function in v.

The Heisenberg Lie algebra $\mathfrak{h}(V, \Omega)$ acts on smooth vectors and these include the coherent states; the Clifford multiplication is defined as $scl(v) = \dot{\rho}_j(v, 0)$ and it splits:

$$(scl(v)f)(z) = \frac{1}{2\hbar}\langle z, v \rangle_j f(z) - (\partial_z f)(v) =: (c(v)f)(z) - (a(v)f)(z)$$

in a creation and an annihilation operators $c(v)$ and $a(v)$ which respectively raises and lowers the degree of a polynomial in z.

To give a description of the kernel of an element in the group Mp^c, we introduce a parametrization of the symplectic group. Writing an element $g \in Sp(V, \Omega)$ as $g = C_g + D_g$ with $C_g = \frac{1}{2}(g - jgj)$ the j-linear part and $D_g = \frac{1}{2}(g + jgj)$ the j-antilinear part, one observes that C_g is invertible and set $Z_g = C_g^{-1} D_g$.

The Berezin kernel $U(z, v) := (Ue_v, e_z)_j$ of the unitary operator U when $(U, g) \in \mathrm{Mp}^c(V, \Omega, j)$ is given by:

$$U(z, v) = \lambda \exp \frac{1}{4\hbar}\{2\langle C_g^{-1}z, v \rangle_j - \langle z, Z_{g^{-1}}z \rangle_j - \langle Z_g v, v \rangle_j\}$$

for some $\lambda \in \mathbb{C}$ with $|\lambda^2 \det C_g| = 1$.

Indeed, since $(e_v)(w) = e^{\frac{1}{2\hbar}\langle w, v \rangle_j} = e^{\frac{1}{4\hbar}\langle v, v \rangle_j}(\rho_j(v, 0)e_0)(w)$, we have

$$U(z, v) = e^{\frac{1}{4\hbar}(\langle z, z \rangle_j + \langle v, v \rangle_j)}(U\rho_j(v, 0)e_0, \rho_j(z, 0)e_0)_j.$$

For $v = g^{-1}z$, $U\rho_j(g^{-1}z, 0) = \rho_j(z, 0)U$ by definition of $\mathrm{Mp}^c(V, \Omega, j)$, and we get $U(z, g^{-1}z) = e^{\frac{1}{4\hbar}(\langle z, z \rangle_j + \langle g^{-1}z, g^{-1}z \rangle_j)}(Ue_0, e_0)_j$.

But $U(z, v)$ is holomorphic in z and antiholomorphic in v, so it is completely determined by its values for $(z, v = g^{-1}z)$. The formula given above follows from the fact that $v = g^{-1}z = C_{g^{-1}}(\mathrm{Id} + Z_{g^{-1}})z$ yields $z = (C_{g^{-1}})^{-1}v - Z_{g^{-1}}z$.

We say that g, λ are the parameters of (U, g) when the kernel of U is given as above.

The product in $\mathrm{Mp}^c(V, \Omega, j)$ of (U_i, g_i) with parameters g_i, λ_i, $i = 1, 2$ has parameters $g_1 g_2, \lambda_{12}$ with

$$\lambda_{12} = \lambda_1 \lambda_2 e^{-\frac{1}{2}a\left(1 - Z_{g_1} Z_{g_2}^{-1}\right)}$$

where $a : GL(V, j)_+ := \{g \in GL(V, j) \mid g + g^* \text{ is positive definite}\} \to \mathbb{C}$ is the unique smooth function defined on the simply connected space $GL(V, j)_+$ such that

$$\det_j g = e^{a(g)} \qquad \text{and} \quad a(I) = 0$$

where $\det_j$ is the complex determinant of a j-linear endomorphism viewed as a complex endomorphism.

This proves that the group $\mathrm{Mp}^c(V, \Omega, j)$ is a Lie group.

2.2. Character and subgroups of $\mathrm{Mp}^c(V, \Omega, j)$

The group $\mathrm{Mp}^c(V, \Omega, j)$ admits a **character** η given by

$$\eta(U, g) = \lambda^2 \det_j C_g \tag{2.7}$$

which is the squaring map on the central $U(1)$. The **metaplectic group** is the kernel of η; it is given by

$$Mp(V, \Omega, j) = \{(U, g) \in \mathrm{Mp}^c(V, \Omega, j) \mid \lambda^2 \det_j C_g = 1\}.$$

It is a double covering of $Sp(V, \Omega)$ and

$$\mathrm{Mp}^c(V, \Omega, j) = (Mp(V, \Omega, j) \times U(1)) / \pm\{\mathrm{Id}\}.$$

We have:

$$r_* X = scl(\nu(\sigma_* X)) + \tfrac{1}{2}\eta_*(X)\mathrm{Id} \qquad \forall X \in \mathfrak{mp}^c. \tag{2.8}$$

• An important fact is that the short exact sequence 2.5 splits over the unitary group. Indeed the unitary group $U(V, \Omega, j)$ injects into $\mathrm{Mp}^c(V, \Omega, j)$:

$$\tilde{\Lambda} : U(V, \Omega, j) \to \mathrm{Mp}^c(V, \Omega, j) : K \mapsto ((U_K, K) \text{ with parameters } K, 1), \tag{2.9}$$

i.e., $(U_K f)(z) = f(K^{-1}z)$. At the level of Lie algebras, we have:

$$\tilde{\Lambda}_*(X) = scl(\nu(X)) + \tfrac{1}{2}\mathrm{Trace}_j(X)\mathrm{Id} \qquad \forall X \in \mathfrak{u}(V, \Omega, j). \tag{2.10}$$

Let $MU^c(V, \Omega, j)$ be the inverse image of $U(V, \Omega, j)$ under σ. It has a character:

$$\lambda : MU^c(V, \Omega, j) \to U(1) : ((U, K) \text{ with parameters } K, \tilde{\lambda}) \mapsto \tilde{\lambda} \tag{2.11}$$

which yields an isomorphism

$$MU^c(V, \Omega, j) \xrightarrow{\sigma \times \lambda} U(V, \Omega, j) \times U(1) \tag{2.12}$$

with inverse $\tilde{\Lambda} \times i$. Observe that the complex determinant defines another character $\det_j \circ \sigma$ on $MU^c(V, \Omega, j)$ and the three characters are related by

$$\eta = \lambda^2 \det_j \circ \sigma. \tag{2.13}$$

• More generally, if $\tilde{j}$ is a compatible complex structure, not necessarily positive, one chooses a positive j commuting with it. The pseudounitary group $U(V, \Omega, \tilde{j})$ of linear endomorphisms injects into $\mathrm{Mp}^c(V, \Omega, j)$:

$$U(V, \Omega, \tilde{j}) \to \mathrm{Mp}^c(V, \Omega, j) : A \mapsto (U, A) \text{ with param. } ((det_j C_A^-)^{-1}, A), \tag{2.14}$$

where C_A^- is the restriction of C_A to $V_- = \{v \in V \,|\, \tilde{j}v = -jv\}$.

In fact, if F is any polarization of (V, Ω), the subgroup of symplectic transformations preserving F injects into $\mathrm{Mp}^c(V, \Omega, j)$.

3. Mp^c structures and Mp^c connections

An Mp^c **structure** on (M, ω) is a principal $\mathrm{Mp}^c(V, \Omega, j)$ bundle $\mathcal{B} \xrightarrow{p_\mathcal{B}} M$ with a fibre-preserving map $\Phi : \mathcal{B} \to \mathcal{B}(M, \omega)$ so that $\Phi(\tilde{f} \cdot \tilde{A}) = \Phi(\tilde{f}) \cdot \sigma(\tilde{A})$ for all $\tilde{f} \in \mathcal{B}$, $\tilde{A} \in \mathrm{Mp}^c(V, \Omega, j)$.

Proposition 3.1 ([10]). *Any symplectic manifold (M, ω) admits Mp^c structures, and the isomorphism classes of Mp^c structures on (M, ω) are parametrized by isomorphism classes of complex line bundles.*

An explicit parametrization of Mp^c structures is obtained as follows: choose a fibre-wise positive ω-compatible complex structure J on TM; this is always possible since the space of compatible complex structures on a given symplectic vector space is contractible. Define $\mathcal{B}(M, \omega, J)$ to be the principal $U(V, \Omega, j)$ bundle of symplectic frames which are complex linear.

If $(\mathcal{B}, \Phi)$ is an Mp^c structure, let

$$\mathcal{B}_J := \Phi^{-1} \mathcal{B}(M, \omega, J).$$

This is a principal $MU^c(V, \Omega, j) \simeq_{\sigma \times \lambda} U(V, \Omega, j) \times U(1)$ bundle and

$$\mathcal{B} \simeq \mathcal{B}_J \times_{MU^c(V, \Omega, j)} \mathrm{Mp}^c(V, \Omega, j).$$

Define

$$\mathcal{B}_J^1(\lambda) := \mathcal{B}_J \times_{MU^c(V, \Omega, j), \lambda} U(1)$$

to be the $U(1)$ principal bundle associated to $\mathcal{B}_J$ and to the character λ of $MU^c(V, \Omega, j)$. The map $\tilde{\lambda} : \mathcal{B}_J \to \mathcal{B}_J^1(\lambda) : \xi \mapsto [(\xi, 1)]$ yields the isomorphism

$$\phi \times \tilde{\lambda} : \mathcal{B}_J \to \mathcal{B}(M, \omega, J) \times_M \mathcal{B}_J^1(\lambda) : \xi \mapsto \phi(\xi), [(\xi, 1)].$$

The line bundle associated to $\mathcal{B}_J^1(\lambda)$ is denoted by $\mathcal{B}_J(\lambda)$; its isomorphism class is independent of the choice of J.

Reciprocally, given any Hermitian line bundle L over M, one defines the Mp^c structure

$$\mathcal{B} := (\mathcal{B}(M, \omega, J) \times_M L^1) \times_{MU^c(V, \Omega, j)} \mathrm{Mp}^c(V, \Omega, j).$$

An Mp^c-**connection** on the Mp^c structure $(\mathcal{B}, \Phi)$ is a principal connection α on $\mathcal{B}$; in particular, it is a 1-form on $\mathcal{B}$ with values in $\mathfrak{mp}^c \simeq_{\sigma_* \times \eta_*} \mathfrak{sp}(V, \Omega)) \oplus \mathfrak{u}(1)$. We decompose it accordingly as

$$\alpha = \alpha_1 + \alpha_0.$$

The character η yields the construction of a $U(1)$ principal bundle

$$\mathcal{B}^1(\eta) := \mathcal{B} \times_{\mathrm{Mp}^c(V, \Omega, j), \eta} U(1)$$

and there exists a map

$$\tilde{\eta} : \mathcal{B} \to \mathcal{B}^1(\eta) : \xi \mapsto [\xi, 1].$$

Then α_0 is the pull-back of a $\mathfrak{u}(1)$-valued 1-form on $\mathcal{B}^1(\eta)$ under the differential of $\tilde{\eta}$ and

$$\alpha_0 = 2\tilde{\eta}^* \beta_0$$

where β_0 is a principal $U(1)$ connection on $\mathcal{B}^1(\eta)$.

   Similarly α_1 is the pull-back under the differential of $\Phi : \mathcal{B} \to \mathcal{B}(M, \omega)$ of a $\mathfrak{sp}(V, \Omega)$-valued 1-form β_1 on $\mathcal{B}(M, \omega)$ and

$$\alpha_1 = \Phi^* \beta_1$$

where β_1 is a principal $Sp(V, \Omega)$ connection on $\mathcal{B}(M, \omega)$, hence corresponding to a linear connection ∇ on M so that $\nabla \omega = 0$.

   Thus an Mp^c-connection on $\mathcal{B}$ induces connections in TM preserving ω and in $\mathcal{B}^1(\eta)$. The converse is true – we pull back and add connection 1-forms in $\mathcal{B}^1(\eta)$ (with a factor 2) and in $\mathcal{B}(M, \omega)$ to get a connection 1-form on $\mathcal{B}$.

   In geometric quantization, a **prequantization structure** is an Mp^c-structure $(\mathcal{B}, \Phi)$ with a connection α so that

$$d\alpha_0 = \pi^* \frac{\omega}{i\hbar}.$$

The prequantization module is the module of sections of symplectic spinors

$$\mathcal{B} \times_{\mathrm{Mp}^c(V, \Omega, j)} \mathcal{H}^{-\infty}(V, \Omega, j)$$

with prequantization operators

$$Q(f) := \nabla^\alpha_{X_f} + \frac{1}{i\hbar} f.$$

We refer to [10] for further development and quantization in this context (which yields automatically half-forms).

   The condition for the existence of such a structure is that

$$[\omega/h] - \tfrac{1}{2} c_1(TM, \omega)^{\mathbb{R}} \tag{3.1}$$

be an integral cohomology class, where $c_1(TM, \omega)$ is the first Chern class of the tangent bundle viewed as a complex bundle via the choice of a compatible complex structure. Indeed $c_1(TM, \omega)$ is the first Chern class of the line bundle associated to $\mathcal{B}(M, \omega, J)$ and the character $\det_{j_o} \sigma$; since $\eta = \lambda^2 \det_{j_o} \sigma$, we have

$$c_1(\mathcal{B}(\eta)) = 2c_1(\mathcal{B}_J(\lambda)) + c_1(TM, \omega),$$

but $c_1(\mathcal{B}(\eta)) = \left[\frac{i}{2\pi} d\beta_0 \right]$ and $c_1(\mathcal{B}_J(\lambda))$ must be an integral class.

Condition (3.1) was obtained in the context of Deformation Quantization by Boris Fedosov [3] in his construction of asymptotic operator representations of a star product on a symplectic manifold. A geometric interpretation of this result using Mp^c-structures is given in [5].

4. Dirac operators

Given a Spin^c structure $(\mathcal{B}, \Phi)$ on an oriented Riemannian manifold (M, g) one considers the spinor bundle

$$\mathcal{S} = \mathcal{B} \times_{(\mathrm{Spin}^c, r)} S.$$

Since the representation r on S is unitary, the spinor bundle carries a natural fiberwise Hermitian inner product, still denoted $\langle , \rangle$.

The smooth sections of $\mathcal{S}$ are called **spinor fields** on M. For two spinor fields ψ, ψ' with compact support on M, one defines their inner product

$$\langle\!\langle \psi, \psi' \rangle\!\rangle := \int_M \langle \psi(x), \psi'(x) \rangle d^g x$$

where $d^g x$ is the measure on M associated to the metric g.

The Clifford multiplication of TM on $\mathcal{S}$ given by

$$Cl([(\tilde{f}, v)]) \left([(\tilde{f}, s)] \right) := [(\tilde{f}, cl(v)s)] \qquad \forall \tilde{f} \in \mathcal{B}, v \in V, s \in S$$

yields a map

$$Cl : \Gamma(M, TM) \times \Gamma(M, \mathcal{S}) \to \Gamma(M, \mathcal{S}) : (X, \psi) \mapsto Cl(X)\psi.$$

A Spin^c **connection** on the Spin^c structure $(\mathcal{B}, \Phi)$ is a principal connection α on $\mathcal{B}$; in particular, it is a 1-form on $\mathcal{B}$ with values in $\mathfrak{spin}^c \simeq \mathfrak{so}(V, \Omega)) \oplus \mathfrak{u}(1)$. We decompose it accordingly as

$$\alpha = \alpha_1 + \alpha_0.$$

The 1-form α_1 is the pull-back under the differential of $\Phi : \mathcal{B} \to \mathcal{B}(M, \tilde{g})$ of a $\mathfrak{so}(V, \tilde{g})$-valued 1-form β_1 on $\mathcal{B}(M, g)$

$$\alpha_1 = \Phi^* \beta_1$$

and β_1 is a principal $SO(V, \tilde{g})$ connection on $\mathcal{B}(M, g)$, hence corresponding to a linear connection ∇ on M such that $\nabla g = 0$. We choose this to be the Levi–Civita connection.

A Spin^c connection α induces a covariant derivative ∇^α of the spinor fields:

$$\nabla^\alpha : \Gamma(M, \mathcal{S}) \to \Gamma(M, T^* M \otimes \mathcal{S}) : \psi \mapsto [X \to \nabla_X^\alpha \psi]$$

and the Clifford multiplication is parallel

$$\nabla_X^\alpha (Cl(Y)\psi) = Cl(\nabla_X Y)\psi + Cl(X)\nabla_X^\alpha \psi.$$

The Spin^c **Dirac operator** is the differential operator of order 1 acting on spinor fields given by the contraction of the covariant derivative and the Clifford multiplication, using concatenation and the identification of TM and T^*M induced by g:

$$D\psi := \sum_i Cl(e_i)\nabla^\alpha_{e^i}\psi = \sum_{ij} g^{ij}Cl(e_i)\nabla^\alpha_{e_j}\psi$$

where e_i is a local frame field for TM and e^i the dual frame field defined by $g(e_i, e^j) = \delta^j_i$.

This Dirac operator D is elliptic and selfadjoint. It acts on the sections of a finite-dimensional bundle. Its square is equal to

$$D^2\psi = -\sum_{jl} g^{jl}\nabla^{\alpha 2}_{jl}\psi + \tfrac{1}{2}\sum_{ijkl} g^{ij}g^{kl}cl(e_i)cl(e_k)\left(R^\alpha(e_j,e_l)\psi - \nabla_{T^\alpha(e_j,e_l)}\psi\right)$$

where R^α is the curvature and T^α the torsion of ∇^α acting on $\mathcal{S}$; with our choice that α_1 corresponds to the Levi–Civita connection on M the torsion T^α vanishes.

The **classical Dirac operator** is defined similarly using a Spin structure $(\mathcal{B}, \Phi)$. Since σ_* yields an isomorphism of Lie algebras between $\mathfrak{spin}$ and $\mathfrak{so}(V, \tilde{g})$, a natural Spin-connection α' on $\mathcal{B}$ is given by

$$\alpha' = \sigma_*^{-1}\Phi^*\beta_1$$

where β_1 is the principal $SO(V, \tilde{g})$ connection on $\mathcal{B}(M, g)$ defined by the Levi–Civita connection on (M, g).

For further results about the Spin^c Dirac operator and its relation to the classical Dirac operator, we refer to [2].

Given an Mp^c structure $(\mathcal{B}, \Phi)$ on a symplectic manifold (M, ω) one considers the symplectic spinor bundle

$$\mathcal{S} = \mathcal{B} \times_{(\mathrm{Mp}^c, r)} \mathcal{H}^{\pm\infty}.$$

Since the representation r on $\mathcal{H}$ is unitary and preserves $\mathcal{H}^\infty$, the spinor bundle carries a fiberwise Hermitian inner product, still denoted $(\ ,\ )_j$.

The smooth sections of $\mathcal{S}$ are called **symplectic spinor fields** on M. For two symplectic spinor fields ψ, ψ' with compact support on M, one defines their inner product

$$\langle\!\langle \psi, \psi'\rangle\!\rangle_j := \int_M (\psi(x), \psi'(x))_j d^\omega x$$

where $d^\omega x$ is the measure on M associated to the symplectic form ω.

The symplectic Clifford multiplication of TM on $\mathcal{S}$ given by

$$sCl([(\tilde{f}, v)])\left([(\tilde{f}, s)]\right) := [(\tilde{f}, scl(v)s)] \qquad \forall \tilde{f} \in \mathcal{B}, v \in V, s \in \mathcal{H}^{\pm\infty}$$

yields a map

$$sCl : \Gamma(M, TM) \times \Gamma(M, \mathcal{S}) \to \Gamma(M, \mathcal{S}) : (X, \psi) \mapsto sCl(X)\psi.$$

An Mp^c connection α induces a covariant derivative ∇^α of the symplectic spinor fields:

$$\nabla^\alpha : \Gamma(M, \mathcal{S}) \to \Gamma(M, T^*M \otimes \mathcal{S}) : \psi \mapsto [X \to \nabla^\alpha_X\psi]$$

and the Clifford multiplication is parallel

$$\nabla_X^\alpha \left(sCl(Y)\psi \right) = sCl(\nabla_X Y)\psi + sCl(X)\nabla_X^\alpha \psi.$$

The Mp^c **symplectic Dirac operator** is the differential operator of order 1 acting on symplectic spinor fields given by the contraction of the covariant derivative and the symplectic Clifford multiplication, using concatenation and the identification of TM and T^*M induced by ω:

$$D\psi := \sum_i sCl(e_i)\nabla_{e^i}^\alpha \psi = -\sum_{ij} \omega^{ij} sCl(e_i)\nabla_{e_j}^\alpha \psi$$

where e_i is a local frame field for TM and e^i is the dual frame field defined by $\omega(e_i, e^j) = \delta_i^j$.

The definition of the Dirac operator in the Mp^c framework given above is a straightforward generalization of the **symplectic Dirac operator** studied by Katharina and Lutz Habermann [6, 7, 8]. They use a **metaplectic structure** $(\mathcal{B}, \Phi)$, i.e., a $Mp(V, \Omega, j)$- principal bundle $\mathcal{B}$ and a fiberpreserving $(Mp(V, \Omega, j), \sigma)$ equivariant map $\Phi : \mathcal{B} \to \mathcal{B}(M, \omega)$. Remark that those do not always exist.

Since σ_* is an isomorphism of Lie algebras between $\mathfrak{mp}(V, \Omega, j)$ and $\mathfrak{sp}(V, \Omega)$, a metaplectic connection α' on $\mathcal{B}$ is given by

$$\alpha' = \sigma_*^{-1} \Phi^* \beta_1$$

where β_1 is a principal $Sp(V, \Omega)$ connection on $\mathcal{B}(M, \omega)$ defined by a linear connection ∇ on (M, ω) so that $\nabla\omega = 0$.

K. Habermann introduces a second symplectic Dirac operator $\tilde{D}$ using an auxiliary compatible almost complex structure J on (M, ω):

$$\tilde{D}\psi := \sum_{ij} g^{ij} Cl(e_i)\nabla_{e_j} \psi \quad \text{with} \quad g(X, Y) := \omega(X, JY)$$

using a connection so that ω and J are parallel.

The commutator of D and $\tilde{D}$ is elliptic and acts on sections of an infinite family of finite-dimensional subbundles of $\mathcal{S}$.

This construction can be performed in the Mp^c framework [1]. It can be further generalized, also in the Riemannian context, using some fields of endomorphisms of the tangent bundle which are covariantly constant. We give a brief description below.

We consider a field A of endomorphisms of the tangent bundle TM of an oriented Riemannian manifold (M, g), such that

$$g(AX, Y) = \epsilon_A g(X, AY) \quad \text{with} \ \epsilon_A = \pm 1, \qquad \forall X, Y \in \Gamma(M, TM),$$

and such that there is a linear connection ∇ preserving g and A, i.e.,

$$\nabla g = 0 \qquad \text{and} \qquad \nabla A = 0.$$

Given a Spin^c structure $(\mathcal{B}, \Phi)$ on (M, g), we consider a Spin^c connection $\alpha = \alpha_1 + \alpha_0$ so that $\alpha_1 = \Phi^* \beta_1$ where β_1 is the connection 1-form on $\mathcal{B}(M, g)$ defined by the linear connection ∇. We define a new Dirac operator D_A:

$$D_A \psi := \sum_i Cl(Ae_i) \nabla_{e^i}^\alpha \psi = \sum_{ij} A_i^k g^{ij} Cl(e_k) \nabla_{e_j}^\alpha \psi, \qquad (4.1)$$

where e_i is a local frame field for TM and e^i is defined by $g(e_i, e^j) = \delta_i^j$. Remark that the Spin^c Dirac operator corresponds to $A = \mathrm{Id}$ with $\epsilon_A = 1$. On a Kähler manifold (M, g, J), one also has D_J with $\epsilon_J = -1$. We have

$$D_A^2 \psi = - \epsilon_A \sum_{jst} (A^2)_s^j g^{st} \nabla_{e_j e_t}^{\alpha 2} \psi$$

$$+ \tfrac{1}{2} \sum_{ijkrst} A_i^k A_s^r g^{ij} g^{st} Cl(e_k) Cl(e_r) \left(R^\alpha(e_j, e_t) \psi - \nabla_{T^\alpha(e_j, e_t)}^\alpha \psi \right). \qquad (4.2)$$

More generally, if A and B are two such fields of endomorphisms, we have

$$(D_A \circ D_B + D_B \circ D_A) \psi = - \sum_{jst} (\epsilon_A (AB)_s^j + \epsilon_B (BA)_s^j) \, g^{st} \nabla_{e_j e_t}^{\alpha 2} \psi$$

$$+ \tfrac{1}{2} \sum_{ijkrst} (A_i^k B_s^r + A_s^r B_i^k) g^{ij} g^{st} Cl(e_k) Cl(e_r) \left(R^\alpha(e_j, e_t) \psi - \nabla_{T^\alpha(e_j, e_t)}^\alpha \psi \right).$$

Similarly we consider a field A of endomorphisms of the tangent bundle TM of a symplectic manifold (M, ω), such that

$$\omega(AX, Y) = \epsilon_A \omega(X, AY) \quad \text{with } \epsilon_A = \pm 1, \qquad \forall X, Y \in \Gamma(M, TM),$$

and such that there is a linear connection ∇ preserving ω and A, i.e.,

$$\nabla \omega = 0 \qquad \text{and} \qquad \nabla A = 0.$$

Given an Mp^c structure $(\mathcal{B}, \Phi)$ on (M, ω), we consider an Mp^c connection $\alpha = \alpha_1 + \alpha_0$ so that $\alpha_1 = \Phi^* \beta_1$ where β_1 is the connection 1-form on $\mathcal{B}(M, \omega)$ defined by the linear connection ∇. We define a new Dirac operator D_A:

$$D_A \psi := \sum_i sCl(Ae_i) \nabla_{e^i}^\alpha \psi = \sum_{ij} A_i^k \omega^{ij} sCl(e_k) \nabla_{e_j}^\alpha \psi \qquad (4.3)$$

where e_i is a local frame field for TM and e^i is defined by $\omega(e_i, e^j) = \delta_i^j$.

The Mp^c Dirac operator corresponds to $A = \mathrm{Id}$ with $\epsilon_A = 1$. The operator $\tilde{D}$ is equal to D_J for J a positive compatible complex structure on (M, ω); then $\epsilon_J = -1$ and there always exists a linear connection preserving ω and J; it may have torsion, but one can always assume that the torsion vector vanishes, i.e., $\sum_i T^\alpha(e_i, e^i) = 0$. If A and B are two such fields of endomorphisms, we have

$$(D_A \circ D_B - D_B \circ D_A) \psi = -\frac{i}{2\hbar} \sum_{jst} (\epsilon_A (AB)_s^j - \epsilon_B (BA)_s^j) \omega^{st} \nabla_{e_j e_t}^{\alpha 2} \psi$$

$$+ \tfrac{1}{2} \sum_{ijkrst} (A_i^k B_s^r - A_s^r B_i^k) \omega^{ij} \omega^{st} sCl(e_k) sCl(e_r) \left(R^\alpha(e_j, e_t) \psi - \nabla_{T^\alpha(e_j, e_t)}^\alpha \psi \right).$$

These generalized symplectic Dirac operators and their commutators are particularly relevant in a homogeneous or symmetric framework, for instance when there is a homogeneous non positive definite compatible almost complex structure on (M, ω) or two homogeneous distributions of supplementary real Lagrangians; this is work in progress with Laurent La Fuente Gravy and John Rawnsley.

References

[1] Michel Cahen, Simone Gutt, John Rawnsley, Symplectic Dirac operators and Mp^c-structures, *Gen Relativ Gravit* (2011) DOI 10.1007/s10714-011-1239-x.

[2] J.J. Duistermaat, *The Heat Kernel Lefschetz Fixed Point Formula for the Spin-c Dirac Operator*, Progress in Nonlinear Diff. Eq. and Their Appl. 18, Birkhäuser, 1995.

[3] B. Fedosov, *Deformation quantization and index theory*. Akademie Verlag, Berlin 1996.

[4] M. Forger and H. Hess, Universal Metaplectic Structures and Geometric Quantization, *Commun. Math. Phys.* 64, (1979) 269–278.

[5] S. Gutt, J. Rawnsley, Mp^c structures, *Math. Phys. Studies* 20, 1997, 103–115.

[6] K. Habermann, The Dirac Operator on Symplectic Spinors *Ann. Global Anal. Geom.* 13 (1995) 155–168.

[7] K. Habermann, Basic Properties of Symplectic Dirac Operators, *Comm. in Math. Phys.* 184 (1997) 629–652.

[8] Katharina Habermann and Lutz Habermann, *Introduction to Symplectic Dirac Operators*, Lecture Notes in Mathematics 1887, Springer-Verlag, 2006.

[9] B. Kostant, Symplectic Spinors. *Symposia Mathematica*, vol. XIV, pp. 139–152, Cambridge University Press, 1974.

[10] P.L. Robinson and J.H. Rawnsley, *The metaplectic representation*, Mp^c *structures and geometric quantization*, Memoirs of the A.M.S. vol. 81, no. 410, 1989.

Michel Cahen
Université Libre de Bruxelles
Campus Plaine, CP 218, Bvd. du triomphe
B-1050 Bruxelles, Belgique
e-mail: mcahen@ulb.ac.be

Simone Gutt
Université Libre de Bruxelles
Campus Plaine, CP 218, Bvd. du triomphe
B-1050 Bruxelles, Belgique
  and
Université de Lorraine à Metz, Ile du Saulcy
F-57045 Metz Cedex 01, France
e-mail: sgutt@ulb.ac.be

Geometric Methods in Physics. XXXI Workshop 2012
Trends in Mathematics, 29–37

$\theta(\hat{x}, \hat{p})$-deformation of the Harmonic Oscillator in a 2D-phase Space

M.N. Hounkonnou, D. Ousmane Samary,
E. Baloïtcha and S. Arjika

Abstract. This work addresses a $\theta(\hat{x}, \hat{p})$-deformation of the harmonic oscillator in a 2D-phase space. Specifically, it concerns a quantum mechanics of the harmonic oscillator based on a noncanonical commutation relation depending on the phase space coordinates. A reformulation of this deformation is considered in terms of a q-deformation allowing to easily deduce the energy spectrum of the induced deformed harmonic oscillator. Then, it is proved that the deformed position and momentum operators admit a one-parameter family of self-adjoint extensions. These operators engender new families of deformed Hermite polynomials generalizing usual q-Hermite polynomials. Relevant matrix elements are computed. Finally, a $su(2)$-algebra representation of the considered deformation is investigated and discussed.

Mathematics Subject Classification (2010). 81Q99; 81S99.

Keywords. Harmonic oscillator, energy spectrum, q-deformation, Hermite polynomials, matrix elements, $su(2)$-algebra.

1. Introduction

Consider a 2D phase space $\mathcal{P} \subset \mathbb{R}^2$. Coordinates of position and momentum are denoted by x and p. Corresponding Hilbert space quantum operators $\hat{x}$ and $\hat{p}$ satisfy the following commutation relation

$$[\hat{x}, \hat{p}] = \hat{x}\hat{p} - \hat{p}\hat{x} = i\theta(\hat{x}, \hat{p}) \tag{1}$$

where θ is the deformation parameter encoding the noncommutativity of the phase space: $\theta(x, p) = 1 + \alpha x^2 + \beta p^2$, $\alpha, \beta \in \mathbb{R}_+$ with the uncertainty relation:

$$\Delta\hat{x}\Delta\hat{p} \geq \frac{1}{2}\left(1 + \alpha(\Delta\hat{x})^2 + \beta(\Delta\hat{p})^2\right)$$

where the parameters α and β are real positive numbers.

The motivations for this study derive from a series of works devoted to the relation (1). Indeed, already in [1] Kempf *et al.* investigated (1) for the particular case $\alpha = 0$ with

$$\Delta\hat{x}\Delta\hat{p} \geq \frac{1}{2}(1 + \beta(\Delta\hat{p})^2)$$

and led to the conclusion that the energy levels of a given system can deviate significantly from the usual quantum mechanical case once energy scales become comparable to the scale $\sqrt{\beta}$. Although the onset of this scale is an empirical question, it is presumably set by quantum gravitational effects. In another work [2], Kempf, for the same model with $\alpha = 0$, led to the conclusion that the anomalies observed with fields over unsharp coordinates might be testable if the onset of strong gravity effects is not too far above the currently experimentally accessible scale about $10^{-18}m$, rather than at the Planck scale of $10^{-35}m$. More recently, in [3], it was shown that similar relation can be applied to discrete models of matter or space-time, including loop quantum cosmology. For more motivations, see [4], [5] and [6], but also [7], [8], [9] and [10] and references therein.

In this work, we investigate how such a deformation may affect main properties, e.g., energy spectrum, of a simple physical system like a harmonic oscillator.

The paper is organized as follows. In Section 2, we introduce a reformulation of the $\theta(\hat{x}, \hat{p})$-deformation in terms of a q-deformation allowing to easily deduce the energy spectrum of the induced deformed harmonic oscillator. Then it is proved that the deformed position and momentum operators admit a one-parameter family of self-adjoint extensions. These operators engender new families of deformed Hermite polynomials generalizing usual q-Hermite polynomials. Section 3 is devoted to the matrix element computation. Finally, in Section 4, we provide a $su(2)$-algebra representation of the considered deformation. Section 5 deals with concluding remarks.

2. q-like realization

It is worth noticing that such a $\theta(x,p)$-deformation (1) admits an interesting q-like realization via the following re-parameterization of deformed creation and annihilation operators:

$$\hat{b}^\dagger = \frac{1}{2}(m_\alpha\hat{x} - im_\beta\hat{p}), \quad \hat{b} = \frac{1}{2}(m_\alpha\hat{x} + im_\beta\hat{p})$$

satisfying the peculiar q-Heisenberg commutation relation:

$$\hat{b}\hat{b}^\dagger - q\hat{b}^\dagger\hat{b} = 1$$

where the parameter q is written in the form

$$q = \frac{1 + \sqrt{\alpha\beta}}{1 - \sqrt{\alpha\beta}} \geq 1$$

and the quantities m_α and m_β are given by

$$m_\alpha = \sqrt{2\alpha \left(\frac{1}{\sqrt{\alpha\beta}} - 1 \right)}, \quad m_\beta = \sqrt{2\beta \left(\frac{1}{\sqrt{\alpha\beta}} - 1 \right)}.$$

With this consideration, the spectrum of the induced harmonic oscillator Hamiltonian $\hat{H} = \hat{b}\hat{b}^\dagger + \hat{b}^\dagger\hat{b}$ is given by

$$\mathcal{E}_n = \frac{1}{2} \left([n]_q + [n+1]_q \right)$$

where the q-number $[n]_q$ is defined by $[n]_q = \frac{1-q^n}{1-q}$. Let $\mathcal{F}$ be a q-deformed Fock space and $\{|n, q\rangle \mid n \in \mathbb{N} \bigcup \{0\}\}$ be its orthonormal basis. The actions of $\hat{b}$, $\hat{b}^\dagger$ on $\mathcal{F}$ are given by

$$\hat{b}|n, q\rangle = \sqrt{[n]_q}|n-1, q\rangle, \quad \text{and} \quad \hat{b}^\dagger|n, q\rangle = \sqrt{[n+1]_q}|n+1, q\rangle,$$

where $|0, q\rangle$ is a normalized vacuum:

$$\hat{b}|0, q\rangle = 0, \qquad \langle q, 0|0, q\rangle = 1.$$

The Hamiltonian operator $\hat{H}$ acts on the states $|n, q\rangle$ to give: $\hat{H}|n, q\rangle = \mathcal{E}_n|n, q\rangle$.

Theorem 1. *The position operator $\hat{x}$ and momentum operator $\hat{p}$, defined on the Fock space $\mathcal{F}$, are not essentially self-adjoint, but have a one-parameter family of self-adjoint extensions.*

Proof. Consider the following matrix elements of the position operator $\hat{x}$ and momentum operator $\hat{p}$:

$$\langle m, q|\hat{x}|n, q\rangle := \left\langle m, q \left| \frac{1}{m_\alpha}(\hat{b}^\dagger + \hat{b}) \right| n, q \right\rangle$$

$$= \frac{1}{m_\alpha} \sqrt{[n+1]_q}\delta_{m,n+1} + \frac{1}{m_\alpha} \sqrt{[n]_q}\delta_{m,n-1}$$

$$\langle m, q|\hat{p}|n, q\rangle := \left\langle m, q \left| \frac{i}{m_\beta}(\hat{b}^\dagger - \hat{b}) \right| n, q \right\rangle$$

$$= \frac{i}{m_\beta} \sqrt{[n+1]_q}\delta_{m,n+1} - \frac{i}{m_\beta} \sqrt{[n]_q}\delta_{m,n-1}.$$

Setting $x_{n,\alpha} = \frac{1}{m_\alpha}\sqrt{[n]_q}$ and $x_{n,\beta} = \frac{1}{m_\beta}\sqrt{[n]_q}$, then the position operator $\hat{x}$ and momentum operator $\hat{p}$ can be represented by the two following symmetric Jacobi matrices, respectively:

$$M_{\hat{x},q,\alpha} = \begin{pmatrix} 0 & x_{1,\alpha} & 0 & 0 & 0 & \cdots \\ x_{1,\alpha} & 0 & x_{2,\alpha} & 0 & 0 & \cdots \\ 0 & x_{2,\alpha} & 0 & x_{3,\alpha} & 0 & \cdots \\ \vdots & \ddots & \ddots & \ddots & \ddots & \ddots \end{pmatrix} \tag{2}$$

and

$$M_{\hat{p},q,\beta} = \begin{pmatrix} 0 & -ix_{1,\beta} & 0 & 0 & 0 & \cdots \\ ix_{1,\beta} & 0 & -ix_{2,\beta} & 0 & 0 & \cdots \\ 0 & ix_{2,\beta} & 0 & -ix_{3,\beta} & 0 & \cdots \\ \vdots & \ddots & \ddots & \ddots & \ddots & \ddots \end{pmatrix}. \tag{3}$$

The quantity $|x_{n,\alpha}| = \frac{1}{m_\alpha}\left|\frac{1-q^n}{1-q}\right|^{1/2}$ is not bounded since $q > 1$ by definition, and $\lim_{n\to\infty} |x_{n,\alpha}| = \infty$. Considering the series $y_\alpha = \sum_{n=0}^\infty 1/x_{n,\alpha}$, we get

$$\overline{\lim_{n\to\infty}} \left(\frac{1/x_{n+1}}{1/x_n}\right) = q^{-1/2} < 1$$

and, hence, the series y_α converges. Besides, as the quantity $q^{-1} + q > 2$,

$$0 < x_{n+1,\alpha} x_{n-1,\alpha} = \frac{1}{m_\alpha^2(1-q)}\left[1 - q^n(q^{-1}+q) + q^{2n}\right]^{1/2} < x_{n,\alpha}^2$$

Hence, the Jacobi matrices in (2) and (3) are not self-adjoint (Theorem 1.5., Chapter VII in Ref. [11]). The deficiency indices of these operators are equal to $(1,1)$. One concludes that the position operator $\hat{x}$ and the momentum operator $\hat{p}$ are no longer essentially self-adjoint but have each a one-parameter family of self-adjoint extensions instead. This means that their deficiency subspaces are one-dimensional. □

Besides, in this case, the deficiency subspaces N_z, $Imz \neq 0$, are defined by the generalized vectors $|z\rangle = \sum_0^\infty P_n(z)||n,q\rangle$ such that [11], [12]:

$$\sqrt{[n]_q}P_{n-1}(z) + \sqrt{[n+1]_q}P_{n+1}(z) = zP_n(z) \tag{4}$$

with the initial conditions $P_{-1}(z) = 0$, $P_0(z) = 1$.

- In the position representation, the states $|x,q>$ such that

$$\hat{x}|x,q\rangle = x|x,q\rangle, \text{ and } |x,q\rangle = \sum_{n=0}^\infty P_{n,q}(x)|n,q\rangle,$$

transforms the relation (4) into

$$m_\alpha x P_{n,q}(x) = \sqrt{[n+1]_q}P_{n+1,q}(x) + \sqrt{[n]_q}P_{n-1,q}(x)$$
$$n = 0,1,\ldots; \; P_{-1,q}(x) = 0, \; P_{0,q}(x) = 1$$

giving

$$2\gamma(x,q)P_{n,q}\left(\frac{2\gamma(x,q)}{(1-q)^{1/2}m_\alpha}\right) = (1-q^{n+1})^{\frac{1}{2}}P_{n+1,q}\left(\frac{2\gamma(x)}{(1-q)^{1/2}m_\alpha}\right)$$
$$+ (1-q^n)^{\frac{1}{2}}P_{n-1,q}\left(\frac{2\gamma(x,q)}{(1-q)^{1/2}m_\alpha}\right) \tag{5}$$

where $2\gamma(x,q) = (1-q)^{1/2}m_\alpha x$.

Setting $\widetilde{P}_{n,q}(\gamma(x,q)) = P_{n,q}\left(\frac{2\gamma(x,q)}{(1-q)^{1/2}m_\alpha}\right)$, the equation (5) can be re-expressed as

$$2\gamma(x,q)\widetilde{P}_{n,q}(\gamma(x,q)) = (1 - q^{n+1})^{\frac{1}{2}}\widetilde{P}_{n+1,q}(\gamma(x,q))$$
$$+ (1 - q^n)^{\frac{1}{2}}\widetilde{P}_{n-1,q}(\gamma(x,q)). \tag{6}$$

Finally, putting $(q;q)_n^{1/2}\widetilde{P}_{n,q}(\gamma(x,q)) = H_{n,q}(x)$, the formula (6) recalls the recurrence relation satisfied by q-Hermite polynomials:

$$2xH_{n,q}(x) = H_{n+1,q}(x) + (1 - q^n)H_{n-1,q}(x)$$

where $(q;q)_n = (1-q)(1-q^2)\cdots(1-q^n)$.

• In the momentum representation, the state $|p,q\rangle$ such that

$$\hat{p}|p,q\rangle = p|p,q\rangle, \quad \text{and} \quad |p,q\rangle = \sum_{n=0}^{\infty} Q_{n,q}(p)|n,q\rangle$$

leads to the following recurrence relation between functions $Q_{n,q}(x)$:

$$im_\beta pQ_{n,q}(p) = \sqrt{[n+1]_q}Q_{n+1,q}(p) - \sqrt{[n]_q}Q_{n-1,q}(p)$$
$$n = 0, 1, \ldots; \; Q_{-1,q}(p) = 0, \; Q_{0,q}(p) = 1.$$

This equation can be also re-expressed as

$$\tilde{\gamma}(\tilde{p},q)Q_{n,q}(p) = (1 - q^{n+1})^{\frac{1}{2}}Q_{n+1,q}(p) - (1 - q^n)^{\frac{1}{2}}Q_{n-1,q}(p) \tag{7}$$

where $\tilde{\gamma}(\tilde{p},q) = (1-q)^{1/2}m_\beta\tilde{p}, \; \tilde{p} = ip$.

Setting $\widetilde{Q}_{n,q}(\tilde{\gamma}(\tilde{p},q)) = Q_{n,q}\left(\frac{2\tilde{\gamma}(\tilde{p},q)}{(1-q)^{1/2}m_\alpha}\right)$, then equation (7) yields

$$2\tilde{\gamma}(\tilde{p},q)\widetilde{Q}_{n,q}(\tilde{\gamma}(\tilde{p},q)) = (1 - q^{n+1})^{\frac{1}{2}}\widetilde{Q}_{n+1,q}(\tilde{\gamma}(\tilde{p},q)) - (1 - q^n)^{\frac{1}{2}}\widetilde{Q}_{n-1,q}(\tilde{\gamma}(\tilde{p},q)).$$

Letting $(q;q)_n^{1/2}\widetilde{Q}_{n,q}(\tilde{\gamma}(\tilde{p},q)) = H_{n,q}(ip)$, we arrive at the recurrence relation satisfied by the complex q-Hermite polynomials given by

$$2ipH_{n,q}(ip) = H_{n+1,q}(ip) - (1 - q^n)H_{n-1,q}(ip).$$

Remark 1. The following is worthy of attention:

(i) In the x-space where the momentum operator is defined by the relation

$$\hat{p} := -i\theta(\hat{x}, \hat{p})\partial_x, \tag{8}$$

any function $\Psi_q(x)$ in x-representation can be expressed in terms of its analog $\Psi_q(p)$ in p-representation by the relation

$$\Psi_q(x) = \int_{-\infty}^{\infty} dp \, \exp\left(\frac{ip}{\alpha\sigma(p)}\arctan\frac{x}{\sigma(p)}\right)\Psi_q(p),$$

where $\sigma(p) = \sqrt{p^2 + \frac{1}{\alpha}}$. Defining the Hilbert space inner product as

$$\langle f, g \rangle = \int \frac{dx}{\theta(\hat{x}, \hat{p})}\bar{f}_q(x)g_q(x)$$

one can readily prove that $\hat{p}$ reverts the property of a Hermitian operator.

(ii) Analogously, in the p-space

$$\hat{x} := i\theta(\hat{x}, \hat{p})\partial_p \tag{9}$$

and

$$\Psi_q(p) = \int_{-\infty}^{\infty} dx \, \exp\left(\frac{-ix}{\alpha\sigma(x)} \arctan \frac{p}{\sigma(x)}\right) \Psi_q(x).$$

The appropriate inner product, in the momentum space, rendering the operator $\hat{x}$ Hermitian is defined as

$$\langle f, g \rangle = \int \frac{dp}{\theta(\hat{x}, \hat{p})} \bar{f}_q(p) g_q(p)$$

with the condition $\lim_{x \to -\infty} \Psi_q(x) = \lim_{x \to \infty} \Psi_q(x) = 0.$

3. Matrix elements

From the natural actions of q-deformed position operator $\hat{x}$ and momentum operator $\hat{p}$ on the basis vectors $|n, q\rangle \in \mathcal{F}$:

$$\hat{x}|n, q\rangle = \frac{1}{m_\alpha}(\hat{b} + \hat{b}^\dagger)|n, q\rangle, \quad \hat{p}|n, q\rangle = \frac{i}{m_\beta}(\hat{b}^\dagger - \hat{b})|n, q\rangle$$

we immediately deduce the matrix elements

$$
\begin{aligned}
\langle m, q | \hat{b}^{\dagger l} \hat{b}^r | n, q \rangle &= \sqrt{\frac{\Gamma_q(n+1)\Gamma_q(n-r+l+1)}{\Gamma_q(n-r+1)\Gamma_q(n-r+1)}} \delta_{m,n-r+l} \\
\langle m, q | \hat{b}^r \hat{b}^{\dagger l} | n, q \rangle &= \sqrt{\frac{\Gamma_q(n+l+1)\Gamma_q(n+l+1)}{\Gamma_q(n+1)\Gamma_q(n-r+l+1)}} \delta_{m,n-r+l}
\end{aligned}
\tag{10}
$$

where $\Gamma_q(.)$ is the q-Gamma function. Denoting by $::$ the normal ordering, then the expectation value of normal ordering product of $\hat{x}^l\hat{p}^r$ can be computed by the following relation:

$$\langle m, q | : \hat{x}^l \hat{p}^r : | n, q \rangle = \frac{i^r}{m_\alpha^l m_\beta^r} \sum_{s=0}^{l} \sum_{t=0}^{r} C_l^s C_r^t \langle m, q | \hat{b}^{\dagger l-s+t} \hat{b}^{s+r-t} | n, q \rangle$$

which can be given explicitly by using relation (10). Then it becomes a matter of computation to determine the basis operators in terms of $\hat{b}$ and $\hat{b}^\dagger$ as follows:

$$|m, q\rangle\langle n, q| =: \frac{\hat{b}^{\dagger m}}{\sqrt{[m]_q!}} e^{-\hat{b}^\dagger \hat{b}} \frac{\hat{b}^n}{\sqrt{[n]_q!}} :.$$

4. $su(2)$-algebra representation

Turning back to the standard expression of the harmonic oscillator Hamiltonian operator, i.e., $\hat{H} = \hat{a}^{\dagger}\hat{a}$, such that

$$\hat{a} = \frac{1}{\sqrt{2}}(\hat{x} + i\hat{p}), \quad \hat{a}^{\dagger} = \frac{1}{\sqrt{2}}(\hat{x} - i\hat{p}),$$

we get explicitly

$$\hat{H} = \frac{1}{2}\left[(1 + \alpha)\hat{x}^2 + (1 + \beta)\hat{p}^2 + 1\right]$$

giving the simpler form $\hat{H} = \frac{1}{2}$ when $\alpha = -1$ and $\beta = -1$.

From (8) and (9), the Hamiltonian H can be considered as non-local and we can define

$$\hat{H}^{\text{loc}} := \hat{H}(\theta, \partial_x\theta, \partial_x^2\theta, \ldots, x, \partial_x, \partial_x^2, \ldots, \alpha, \beta)$$

with

$$\theta, \partial_x\theta, \cdots = f(\theta, \partial_x\theta, \partial_x^2\theta, \ldots, x, \partial_x, \partial_x^2, \ldots, \alpha, \beta)$$

adding some nonlinearity to the Hamiltonian operator nonlocality.

Assume the parameters α and β satisfy the condition: $|\alpha| \ll 1, |\beta| \ll 1$ and put $\tilde{\alpha} = \alpha$ and $\tilde{\beta} = -\beta$. Then $\hat{x}, \hat{p}, \theta$ can be viewed as the elements of $su(2)$-algebra, i.e.,

$$[\hat{x}, \hat{p}] = i\theta, \quad [\hat{p}, \theta] = i\alpha\{\hat{x}, \theta\} = i\tilde{\alpha}\hat{x}, \quad [\theta, \hat{x}] = -i\beta\{\hat{p}, \theta\} = i\tilde{\beta}\hat{p}.$$

Let $\overrightarrow{J} := (\hat{x}, \hat{p}, \theta)$ be the angular momentum such that there exist states $|j, m\rangle \in \mathcal{F}$ satisfying the condition $\langle j, m|j, m'\rangle = \delta_{mm'}$. Define the operators $\hat{J}_+$ and $\hat{J}_-$ by

$$\hat{J}_+ := \frac{1}{\beta}\hat{x} + \frac{i}{\alpha}\hat{p}, \quad \hat{J}_- := \frac{1}{\beta}\hat{x} - \frac{i}{\alpha}\hat{p}.$$

Proposition 1. *There exists an arbitrary number ν such that*

$$\hat{J}_-|j, m\rangle = C_-(m, j)|j, m - \nu\rangle, \quad \hat{J}_+|j, m\rangle = C_+(m, j)|j, m + \nu\rangle,$$
$$\theta|j, m\rangle = f(m, j)|j, m\rangle \tag{11}$$

where $C_-(m, j)$, $C_+(m, j)$ and $f(m, j)$ are three constants depending on j and m.

The parameters j and m depend on α and β. The unitary representation of $su(2)$-algebra, $\{|j, m\rangle, \ j \in \mathbb{N}, \ -j \le m \le j\}$, is infinite dimensional. The operators $\{\hat{x}, \hat{p}, \theta\}$ act on the Fock space $\mathcal{H} = \{|j, m\rangle / m \in \mathbb{N} \cup \{0\}\}$ following (11). Note that θ and $\overrightarrow{J}^2 = (1 + 2\alpha)\hat{x}^2 + (1 + 2\beta)\hat{p}^2 + 1$ commute. Therefore, $\overrightarrow{J}^2$ and $\hat{H}$ commute too. Besides, the following commutation relations are in order:

$$[\theta, \hat{J}_+] = \hat{x} + i\hat{p}, \quad [\theta, \hat{J}_-] = -(\hat{x} - i\hat{p}). \tag{12}$$

In the interesting particular case where $\alpha = \beta$, the relations (12) are reduced to

$$[\theta, \hat{J}_+] = \alpha\hat{J}_+, \quad [\theta, \hat{J}_-] = -\alpha\hat{J}_-, \quad [\hat{J}_+, \hat{J}_-] = 2\alpha^{-2}\theta.$$

Taking $f(m, j) = m$ yields the condition

$$\theta\hat{J}_+|j, m\rangle = (m + \alpha)\hat{J}_+|j, m\rangle, \quad \theta\hat{J}_-|j, m\rangle = (m - \alpha)\hat{J}_+|j, m\rangle.$$

Besides, we have

$$\hat{J}_+|j,m\rangle = C_+|j,m+\alpha\rangle, \quad \hat{J}_-|j,m\rangle = C_-|j,m-\alpha\rangle$$

where

$$C_+ = \sqrt{(j-m)(j+m+\alpha)}, \quad C_- = \sqrt{(j+m)(j-m+\alpha)}.$$

The eigenfunctions of the Hamiltonian $\hat{H}$ in the position and momentum representations are given, respectively, by

$$\Psi_{j,m}(x) = \langle x|j,m\rangle, \quad \Psi_{j,m}(p) = \langle p|j,m\rangle$$

solution of the equation

$$\hat{H}\Psi_{j,m}(x) = \frac{\alpha^2}{2}\sqrt{(j+m)(j-m)(j+m+\alpha)(j-m+\alpha)}\Psi_{j,m}(x)$$

easily obtainable by solving the eigenvalue problem for the Casimir operator $\hat{J}_+\hat{J}_-$. Furthermore, we get

$$\hat{J}^2\Psi_{j,m}(x) = 2\alpha^2\sqrt{(j+m)(j-m)(j+m+2\alpha)(j-m+2\alpha)}\Psi_{j,m}(x).$$

5. Concluding remarks

In work, we have introduced a reformulation of the $\theta(\hat{x},\hat{p})$-deformation in terms of a q-deformation allowing to easily deduce the energy spectrum of the induced deformed harmonic oscillator. Then we have proved that the deformed position and momentum operators admit a one-parameter family of self-adjoint extensions. These operators have engendered new families of deformed Hermite polynomials generalizing usual q-Hermite polynomials. We have also computed relevant matrix elements. Finally, a $su(2)$-algebra representation of the considered deformation has been investigated and discussed.

Acknowledgment

This work is partially supported by the ICTP through the OEA-ICMPA-Prj-15. The ICMPA is in partnership with the Daniel Iagolnitzer Foundation (DIF), France. MNH expresses his gratefulness to Professor A. Odzijewicz and all his staff for their hospitality and the good organization of the Workshops in Geometric Methods in Physics.

References

[1] Kempf, A. Mangano, G. and Mann, R.B.: (1995), *Hilbert Space Representation of the Minimal Length Uncertainty Relation*, J. Phys. D. 52, pp. 1108.

[2] Kempf, A.: (2000), *A Generalized Shannon Sampling Theorem, Fields at the Planck Scale as Bandlimited Signals*, Phys. Rev. Lett. 85, pp. 2873 *[e-print hep-th/9905114]*.

[3] Kempf, A.: (2011), *Generalized uncertainty principles and localization in discrete space, [e-print hep-th/1112.0994]*.

[4] D.J. Gross, P.F. Mende.: (1988), *The Minimal Length in String Theory*, Nucl. Phys. B303, 407.

[5] D. Amati, M. Ciafaloni, G. Veneziano.: (1989), *Can spacetime be probed below the string size* Phys. Lett. B 216 41.

[6] L.J. Garay.: (1995) *Models of neutrino masses and mixings*, Int. J. Mod. Phys. **A10**, 145.

[7] Hirshfeld, A.C. and Henselder, P.: (2002), *Deformation quantization in the teaching of quantum mechanics*, American Journal of Physics, 70 (5) pp. 537–547, May.

[8] Quesne C. and Tkachuk V.M.: *Lorentz-covariant deformed algebra with minimal length*, [arXiv:hep-th/0612093].

[9] Lay Nam Chang, Dlordje Minic, Naotoshi Okamura and Tatsu Takeuchi: *Exact solutions of the harmonic oscillator in arbitrary dimensions with minimal length uncertainty relations*, [arXiv:hep-th/0111181].

[10] Sketesko M.M and Tkachuk V.M.: *Perturbative hydrogen-atom spectrum in deformed space with minimal length*, [arXiv:hep-th/0603042].

[11] Ju.M. Berezanskiĭ, *Expansions in Eigenfunctions of Selfadjoint Operators*, (Amer. Math. Soc., Providence, Rhode Island, 1968).

[12] Burban I.M.: *Generalized q-deformed oscillators, q-Hermite polynomials, generalized coherent states.*

M.N. Hounkonnou, D. Ousmane Samary,
E. Baloïtcha and S. Arjika
University of Abomey-Calavi
International Chair in Mathematical Physics and Applications
ICMPA-UNESCO CHAIR
072B.P.:50, Cotonou, Rep. of Benin
e-mail: norbert.hounkonnou@cipma.uac.bj; hounkonnou@yahoo.fr
        dine.ousmanesamary@cipma.uac.bj
        ezinvi.baloitcha@cipma.uac.bj
        rjksama2008@gmail.com

Geometric Methods in Physics. XXXI Workshop 2012
Trends in Mathematics, 39–45

Star Products and Certain Star Product Functions

Mari Iida, Chishu Tsukamoto and Akira Yoshioka

To the memory of Boris Fedosov

Abstract. The aim of this note is to provide a short introduction to non-formal star products with some concrete examples. We discuss star exponentials, and their application to trigonometric functions following Eisenstein.

Mathematics Subject Classification (2010). Primary 53D55; Secondary 11M36.

Keywords. Star product, deformation quantization.

1. Introduction

The Moyal product is a typical example of a star product. The Moyal product algebra is isomorphic to the Weyl algebra and is regarded as a polynomial expression of the Weyl algebra. In this note, we give a family of star products, parametrized by complex symmetric matrices, in which the Moyal product is contained.

We consider extensions of the usual point-wise product of polynomials to the space of smooth functions in two different directions. The first one is to extend the product by regarding the Planck constant as a formal parameter. This formal extension leads to the concept of deformation quantization on manifolds.

The second one is a non-formal extension, i.e., star products converging with respect to the Planck constant. This will be the subject of this contribution. We will recall certain classes of functions for which the star products are well defined as convergent series. We discuss the star exponential functions. Using star exponential functions we can define several star functions. In this note, we introduce trigonometric functions following the discussion in Weil's book [1].

The third author is supported by Grant-in-Aid for Scientific Research (# 24540097).

2. Star products

Let Λ be an arbitrary $n \times n$ complex matrix, we define a product on complex polynomials $f(u_1, \ldots, u_n)$, $g(u_1, \ldots, u_n) \in \mathcal{P}(\mathbb{C}^n)$ such that (OMMY [2])

$$f *_\Lambda g = f \exp\left(\tfrac{i\hbar}{2} \overleftarrow{\partial} \Lambda \overrightarrow{\partial}\right) g$$

$$= fg + \tfrac{i\hbar}{2} f \left(\overleftarrow{\partial} \Lambda \overrightarrow{\partial}\right) g + \cdots + \tfrac{1}{k!} \left(\tfrac{i\hbar}{2}\right)^k f \left(\overleftarrow{\partial} \Lambda \overrightarrow{\partial}\right)^k g + \cdots$$

where $\overleftarrow{\partial} \Lambda \overrightarrow{\partial}$ is the biderivation given by

$$f \overleftarrow{\partial} \Lambda \overrightarrow{\partial} g = f \left(\sum_{ij} \Lambda^{ij} \overleftarrow{\partial_{u_i}} \overrightarrow{\partial_{u_j}}\right) g = \sum_{ij} \Lambda^{ij} \partial_{u_i} f \, \partial_{u_j} g.$$

We remark here that the product is well defined on polynomials. It is easy to see that

Proposition 1. *For any Λ, the product $*_\Lambda$ is associative.*

We call $*_\Lambda$ a star product and the algebra $(\mathcal{P}(\mathbb{C}^n), *_\Lambda)$ a star product algebra.

Remark 1. We note that

1. when $\Lambda = 0$, $*_\Lambda$ is the usual multiplication of polynomials,
2. when Λ is symmetric, $*_\Lambda$ is commutative.

2.1. A family of star products and their equivalence

Let us consider a family of complex matrices Λ with common skew symmetric part. For these Λ's, we consider a family of star product algebras $\{(\mathcal{P}(\mathbb{C}^n), *_\Lambda)\}$.

We denote the difference of these by $K = \Lambda' - \Lambda$ and we define a linear isomorphism

$$T_K f = \exp\left(\tfrac{i\hbar}{4} \partial K \partial\right) f = \sum_{n \geq 0} \tfrac{1}{n!} \left(\tfrac{i\hbar}{4}\right)^n (\partial K \partial)^n f$$

where $\partial K \partial = \sum_{ij} K_{ij} \partial_{u_i} \partial_{u_j}$. Then we have

Proposition 2. *T_K is an intertwiner between the products $*_{\Lambda'}$ and $*_\Lambda$;*

$$(T_K f *_{\Lambda'} T_K g) = T_K (f *_\Lambda g).$$

Hence, the algebraic structure of $*_\Lambda$ depends only on the skew symmetric part of Λ.

When $n = 2m$ and the common skew symmetric part is $J = \left(\begin{smallmatrix} 0 & -1_m \\ 1_m & 0 \end{smallmatrix}\right)$ then each $(\mathcal{P}(\mathbb{C}^n), *_\Lambda)$ is isomorphic to the Weyl algebra. Hence each element of the family $\{(\mathcal{P}(\mathbb{C}^n), *_\Lambda)\}$ with skew symmetric part J is regarded as a polynomial expression of the Weyl algebra W_{2m}.

3. Extension of star product

We want to extend the star product to some more general spaces of functions.

We study the following two directions.

1. The first directions are formal star products – i.e., star products on the space of all formal power series of $\hbar$ with coefficients in smooth functions.
2. The second direction are non-formal deformations.

3.1. Formal extension

We extend the star product $*_\Lambda$ to the space of all formal power series with coefficients in the smooth functions on $\mathbb{R}^n$.

Let us consider the space of all formal power series

$$\mathcal{A}_\hbar = C^\infty(\mathbb{R}^n)[[\hbar]].$$

Then we have

Proposition 3. *The star product $*_\Lambda$ is well defined on $\mathcal{A}_\hbar$ such that*

$$f *_\Lambda g = fg + \tfrac{i\hbar}{2}\{f,g\} + \cdots + \hbar^n C_n(f,g) + \cdots$$

where $\{f,g\}$ is the Poisson bracket and C_n is a bidifferential operator. By this we obtain an associative algebra $(\mathcal{A}_\hbar, *_\Lambda)$.

3.1.1. Deformation quantization on manifolds.
The concept of a formal star product leads to deformation quantization of Poisson manifolds ([3, 4]).

Let us consider a Poisson manifold $(M, \{\ ,\ \})$, and put $\mathcal{A}_\hbar(M) = C^\infty(M)[[\hbar]]$.

Definition 1. An associative product $*$ on $\mathcal{A}_\hbar(M)$ is called a deformation quantization on M when it has an expansion

$$f *_\Lambda g = fg + \tfrac{i\hbar}{2}C_1(f,g) + \cdots + \hbar^n C_n(f,g) + \cdots$$

for any $f, g \in \mathcal{A}_\hbar(M)$, where C_n is a bidifferential operator on M and

$$C_1(f,g) - C_1(g,f) = 2\{f,g\}.$$

Note that $*$ can be localized to any arbitrary domain $U \subset M$, that is, we have a star product $f * g$ for any $f, g \in \mathcal{A}_\hbar(U)$.

3.1.2. Existence of deformation quantization.
When $(M, \{\ ,\ \})$ is symplectic, the deformation quantization $*$ has a nice property.

On a Darboux chart $(U, (u_1, \ldots, u_n, v_1, \ldots, v_n))$, the Poisson bracket can be expressed in the form

$$\{f,g\} = \sum_i \frac{\partial f}{\partial u_i}\frac{\partial g}{\partial v_i} - \frac{\partial f}{\partial v_i}\frac{\partial g}{\partial u_i} = f\overset{\leftarrow}{\partial}\Lambda\overset{\rightarrow}{\partial}g,$$

where $\Lambda = \left(\begin{smallmatrix} 0 & -1_n \\ 1_n & 0 \end{smallmatrix}\right)$. Then we have ([5–7])

Proposition 4 (Quantized Darboux theorem). *For any deformation quantization* $*$ *on a symplectic manifold* $(M, \{\ ,\ \})$, *locally the product* $*$ *is isomorphic to the Moyal product on* $C^\infty(U)[[\hbar]]$.

Vice versa, by gluing local Moyal algebras we obtain a deformation quantization.

Theorem 1 (Existence [6–8]). *For any symplectic manifold* $(M, \{\ ,\ \})$, *there exists a deformation quantization.*

Furthermore,

Theorem 2 ([9]). *For any Poisson manifold, there exists a deformation quantization.*

3.2. Non-formal extension

Now we consider non-formal extensions of star products. Here the situation is quite different from the formal extensions. For instance, we see

- The expansion

$$f *_\Lambda g = f \exp\left(\tfrac{i\hbar}{2}\overleftarrow{\partial} \Lambda \overrightarrow{\partial}\right) g$$

$$= fg + \tfrac{i\hbar}{2} f \left(\overleftarrow{\partial} \Lambda \overrightarrow{\partial}\right) g + \cdots + \tfrac{1}{n!}\left(\tfrac{i\hbar}{2}\right)^n f \left(\overleftarrow{\partial} \Lambda \overrightarrow{\partial}\right)^n g + \cdots$$

is not always convergent for functions f, g.
- Gluing of local star product algebra is not possible in general. So, we cannot consider a star product on a general Poisson manifold.

3.2.1. Star products on certain holomorphic function spaces.
We consider the star products on holomorphic functions on $\mathbb{C}^n$ ([10, 11]). Namely, for every positive number p we put

Definition 2.
$$\mathcal{E}_p = \{f \in Hol(\mathbb{C}^n) \,|\, |f|_{p,s} < \infty, \quad \forall s > 0\,\}$$
where $|f|_{p,s}$ is the semi-norm given by
$$|f|_{p,s} = \sup_{z \in \mathbb{C}^n} |f(z)| \exp\left(-s|z|^p\right).$$

Then it is easy to see $\mathcal{E}_p \subset \mathcal{E}_{p'}$, for $p < p'$. The space $\mathcal{E}_p$ is a commutative Fréchet algebra under the usual multiplication of functions. Moreover, the star product and the intertwiner are convergent for certain p. Namely, we have

Theorem 3.
1. *For* $0 < p \le 2$, $(\mathcal{E}_p, *_\Lambda)$ *is a Fréchet algebra. Moreover, for any* Λ' *having the same skew symmetric part as* Λ, $I_\Lambda^{\Lambda'} = \exp(\tfrac{i\hbar}{4}\partial K \partial)$ *with* $K = \Lambda' - \Lambda$ *is well defined intertwiner from* $(\mathcal{E}_p, *_\Lambda)$ *to* $(\mathcal{E}_p, *_{\Lambda'})$.
2. *For* $p > 2$, *the multiplication* $*_\Lambda : \mathcal{E}_p \times \mathcal{E}_{p'} \to \mathcal{E}_p$ *is well defined for* p' *such that* $\tfrac{1}{p} + \tfrac{1}{p'} = 2$, *and* $(\mathcal{E}_p, *_\Lambda)$ *is a* $\mathcal{E}_{p'}$*-bimodule.*

4. Star exponentials

Exponentials are very important in analysis, however they are hard to deal with in star product algebras in general.

For a star polynomial H_*, we want to define a star exponential $e_*^{t\frac{H_*}{i\hbar}}$. However, except special cases, the expansion $\sum_n \frac{t^n}{n!} \left(\frac{H_*}{i\hbar}\right)^n$ is not convergent, so we define a star exponential by means of a differential equation.

Definition 3. The star exponential $e_*^{t\frac{H_*}{i\hbar}}$ is given as solution of the following differential equation
$$\tfrac{d}{dt} F_t = H_* *_\Lambda F_t, \quad F_0 = 1.$$

4.1. Examples

We are interested in the star exponentials of linear, and quadratic polynomials. For these, we can solve the differential equation in explicit form. For simplicity, we consider matrices Λ with skew symmetric part $J_0 = \left(\begin{smallmatrix} 0 & -1 \\ 1 & 0 \end{smallmatrix}\right)$. We write $\Lambda = K + J_0$ where K is a complex symmetric matrix.

Linear case.

Proposition 5. *For* $l = \sum_j a_j u_j = \langle \boldsymbol{a}, \boldsymbol{u} \rangle$
$$e_*^{t(l/i\hbar)} = e^{t^2 \boldsymbol{a} K \boldsymbol{a}/4i\hbar} e^{t(l/i\hbar)}.$$

Quadratic case.

Proposition 6. *For* $Q_* = \langle \boldsymbol{u} A, \boldsymbol{u} \rangle_*$ *where* A *is a* $2m \times 2m$ *complex symmetric matrix,*
$$e_*^{t(Q_*/i\hbar)} = \frac{2^m}{\sqrt{\det(I - \kappa + e^{-2t\alpha}(I + \kappa))}} e^{\frac{1}{i\hbar} \langle \boldsymbol{u} \frac{1}{I - \kappa + e^{-2t\alpha}(I+\kappa)} (I - e^{-2t\alpha}) J, \boldsymbol{u} \rangle}$$
where $\kappa = K J_0$ *and* $\alpha = A J_0$.

5. Star trigonometric functions

In this section, we discuss a star product version of trigonometric functions according to Eisenstein (cf. Veil [1]).

We consider the star product for the simple case where the matrix is of the form $\Lambda = \left(\begin{smallmatrix} \rho & 0 \\ 0 & 0 \end{smallmatrix}\right)$ (cf. [11]). In this case, the product is essentially given on the space of polynomials of one variable. Namely, we consider functions $f(w), g(w)$ of complex variable w and the commutative star product $*_\tau$ with complex parameter τ such that
$$f(w) *_\tau g(w) = f(w) e^{\frac{\tau}{2} \overleftarrow{\partial}_w \overrightarrow{\partial}_w} g(w).$$

A direct calculation gives
$$\exp_{*_\tau} itw = \exp(itw - (\tau/4)t^2).$$

5.1. Star trigonometric functions ε_{k*}

Let us consider $\varepsilon_1(x) = \sum_e \frac{1}{x+\mu}$ where the symbol $\sum_e$ means the Eisenstein summation $\lim_{N\to+\infty} \sum_{\mu=-N}^{N}$, and $\varepsilon_k(x) = \sum_{\mu=-\infty}^{+\infty} \frac{1}{(x+\mu)^k}$ $(k \geq 2)$ (cf. Chapter II, [1]). These functions are periodic with respect to the real part of x, hence we have Fourier series expansions. For example, for x with positive imaginary part we see

$$\varepsilon_1(x) = \sum_{\ell=-\infty}^{+\infty} A_\ell e^{2\ell\pi i x} = \sum_{\ell=1}^{\infty} -2\pi i e^{2\ell\pi i x} + \pi i.$$

By means of the Fourier expansion, we define the star function ε_{1*} by

$$\varepsilon_{1*}(x) = \sum_{\ell=-\infty}^{+\infty} A_\ell \exp_{*_\tau} 2\ell\pi i(w + x).$$

We also define $\varepsilon_{k*}(x)$ $(k \geq 2)$ by the same way. The basic identities for $\varepsilon_k(x)$ are transferred into relations between the Fourier coefficients. Hence, the exponential law

$$\exp_{*_\tau} 2\ell\pi i(w + x) *_\tau \exp_{*_\tau} 2\ell\pi i(w + y) = \exp_{*_\tau} 2\ell\pi i(w + x + w + y)$$

gives that ε_{k*} satisfies the star version of the basic relations given in Chapter II of Weil's book [1].

5.2. Discussion

Using the star exponential, we have the inverse element of $(w + x + \mu)$ and hence we can define star versions of ε_k by replacing $\frac{1}{(x+\mu)^k}$ with $(w+x+\mu)_*^{-k}$ in $\varepsilon_k(x) = \sum_{\mu=-\infty}^{+\infty} \frac{1}{(x+\mu)^k}$ $(k \geq 2)$. We remark that for $\varepsilon_k(x)$ $(k \geq 2)$ the summation can be also taken as Eisenstein summation because of the strong convergence.

For $\Re\tau > 0$, the star exponential $\exp_{*_\tau} it(w + x + \mu)$ is rapidly decreasing with respect to the parameter t. Hence, it is easy to see the following: For complex parameter x and for every integer μ, the following integral converges to give the inverse element of $(w + x + \mu) *_\tau \cdots *_\tau (w + x + \mu)$ (k times) with respect to the product $*_\tau$

$$(w + x + \mu)_*^{-k} = \int_0^\infty (-i)^k \frac{t^{k-2}}{(k-1)!} \exp_{*_\tau} it(w + x + \mu)dt.$$

Now we define a star version of $\varepsilon_k(x)$ for $k \geq 1$ by

$$\varepsilon_{k*}(x) = \sum_e (w + x + \mu)_*^{-k}, \ (k \geq 1).$$

If we put $E_\mu = \int_0^{+\infty} \exp_{*_\tau} it(w + x + \mu)dt$, integration by parts yields

$$E_\mu = \frac{i}{x + \mu} - \frac{1}{x + \mu} w *_\tau E_\mu.$$

Then the inequality $|E_\mu| \leq \int_0^{+\infty} \exp\left(-\frac{\tau}{4}t^2 - t(|w| + |x|)\right)dt$ shows the convergence of $\varepsilon_{1*}(x) = \sum_e (w + x + \mu)_*^{-1} = \sum_e E_\mu$. As to $\varepsilon_{k*}(x)$ $(k \geq 2)$ we have the convergence by a similar manner.

The definitions of star functions for $\varepsilon_k(x)$ are given differently from the previous subsection. The relation between these are still unknown.

Acknowledgment

The third author expresses his great thanks to the organizers of the XXXI Workshop on the Geometric Methods in Physics in Białowieza for their hospitality.

References

[1] Weil, André, *Elliptic functions according to Eisenstein and Kronecker*. Reprint of the 1976 original. Classics in Mathematics, Springer-Verlag, Berlin, 1999.

[2] Omori, Hideki; Maeda, Yoshiaki; Miyazaki, Naoya; Yoshioka, Akira, *Orderings and Non-formal Deformation quantization*, Lett. Math. Phys. **82** (2007) 153–175.

[3] Bayen, F., Flato, M., Fronsdal, C., Lichnerowicz, A., Sternheimer, D.: *Deformation Theory and Quantization: I. Deformations of Symplectic Structures*. Ann. Phys. 111, 61–110 (1978).

[4] Bieliavsky, P.; Cahen, M.; Gutt, S. *Symmetric symplectic manifolds and deformation quantization*. Modern group theoretical methods in physics (Paris, 1995), 63–73, Math. Phys. Stud., 18, Kluwer, Acad. Publ., Dordrecht, 1995.

[5] Gutt, Simone. *Equivalence of deformations and associated *-products*. Lett. Math. Phys. 3 (1979), no. 4, 297–309.

[6] Fedosov, Boris V. *A simple geometrical construction of deformation quantization*. J. Differential Geom. 40 (1994), no. 2, 213–238.

[7] Omori, Hideki; Maeda, Yoshiaki; Yoshioka, Akira, *Weyl manifolds and deformation quantization*. Adv. Math. 85 (1991), no. 2, 224–255.

[8] De Wilde, Marc; Lecomte, Pierre B. A. *Existence of star-products and of formal deformations of the Poisson Lie algebra of arbitrary symplectic manifolds*. Lett. Math. Phys. 7 (1983), no. 6, 487–496.

[9] Kontsevich, Maxim. *Deformation quantization of Poisson manifolds*. Lett. Math. Phys. 66 (2003), no. 3, 157–216.

[10] Omori, Hideki; Maeda, Yoshiaki; Miyazaki, Naoya; Yoshioka, Akira, *Convergent star products on Fréchet linear Poisson algebras of Heisenberg type*. Global differential geometry: the mathematical legacy of Alfred Gray (Bilbao, 2000), 391–395, Contemp. Math., 288, Amer. Math. Soc., Providence, RI, 2001.

[11] Iida, Mari; Yoshioka, Akira, *Star products and applications*. J. Geom. Symmetry Phys. 20 (2010), 49–56.

Mari Iida, Chishu Tsukamoto and Akira Yoshioka
Department of Mathematics, Tokyo University of Science
Kagurazaka 1–3
Shinjyuku-ku
Tokyo 162 8601, Japan
e-mail: j1111701@ed.kagu.tus.ac.jp
        j1111609@ed.kagu.tus.ac.jp
        yoshioka@rs.kagu.tus.ac.jp

Geometric Methods in Physics. XXXI Workshop 2012
Trends in Mathematics, 47–63

Paragrassmann Algebras as Quantum Spaces Part I: Reproducing Kernels

Stephen Bruce Sontz

Abstract. Paragrassmann algebras are given a sesquilinear form for which one subalgebra becomes a Hilbert space known as the Segal–Bargmann space. This Hilbert space as well as the ambient space of the paragrassmann algebra itself are shown to have reproducing kernels. These algebras are not isomorphic to algebras of functions so some care must be taken in defining what "evaluation at a point" corresponds to in this context. The reproducing kernel in the Segal–Bargmann space is shown to have most, though not all, of the standard properties. These quantum spaces provide non-trivial examples of spaces which have a reproducing kernel but which are not spaces of functions.

Mathematics Subject Classification (2010). 46E22, 81R05.

Keywords. Reproducing kernel, quantum space, paragrassmann algebra.

1. Introduction

This paper is inspired in large measure by the work in [1] on paragrassmann algebras. We begin in Sections 1–4 by reviewing some of the material in [1], though sometimes re-working that presentation by using our own notation and sometimes by making mild generalizations. Please consult [1] for references to previous works on this topic in mathematics and physics. We note that the deformation parameter q in this paper is non-zero and complex, while in [1] it lies on the unit circle in the complex plane. But more importantly, the conjugation used here is different from that in [1]. So strictly speaking this paper treats topics not discussed in [1], though there are ideas in common. The core material of the paper starts in Section 5 where reproducing kernels are defined and discussed in the context of a Segal–Bargmann space that we define as a subalgebra of a paragrassmann algebra. The Segal–Bargmann (or coherent state) transform is introduced in Section 6, and its relation to the reproducing kernel is proved. We follow in Section 7 with a proof of the existence of the reproducing kernel in the full space of paragrassmann variables. In the last section we discuss some possible avenues for future research.

One of these possibilities, the definition and study of Toeplitz operators in this context, is the topic of a companion paper [2].

2. Preliminaries

In this article we take l to be an integer with $l \geq 2$. (N.B. Our parameter l corresponds to k' in [1].) We take the set $\{\theta, \overline{\theta}\}$ of two elements and consider the free algebra over the field of complex numbers $\mathbb{C}$ generated by this set. It is denoted by $\mathbb{C}\{\theta, \overline{\theta}\}$. In this paper all spaces are vector spaces over the field $\mathbb{C}$, and all algebras are *unital*, that is, have an identity element 1. Moreover, algebra morphisms map the identity 1 in the domain to 1 in the codomain.

Definition 1. *Let $l \geq 2$ be an integer and $q \in \mathbb{C} \setminus \{0\}$. The* paragrassmann algebra *$PG_{l,q}$ with* paragrassmann variables *θ and $\overline{\theta}$ is defined by*

$$PG_{l,q} = PG_{l,q}(\theta, \overline{\theta}) := \mathbb{C}\{\theta, \overline{\theta}\}/\langle \theta^l, \overline{\theta}^l, \theta\overline{\theta} - q\overline{\theta}\theta \rangle.$$

Here as usual the notation $\langle \cdot \rangle$ means the two-sided ideal generated by the elements listed inside the braces. In [1] only the particular value $q = e^{2\pi i/l}$ is considered. We let $\theta, \overline{\theta}$ also denote the quotients (i.e., equivalence classes) of these two elements in PG_l. We will be studying $PG_{l,q}$. The equation $\theta\overline{\theta} - q\overline{\theta}\theta = 0$ in $PG_{l,q}$ is called the *q-commutation relation*, while $\theta^l = 0$ and $\overline{\theta}^l = 0$ in $PG_{l,q}$ are called the *nilpotency conditions*. The algebra $PG_{l,q}$ could be viewed as 'classical' object in some sense even though it is also a 'quantum' object, that is, it is not commutative. Viewed as a classical space there is a quantization of it given by Toeplitz operators. (See [2].)

We will be using this index set throughout: $I_l = \{0, 1, \ldots, l-1\}$. When an index, say i, is given without an explicit index set, we assume $i \in I_l$. The *Segal–Bargmann (or holomorphic) space* is $\mathcal{B}_H = \mathcal{B}_H(\theta) := \mathrm{span}_{\mathbb{C}} \{\theta^i \,|\, i \in I_l\}$. Similarly, the *anti-Segal–Bargmann (or anti-holomorphic) space* is defined to be $\mathcal{B}_{AH} = \mathcal{B}_{AH}(\overline{\theta}) := \mathrm{span}_{\mathbb{C}} \{\overline{\theta}^i \,|\, i \in I_l\}$.

The Segal–Bargmann space is not only a vector subspace of $PG_{l,q}(\theta, \overline{\theta})$; it is also a subalgebra. Actually, it is a commutative subalgebra isomorphic to the truncated polynomial algebra $\mathbb{C}[\theta]/\langle \theta^l \rangle$. (Similarly, for the anti-Segal–Bargmann space.) The subspace $\mathcal{S} = \mathcal{B}_H + \mathcal{B}_{AH}$ of $PG_{l,q}$ plays a special role. Note that this is not a subalgebra.

We have two canonical bases of $PG_{l,q}$: The *Wick basis* $W = \{\overline{\theta}^i\theta^j \,|\, i, j \in I_l\}$ and the *anti-Wick basis* $AW = \{\theta^i\overline{\theta}^j \,|\, i, j \in I_l\}$. Here we follow [1] by saying that an expression with all factors of θ to the right (resp., left) of all factors of $\overline{\theta}$ is in *Wick* (resp., *anti-Wick*) order. In quantum physics (e.g., see [3]) the original definition is that an expression with all annihilation operators to the right of all creation operators is in Wick order. Since we have no way for identifying at this classical level which variable corresponds to annihilation, our present definition is a rather arbitrary choice whose only virtue is that it agrees with [1]. Clearly, we have $\dim_{\mathbb{C}} PG_{l,q} = l^2$.

The integral is a linear map $PG_{l,q} \to \mathbb{C}$ defined on the basis AW by

$$\iint d\theta \; (\theta^i \overline{\theta}^j) \; d\overline{\theta} := \delta_{i,l-1} \delta_{j,l-1},$$

where δ_{ab} is the Kronecker delta of the integers a, b. This is a Berezin type integral, by which is meant the only basis element with non-zero integral is the highest power element $\theta^{l-1} \overline{\theta}^{l-1}$.

3. Conjugation

We next introduce a conjugation (or $*$-operation) in $PG_{l,q}$ by expanding an arbitrary $f \in PG_{l,q}$ in the basis AW as $f = \sum_{i,j} f_{ij} \theta^i \overline{\theta}^j$, where the coefficients $f_{ij} \in \mathbb{C}$ are uniquely determined. Then we define the *conjugation* of f by $f^* := \sum_{i,j} f_{ij}^* \theta^j \overline{\theta}^i$. (The usual complex conjugate of $\lambda \in \mathbb{C}$ is denoted by λ^*.) This gives the expansion of f^* in the same basis AW. This is an anti-linear operation. It also immediately follows that $f^{**} = f$, that is, this operation is an involution. Clearly, $(\theta^i \overline{\theta}^j)^* = \theta^j \overline{\theta}^i$.

We note, without giving the proof, that the algebra $PG_{l,q}$ is a $*$-algebra, that is we have $(fg)^* = g^* f^*$ for all $f, g \in PG_{l,q}$, if and only if $q \in \mathbb{R} \setminus \{0\}$.

Remarks. In earlier preprint versions of this paper, I assumed that the conjugation made $PG_{l,q}$ into a $*$-algebra. So I only considered the case $q \in \mathbb{R} \setminus \{0\}$. This is an unnecessary restriction. Here we consider the more general case $q \in \mathbb{C} \setminus \{0\}$. The only way that the conjugation enters into the subsequent theory is through the definition of the sesquilinear form given in the next section. And the properties that we will use of that sesquilinear form do not require that the conjugation gives $PG_{l,q}$ a $*$-algebra structure.

I thank R. Fresneda [4] for clarifying for me that the definition of the conjugation used in [1] is the anti-linear extension of $(\theta^i \overline{\theta}^j)^* = \overline{\theta}^i \theta^j$. (This does not give a $*$-algebra.) The fact that we will be using a different definition of the conjugation means that we are considering structures that are strictly speaking distinct from those discussed in [1]. Nonetheless, there will still be things in common with the approach in [1]. What is behind these different definitions of conjugation are different ways of dealing with an ordering problem.

Note that θ and $\overline{\theta}$ are a pair of conjugate complex variables, that is, $\theta^* = \overline{\theta}$ and $\overline{\theta}^* = \theta$ and the intersection of the two subalgebras generated by θ and by $\overline{\theta}$, respectively, is simply the smallest it could possibly be: $\mathbb{C}1$. This is in close analogy with the pair of conjugate complex variables z and $\overline{z}$ (which are *functions* $\mathbb{C} \to \mathbb{C}$ and *not* points in $\mathbb{C}$) as studied in complex analysis, where z generates the algebra of holomorphic functions on $\mathbb{C}$ while $\overline{z}$ generates the algebra of anti-holomorphic functions. Also the intersection of these algebras consists of the constant functions. Notice how the non-commutative geometry approach of viewing elements of algebras (in this case functions) as the primary objects of study clarifies a common confusion even in this commutative example where one might not otherwise understand how the complex plane (whose complex dimension is one) can support two independent complex variables, neither of which is more 'fundamental' than the other. This short discussion motivates the definition of a *variable* as any element in a unital algebra that is not a scalar multiple of the identity element. And then a pair of *complex variables* in any unital ∗-algebra is defined as any pair of conjugate variables which generate subalgebras with intersection $\mathbb{C}1$.

4. Sesquilinear form

Much of the material in this section comes from the paper [1], though we define a more general sesquilinear form. We want to introduce a sesquilinear form on $PG_{l,q}$ in order to turn it into something like an L^2 space. We start with any element (a 'positive weight') in $PG_{l,q}$ of the form

$$w = w(\theta, \overline{\theta}) = \sum_{m \in I_l} w_{l-1-m} \theta^m \overline{\theta}^m, \tag{1}$$

where $w_n > 0$ for all $n \in I_l$. The strange looking way of writing the sub-index on the right side of this equation will be justified later on. In [1] the authors take $w_n = [n]_q!$ which is a q-deformed factorial of the integer n. In any case, this couples the 'weight' factors w_n with the deformation parameter q, which itself is coupled in [1] with the nilpotency power l. We have preferred to keep all of these parameters decoupled from one and other.

Take $f = f(\theta, \overline{\theta})$ and $g = g(\theta, \overline{\theta})$ in $PG_{l,q}$. Informally, we would like to define the *sesquilinear form* or *inner product* as in [1] by

$$\langle f, g \rangle_w := \iint d\theta : f(\theta, \overline{\theta})^* g(\theta, \overline{\theta}) w(\theta, \overline{\theta}) : d\overline{\theta}, \tag{2}$$

where $:\,:$ is the *anti-Wick* (or *anti-normal*) *ordering*, that is, put all θ's to the left and all $\overline{\theta}$'s to the right. However, the anti-Wick ordering is only well defined on the space $\mathbb{C}\{\theta, \overline{\theta}\}$ and does not pass to its quotient space $PG_{l,q}(\theta, \overline{\theta})$. By the formal expression (2) we really mean

$$\langle f, g \rangle_w := \sum_{m \in I_l} w_{l-1-m} \iint d\theta \, \theta^m : f(\theta, \overline{\theta})^* :: g(\theta, \overline{\theta}) : \overline{\theta}^m \, d\overline{\theta} \in \mathbb{C}, \tag{3}$$

where the *anti-Wick product* $: \cdot :: \cdot :$ is defined as the $\mathbb{C}$-bilinear extension of $: \theta^a \overline{\theta}^b :: \theta^c \overline{\theta}^d : \equiv \theta^{a+c} \overline{\theta}^{b+d}$ for pairs of basis elements in AW. Clearly the expression in (3) is anti-linear in f and linear in g. As the reader can show, the form (3) is complex symmetric, that is we have that $\langle f, g \rangle_w^* = \langle g, f \rangle_w$. We also define $\|f\|_w^2 := \langle f, f \rangle_w$. Now $\langle f, f \rangle_w = \langle f, f \rangle_w^*$ is real, but negative for some f's. So, in general, we are not defining $\|f\|_w$. Let $\theta^a \overline{\theta}^b$ and $\theta^c \overline{\theta}^d$ be arbitrary elements in the basis AW. Then:

$$\langle \theta^a \bar{\theta}^b, \theta^c \bar{\theta}^d \rangle_w = \iint d\theta : (\theta^a \bar{\theta}^b)^* \theta^c \bar{\theta}^d w(\theta, \bar{\theta}) : d\bar{\theta}$$

$$= \sum_{n \in I_l} w_{l-1-n} \iint d\theta\; \theta^n : \theta^b \bar{\theta}^a :: \theta^c \bar{\theta}^d : \bar{\theta}^n\; d\bar{\theta}$$

$$= \sum_{n} w_{l-1-n} \iint d\theta\; \theta^{b+c+n} \bar{\theta}^{a+d+n} d\bar{\theta}$$

$$= \sum_{n} w_{l-1-n}\, \delta_{b+c+n,l-1}\, \delta_{a+d+n,l-1} \quad \text{(Kronecker deltas)}$$

$$= \begin{cases} w_{b+c} = w_{a+d} & \text{if } b + c = a + d \leq l - 1 \\ 0 & \text{otherwise} \end{cases}$$

$$= \delta_{a+d,b+c}\, \chi_l(a+d)\, w_{a+d} \in \mathbb{R}, \tag{4}$$

where χ_l is the characteristic function of I_l.

Now there are always pairs such that $(a,b) \neq (c,d)$ but $b + c = a + d \leq l - 1$. So the basis AW is not orthogonal. But $\langle \theta^a \bar{\theta}^b, \theta^c \bar{\theta}^d \rangle_w = 0$ if $a - b \neq c - d$. It follows that $\{1, \theta, \theta^2, \ldots, \theta^{l-1}, \bar{\theta}, \bar{\theta}^2, \ldots, \bar{\theta}^{l-1}\}$ is orthogonal. Also, from (4) we immediately get $\|\theta^a \bar{\theta}^b\|_w^2 = \chi_l(a+b)\, w_{a+b}$. So there exists $f \in PG_{l,q}$ with $f \neq 0$ and $\|f\|_w^2 = 0$. We also get $\|\theta^a\|_w^2 = w_a$ and $\|\bar{\theta}^b\|_w^2 = w_b$. The 'nice' subindices of the w's in these identities are the reason we took the unusual looking convention for the subindices in the definition (1) of $w(\theta, \bar{\theta})$.

We now define $\phi_n(\theta, \bar{\theta}) := w_n^{-1/2} \theta^n$ for every $n \in I_l$. We also write $\phi_n(\theta) = w_n^{-1/2} \theta^n$, since this element does not 'depend' on $\bar{\theta}$, that is, it lies in the subalgebra generated by θ alone. These vectors clearly form an orthonormal basis of $\mathcal{B}_H$. Note that $\phi_n^*(\theta, \bar{\theta}) = (\phi_n(\theta, \bar{\theta}))^* = w_n^{-1/2} \bar{\theta}^n$ using $w_n > 0$. We also write this element as $\phi_n^*(\bar{\theta})$.

5. Reproducing kernel for the Segal–Bargmann space

It is not possible to find an algebra of complex-valued functions which is an iso-morphic copy of the commutative algebra $\mathcal{B}_H$. We can see this is so by simply noting that $\theta \in \mathcal{B}_H$ satisfies $\theta \neq 0$ since $l \geq 2$ and is nilpotent, namely, $\theta^l = 0$. But no non-zero complex-valued function is nilpotent. Similarly, the commutative algebra $\mathcal{B}_{AH}$ is not isomorphic to an algebra of functions.

Nonetheless $\mathcal{B}_H$ and $\mathcal{B}_{AH}$ are reproducing kernel Hilbert spaces, properly un-derstood. The classical theory of reproducing kernel Hilbert spaces whose elements are functions goes back to the seminal works of Aronszajn and Bergman in the 20th century. (See [5] and [6].) Here we start with $\mathcal{B}_H$, the Segal–Bargmann space, which we will now write as $\mathcal{B}_H(\theta)$ to indicate that the paragrassmann variable in this space is θ. First, let us note that when we write an arbitrary element in this

Segal–Bargmann space uniquely as

$$f(\theta) = \sum \lambda_j \theta^j,$$

where $\lambda_j \in \mathbb{C}$ for all $j \in I_l$, this really can be interpreted as a function of θ. In fact, if we let $f(x) = \sum_{j=0}^{N} \beta_j x^j \in \mathbb{C}[x]$ be an arbitrary polynomial in x, an indeterminant, then we can define a *functional calculus* (where the 'functions' are polynomials) of any element $a \in \mathcal{B}_H(\theta)$ precisely by *defining* $f(a)$ to be $\sum_{j=0}^{N} \beta_j a^j$. Even if $\deg f < l - 1$, we will take $N \geq l - 1$. If we now take $\beta_j = \lambda_j$ for all $j \in I_l$ and any value whatsoever for β_j for $j \geq l$, then $f(\theta)$ is simply the arbitrary element $\sum_j \lambda_j \theta^j$ considered above. The mapping from $\mathbb{C}[x]$, the algebra of polynomials $f(x)$ in x, to $\mathcal{B}_H(\theta)$ given by $f(x) \mapsto f(\theta)$ is clearly an algebra morphism that is surjective. This is standard material, but it aids us in considering $f(\theta) = \sum_{j \in I_l} \lambda_j \theta^j$ in $\mathcal{B}_H(\theta)$ and its corresponding element $f(\eta) = \sum_{j \in I_l} \lambda_j \eta^j$ in $\mathcal{B}_H(\eta)$ for another paragrassmann variable η. We provide $PG_{l,q}(\eta, \overline{\eta})$ and its subspace $\mathcal{B}_H(\eta)$ with essentially the same inner product as above, simply replacing $\theta, \overline{\theta}$ with $\eta, \overline{\eta}$ everywhere.

Also, we wish to emphasize that this functional calculus is what replaces in this context the concept of 'evaluation at a point' in the usual theory of reproducing kernel Hilbert spaces of functions.

We would like to establish for every $f(x) \in \mathbb{C}[x]$ the *reproducing formula*

$$f(0) = \langle K(\theta, \eta), f(\eta) \rangle_w, \tag{5}$$

where the inner product here is roughly speaking taken with respect to the variable η (that is, basically in $\mathcal{B}_H(\eta)$; more on this later) and where

$$K(\theta, \eta) \in \mathcal{B}_{AH}(\overline{\theta}) \otimes \mathcal{B}_H(\eta).$$

This last condition says that the *reproducing kernel* $K(\theta, \eta)$ is holomorphic in η and anti-holomorphic in θ. This condition is in analogy with the theory of reproducing kernel functions in holomorphic function spaces. (See [7], for example.) The reader should note that we are using the notation $K(\theta, \eta)$, which is analogous to the notation in the classical theory of reproducing kernel Hilbert spaces. If we had been consistent with our own conventions, we would have denoted this as $K(\overline{\theta}, \eta)$.

Now the unknown we have to solve for is the kernel $K(\theta, \eta)$. We write

$$K(\theta, \eta) = \sum_{i,j} a_{ij} \overline{\theta}^i \otimes \eta^j, \tag{6}$$

an arbitrary element in $\mathcal{B}_{AH}(\overline{\theta}) \otimes \mathcal{B}_H(\eta)$, and see what are the conditions that the reproducing formula (5) imposes on the coefficients $a_{ij} \in \mathbb{C}$. We take an arbitrary element $f(\theta) = \sum_k \lambda_k \theta^k \in \mathcal{B}_H(\theta)$. So we have the corresponding element $f(\eta) = \sum_k \lambda_k \eta^k \in \mathcal{B}_H(\eta)$. We then calculate out the right side of equation (5) and get

$$\langle K(\theta, \eta), f(\eta) \rangle_w = \left\langle \sum_{i,j} a_{ij} \, \bar{\theta}^i \otimes \eta^j, f(\eta) \right\rangle_w = \sum_{i,j} a_{ij}^* \, \langle \eta^j, f(\eta) \rangle_w \, \theta^i$$

$$= \sum_{i,j} a_{ij}^* \, \langle \eta^j, \sum_k \lambda_k \eta^k \rangle_w \, \theta^i = \sum_{i,j} a_{ij}^* \sum_k \lambda_k \langle \eta^j, \eta^k \rangle_w \, \theta^i$$

$$= \sum_{i,j} a_{ij}^* \sum_k \lambda_k \delta_{j,k} w_j \, \theta^i = \sum_{i,j} a_{ij}^* \, \lambda_j w_j \, \theta^i = \sum_i \left(\sum_j w_j a_{ij}^* \, \lambda_j \right) \theta^i.$$

(Note that the second equality is nothing other than the promised, rigorous definition of the inner product in (5).) Now we want this to be equal to $f(\theta) = \sum_i \lambda_i \theta^i$ for *all* $f(\theta)$ in $\mathcal{B}_H(\theta)$, that is, for all vectors $\{\lambda_i | i \in I_l\}$ in $\mathbb{C}^l$. So the matrix $(w_j a_{ij}^*)$ has to act as the identity on $\mathbb{C}^l$ and thus has to be the identity matrix (δ_{ij}), where δ_{ij} is the Kronecker delta. The upshot is that $a_{ij} = \delta_{ij}/w_j$ does the job, and nothing else does. So, substituting in equation (6) we see that

$$K(\theta, \eta) = \sum_j \frac{1}{w_j} \bar{\theta}^j \otimes \eta^j \tag{7}$$

is the unique reproducing kernel 'function.'

And, as one might expect, an abstract argument also shows that reproducing kernels are unique. For suppose that $K_1(\theta, \eta) \in \mathcal{B}_{AH}(\bar{\theta}) \otimes \mathcal{B}_H(\eta)$ is also a reproducing kernel. Then the standard argument makes sense in this context, namely,

$$K_1(\rho, \eta) = \langle K(\eta, \theta), K_1(\rho, \theta) \rangle_w = \langle K_1(\rho, \theta), K(\eta, \theta) \rangle_w^*$$
$$= K(\eta, \rho)^* = K(\rho, \eta), \tag{8}$$

where ρ is another paragrassmann variable. The astute reader will have realized that these innocent looking formulas require a bit of justification, including a rigorous definition of the inner product in this context. We leave most of these details to the reader. But, for example, in the last equality we are using the standard natural isomorphisms

$$(\mathcal{B}_{AH}(\bar{\eta}) \otimes \mathcal{B}_H(\rho))^* \cong \mathcal{B}_H(\eta) \otimes \mathcal{B}_{AH}(\bar{\rho}) \cong \mathcal{B}_{AH}(\bar{\rho}) \otimes \mathcal{B}_H(\eta).$$

The relation $K(\eta, \rho)^* = K(\rho, \eta)$ is also well known in the classical theory. For example, Eq. (1.9a) in [8] is an analogous result. Also, by putting K_1 equal to K in the first equality of (8) we get another result that is analogous to a result in the classical case. This is

$$K(\rho, \eta) = \langle K(\eta, \theta), K(\rho, \theta) \rangle_w, \tag{9}$$

which is usually read as saying that two 'evaluations' of the reproducing kernel (with the holomorphic variable, here θ, being the same in the two) have an inner product with respect to that holomorphic variable that is itself an 'evaluation' of the reproducing kernel. See [7], Theorem 2.3, part 4, for the corresponding identity in the context of holomorphic function spaces. Next we put $\eta = \rho$ to get

$$K(\rho, \rho) = \langle K(\rho, \theta), K(\rho, \theta) \rangle_w = ||K(\rho, \theta)||_w^2 = ||K(\rho, \cdot)||_w^2. \tag{10}$$

And again this last formula is analogous to a result in the classical theory. Later on, we will discuss the positivity of $K(\rho, \rho)$.

We note that this theory is consistent with many expectations coming from the usual theory of reproducing kernels. As we have seen, $K(\theta, \eta)$ is the only element in the appropriate space with the reproducing property. Also, we clearly have the following well-known relation with the elements of the standard orthonormal basis, namely that

$$K(\theta, \eta) = \sum_j \phi_j(\theta)^* \otimes \phi_j(\eta). \tag{11}$$

This follows from (7) and the definition of ϕ_j. Suppose that $\psi_j(\eta)$ for $j \in I_l$ is another orthonormal basis of $\mathcal{B}_H(\eta)$. It then follows from $\langle f^*, g^* \rangle_w = \langle f, g \rangle_w^*$ for all $f, g \in PG_{l,q}$ (which we leave to the reader as another exercise) that $\psi_j(\theta)^*$ is an orthonormal basis of $\mathcal{B}_{AH}(\theta)$. Then by a standard argument in linear algebra we obtain from (11) that

$$K(\theta, \eta) = \sum_j \psi_j(\theta)^* \otimes \psi_j(\eta).$$

This formula is the analogue of an identity in the classical theory. (For example see [8], Eq. (1.9b) or [7], Theorem 2.4.)

One result from the classical theory of reproducing kernel Hilbert spaces of functions seems to have no analogue in this context: the point-wise estimate that one gets by applying the Cauchy–Schwarz inequality to the reproducing formula. See [7], Theorem 2.3, part 5 for this result in the holomorphic function setting.

The reproducing formula in the usual theory of reproducing kernel Hilbert spaces is interpreted as saying that the Dirac delta function is realized as integration against a smooth kernel function. Since the inner product in the reproducing formula (5) is a Berezin integral, we can say that the Dirac delta function in this context is realized as Berezin integration against the 'smooth function' $K(\theta, \eta)$. But what is the Dirac delta function in this context? We recall that $f(\theta)$ in this paper is merely convenient notation for an element in a Hilbert space. We are not evaluating f at a point θ in its domain. It seems that the simplest interpretation for the Dirac delta $\delta_{\eta \to \theta}$ in the present context is that it acts on $f(\eta)$ to produce $f(\theta)$, namely that it is a *substitution* operator. We use a slightly different notation for this Dirac delta in part because it is different from the usual Dirac delta and also to distinguish it from the Kronecker delta function that we have been using. An appropriate definition and notation would be $\delta_{\eta \to \theta}[f(\eta)] := f(\theta)$ so that $\delta_{\eta \to \theta} : \mathcal{B}_H(\eta) \to \mathcal{B}_H(\theta)$ is an isomorphism of Hilbert spaces and of algebras. In particular, with this way of defining the Dirac delta we do not get a functional acting on a space of test functions. However, the left side of equation (5) is $\delta_{\eta \to \theta}[f(\eta)]$.

Notice that in this approach

$$\delta_{\eta\to\theta} \in \mathrm{Hom}_{\mathrm{Vect}_{\mathbb{C}}}(\mathcal{B}_H(\eta), \mathcal{B}_H(\theta)) \cong \mathcal{B}_H(\theta) \otimes \mathcal{B}_H(\eta)' \cong \mathcal{B}_H(\theta) \otimes \mathcal{B}_{AH}(\overline{\eta})$$

using standard notation from category theory and viewing the space $\mathcal{B}_{AH}(\overline{\eta})$ as the dual space $\mathcal{B}_H(\eta)'$. This does agree with our previous analysis where we had that $K(\theta,\eta) \in \mathcal{B}_{AH}(\overline{\theta}) \otimes \mathcal{B}_H(\eta)$, because the inner product in equation (5) is anti-linear in its first argument. So that previous analysis simply identifies which element in $\mathcal{B}_H(\theta) \otimes \mathcal{B}_{AH}(\overline{\eta})$ is the Dirac delta, namely

$$\delta_{\eta\to\theta} = \sum_j \frac{1}{w_j}\theta^j \otimes \overline{\eta}^j = \sum_j \phi_j(\theta) \otimes \phi_j^*(\overline{\eta}),$$

which is analogous to a standard formula for the Dirac delta.

We now try to see to what extent this generalized notion of a reproducing kernel has the positivity properties of a usual reproducing kernel. First, we 'evaluate' the diagonal 'elements' $K(\theta,\theta)$ in $\mathcal{B}_{AH}(\overline{\theta}) \otimes \mathcal{B}_H(\theta) \cong \mathcal{B}_H(\theta)' \otimes \mathcal{B}_H(\theta)$ getting

$$K(\theta,\theta) = \sum_j \frac{1}{w_j}\overline{\theta}^j \otimes \theta^j = \sum_j \frac{1}{w_j}(\theta^j)^* \otimes \theta^j.$$

This element is positive by using the usual definition of a positive element in a $*$-algebra, *provided* we adequately define the $*$-operation in $\mathcal{B}_{AH}(\overline{\theta}) \otimes \mathcal{B}_H(\theta) \cong \mathcal{B}_H(\theta)' \otimes \mathcal{B}_H(\theta)$. Given that this exercise has been done and without going into further details we merely comment that we can identify $\mathcal{B}_H(\theta)' \otimes \mathcal{B}_H(\theta)$ with $\mathcal{L}(\mathcal{B}_H(\theta))$, the vector space (and $*$-algebra) of all of the linear operators from the Hilbert space $\mathcal{B}_H(\theta)$ to itself. Under this identification the positive elements of $\mathcal{B}_H(\theta)' \otimes \mathcal{B}_H(\theta)$ correspond exactly to the positive operators in $\mathcal{L}(\mathcal{B}_H(\theta))$, and we have that

$$K(\theta,\theta) = \sum_j \frac{1}{w_j}|\theta^j\rangle\langle\theta^j| = \sum_j |\phi_j(\theta)\rangle\langle\phi_j(\theta)| \in \mathcal{L}(\mathcal{B}_H(\theta)),$$

using the Dirac bra and ket notation. But this is clearly a positive linear operator since

$$\sum_j |\phi_j(\theta)\rangle\langle\phi_j(\theta)| = I_{\mathcal{B}_H(\theta)} \equiv I \geq 0,$$

the identity operator on $\mathcal{B}_H(\theta)$. The upshot is that $K(\theta,\theta) = I$. But $\|K(\theta,\cdot)\|_w^2 = K(\theta,\theta)$ as we showed in (10). So, $\|K(\theta,\cdot)\|_w^2 = I$ as well.

Next, we take a finite number of pairs of paragrassmann variables $\theta_n, \overline{\theta}_n$ and a finite sequence of complex numbers λ_n, where $n = 1, \ldots, N$. Then we investigate the positivity of the usual expression, that is, we consider

$$\sum_{n,m=1}^N \lambda_n^* \lambda_m K(\theta_n, \theta_m) = \sum_{n,m=1}^N \lambda_n^* \lambda_m \sum_{j\in I_l} \frac{1}{w_j}\overline{\theta}_n^j \otimes \theta_m^j$$

$$= \sum_{j\in I_l} \frac{1}{w_j} \sum_{n,m=1}^N \lambda_n^* \lambda_m (\theta_n^j)^* \otimes \theta_m^j = \sum_{j\in I_l} \frac{1}{w_j} \left(\sum_{n=1}^N \lambda_n \theta_n^j\right)^* \otimes \left(\sum_{m=1}^N \lambda_m \theta_m^j\right),$$

which is a positive element in the appropriate $*$-algebra and therefore also corresponds to a positive linear map. A detail here is that one has to define a $*$-algebra where the sums $\sum_n \lambda_n \theta_n^j$ makes sense for all $j \in I_l$. But this is a straightforward exercise left to the reader.

It now is natural to ask whether this procedure can be reversed, as we know is the case with the usual theory of reproducing kernel functions. That is to say, can we start with a mathematical object, call it K, that has the properties (in particular, the positivity) of a reproducing kernel in this context and produce from it a reproducing kernel Hilbert space that has that given object K as its reproducing kernel? This seems not to be possible, at least not using an argument based on the identity (9) as is done in the classical case. It turns out that (9) is only analogous to the identity in the classical case, since it says something decidedly different given that the left side of it is not a complex number. In the classical theory the operation of evaluation at a point gives a complex number. But in this context the evaluation of a function at a variable in an algebra gives another element in that same algebra. In the argument in the classical case, one uses the analogue of (9) to define an inner product (on a set of functions) having available only the candidate mathematical object K (in that case a function of two variables). But one can not use (9) directly to define in this context a complex-valued inner product. Perhaps an inverse procedure can be found, but it will have to differ somewhat from the procedure in the classical case.

We now come back to the question of finding an analogy to a point-wise bound for $f(\theta) \in \mathcal{B}_H(\theta)$. Suppose that $f(\theta) \neq 0$ and put $u_0 := \|f(\theta)\|_w^{-1} f(\theta)$, a unit vector in $\mathcal{B}_H(\theta)$. Extend this to an orthonormal basis u_j of $\mathcal{B}_H(\theta)$ for $j \in I_l$. So we get the operator inequality $|u_0\rangle\langle u_0| \leq \sum_j |u_j\rangle\langle u_j| = I_{\mathcal{B}_H(\theta)}$. Then we obtain

$$|f(\theta)\rangle\langle f(\theta)| = \|f(\theta)\|_w^2 \, |u_0\rangle\langle u_0| \leq \|f(\theta)\|_w^2 \, I_{\mathcal{B}_H(\theta)} = \|f(\theta)\|_w^2 \, \|K(\theta,\cdot)\|_w^2 \quad (12)$$

which is an operator inequality involving positive operators. For $f(\theta) = 0$ this inequality is trivially true (and is actually an equality). The point here is that (12) has some resemblance to the point-wise estimate in the classical case. (See [7].) The left side of (12) can be considered a type of 'outer product' of $f(\theta)$ with itself.

An entirely analogous argument shows that $\mathcal{B}_{AH}$ is a reproducing kernel Hilbert space with reproducing kernel K_{AH} given by $K_{AH}(\theta, \eta) = K(\theta, \eta)^*$. Finally, we note that the results of this section depend on the weight $w(\theta, \overline{\theta})$ and are independent of the value of the parameter $q \in \mathbb{C} \setminus \{0\}$.

6. Coherent states and the Segal–Bargmann transform

We now introduce the coherent state quantization of Gazeau. (See [9] for example.) We let $\mathcal{H}$ be any complex Hilbert space of dimension l and choose any orthonormal basis of $\mathcal{H}$, which we will denote as e_n for $n \in I_l$. While θ is a complex variable, it does not 'run' over a domain of values, say in some phase space. So the coherent state $|\theta\rangle$ we are about to define is one object, not a parameterized family of objects.

Actually, we define two *coherent states* corresponding to the variable θ:

$$|\theta\rangle := \sum_{n \in I_l} \phi_n(\theta) \otimes e_n \in \mathcal{B}_H(\theta) \otimes \mathcal{H}, \quad \langle\theta| := \sum_{n \in I_l} \phi_n^*(\bar{\theta}) \otimes e_n' \in \mathcal{B}_{AH}(\theta) \otimes \mathcal{H}'.$$

Here $\mathcal{H}'$ denotes the dual space of all linear functionals on $\mathcal{H}$, and e_n' is its orthonormal basis that is dual to e_n, where $n \in I_l$. We refrain from using the super-script $*$ for the dual objects in order to avoid confusion in general with the conjugation. We are following Gazeau's conventions here. (See [9].) As noted in [1] these objects give a resolution of the identity using the Berezin integration theory, and so this justifies calling them coherent states. Now we have

$$\langle\theta\,|\,\eta\rangle = \sum_{j,k} \phi_j^*(\bar{\theta}) \otimes \phi_k(\eta) \langle e_j', e_k\rangle = \sum_{j,k} \phi_j^*(\bar{\theta}) \otimes \phi_k(\eta)\, \delta_{j,k}$$

$$= \sum_j \phi_j^*(\bar{\theta}) \otimes \phi_j(\eta) = K(\theta, \eta).$$

Here the inner product (or pairing) of the coherent states $\langle\theta|$ and $|\eta\rangle$ is simply *defined* by the first equality, which is a quite natural definition. So the inner product of two coherent states gives us the reproducing kernel. This is a result that already appears in the classical theory. (See [8], Eq. (1.9a) for a formula of this type.)

Since we have coherent states, we can define the corresponding *Segal–Bargmann* (or *coherent state*) transform in the usual way. Essentially the same transform was discussed in [1], where it is denoted by $\mathcal{W}$. This will be a unitary isomorphism $C : \mathcal{H} \to \mathcal{B}_H(\theta)$. (Recall that we are considering an abstract Hilbert space $\mathcal{H}$ of dimension l with orthonormal basis e_n.) We define C for all $\psi \in \mathcal{H}$ by

$$C\psi(\theta) := \langle\,\langle\theta|, \psi\rangle.$$

Note that $\langle\theta| \in \mathcal{B}_{AH}(\theta)\otimes\mathcal{H}'$. So the outer bracket $\langle\cdot, \cdot\rangle$ here refers to the pairing of the dual space $\mathcal{H}'$ with $\mathcal{H}$. As noted earlier $\langle\theta|$ is one object, not a parameterized family of objects. So $C\psi(\theta)$ is one element in the Hilbert space $\mathcal{B}_H(\theta)$. And as usual the θ in the notation is a convenience for reminding us what the variable is; it is not a point at which we are evaluating $C\psi$. Actually, in this context where it is understood that the variable under consideration is θ, $C\psi(\theta)$ and $C\psi$ are two notations for one and the same object.

Substituting the definition for $\langle\theta|$ gives

$$C\psi(\theta) = \langle\,\langle\theta|, \psi\rangle = \left\langle \sum_{n \in I_l} \phi_n^*(\bar{\theta}) \otimes e_n', \psi \right\rangle$$

$$= \sum_{n \in I_l} \langle\phi_n^*(\bar{\theta}) \otimes e_n', \psi\rangle = \sum_{n \in I_l} \langle e_n', \psi\rangle\, \phi_n(\theta).$$

So by taking ψ to be e_j we see that the Segal–Bargmann transform $C : \mathcal{H} \to \mathcal{B}$ is the (unique) linear transformation mapping the orthonormal basis e_j to the orthonormal basis $\phi_j(\theta)$. This shows that C is indeed a unitary isomorphism. Also the above shows that this is an 'integral kernel' operator with kernel given by the coherent state $\langle\theta| = \sum_j \phi_j^*(\bar{\theta}) \otimes e_j'$. This is also a well-known relation in the Segal–Bargmann theory. Equation (2.10b) in [8] is this sort of formula.

The Segal–Bargmann transform of the coherent state $|\eta\rangle$ is given by

$$C|\eta\rangle(\theta) = \langle\langle\theta|, |\eta\rangle\rangle = \langle\theta|\eta\rangle = K(\theta,\eta).$$

So, this says that the Segal–Bargmann transform of a coherent state is the reproducing kernel. This is also a type of relation known in the classical context of [8], where it appears as Eq. (2.8).

There are also two coherent states corresponding to the other variable $\bar{\theta}$, which also give a resolution of the identity. And $\langle\bar{\theta}|\bar{\eta}\rangle$ gives the reproducing kernel for the anti-holomorphic Segal–Bargmann space, and so forth.

7. Reproducing kernel for the paragrassmann space

Consider $PG_{l,q}(\theta,\bar{\theta})$ with $l = 2$, a 'fermionic' case. This is a non-commutative algebra of dimension 4. The basis AW is $\{1, \theta, \bar{\theta}, \theta\bar{\theta}\}$. The weight 'function' in this case is

$$w(\theta,\bar{\theta}) = w_1 + w_0\theta\bar{\theta}.$$

(We are using our convention for the sub-indices. See equation (1).) So if this space has a reproducing kernel $K_{PG}(\theta,\bar{\theta},\eta,\bar{\eta})$, it must satisfy

$$f(\theta,\bar{\theta}) = \langle K_{PG}(\theta,\bar{\theta},\eta,\bar{\eta}), f(\eta,\bar{\eta})\rangle_w$$

for all $f = f(\theta,\bar{\theta}) \in PG_{2,q}$. (Notice that here, and throughout this section, we use a functional calculus of a pair of *non-commuting* variables in place of evaluation at a point. Actually, we can define a functional calculus for all $f \in \mathbb{C}\{\theta,\bar{\theta}\}$.) This last equation in turn is equivalent to

$$\theta^i\bar{\theta}^j = \langle K_{PG}(\theta,\bar{\theta},\eta,\bar{\eta}), \eta^i\bar{\eta}^j\rangle_w$$

for all $i, j \in I_2 = \{0, 1\}$. Substituting the general element

$$K_{PG}(\theta,\bar{\theta},\eta,\bar{\eta}) = \sum_{abcd} k_{abcd}\theta^a\bar{\theta}^b \otimes \eta^c\bar{\eta}^d$$

(which lies in a space of dimension 2^4) into the previous equation gives

$$\theta^i\bar{\theta}^j = \sum_{abcd} k_{abcd}^*\theta^b\bar{\theta}^a\langle\eta^c\bar{\eta}^d, \eta^i\bar{\eta}^j\rangle_w = \sum_{ab}\left(\sum_{cd} k_{abcd}^*G_{cdij}\right)\theta^b\bar{\theta}^a,$$

where $G_{cdij} = \langle\eta^c\bar{\eta}^d, \eta^i\bar{\eta}^j\rangle_w$. So the unknown coefficients $k_{abcd} \in \mathbb{C}$ must satisfy

$$\sum_{cd} k_{abcd}^*G_{cdij} = \delta_{(b,a),(i,j)},$$

where this Kronecker delta is 1 if the ordered pair (b, a) is equal to the ordered pair (i, j) and otherwise is 0. So everything comes down to showing the invertibility of the matrix $G = (G_{cdij})$, whose rows (resp., columns) are labelled by $\eta^c\bar{\eta}^d$ (resp.,

$\eta^i \overline{\eta}^j \rangle_w)$ where the ordered pairs $(c,d), (i,j) \in I_2 \times I_2$. So G is a 4×4 matrix. To calculate it we note that the pertinent identities are as follows:

$$\langle 1, 1 \rangle_w = w_0,$$
$$\langle 1, \eta\overline{\eta} \rangle_w = \langle \eta\overline{\eta}, 1 \rangle_w = w_1,$$
$$\langle \eta, \eta \rangle_w = \langle \overline{\eta}, \overline{\eta} \rangle_w = w_1,$$
$$\langle \eta\overline{\eta}, \eta\overline{\eta} \rangle_w = 0.$$

All other inner products of pairs of elements in AW are zero. Hence, the matrix G we are dealing with here in the case $l = 2$ is

$$G = \begin{pmatrix} w_0 & 0 & 0 & w_1 \\ 0 & w_1 & 0 & 0 \\ 0 & 0 & w_1 & 0 \\ w_1 & 0 & 0 & 0 \end{pmatrix}$$

with respect to the ordered basis $\{1, \eta, \overline{\eta}, \eta\overline{\eta}\}$. Then $\det G = -(w_1)^4 \neq 0$ and so

$$G^{-1} = \begin{pmatrix} 0 & 0 & 0 & 1/w_1 \\ 0 & 1/w_1 & 0 & 0 \\ 0 & 0 & 1/w_1 & 0 \\ 1/w_1 & 0 & 0 & -w_0/w_1^2 \end{pmatrix}$$

by using standard linear algebra. It follows that the (unique!) reproducing kernel for $PG_{2,q}$ is given by

$$K_{PG}(\theta, \overline{\theta}, \eta, \overline{\eta}) = \frac{1}{w_1} \theta\overline{\theta} \otimes 1 + \frac{1}{w_1} \overline{\theta} \otimes \eta + \frac{1}{w_1} \theta \otimes \overline{\eta} + \frac{1}{w_1} 1 \otimes \eta\overline{\eta} - \frac{w_0}{w_1^2} \theta\overline{\theta} \otimes \eta\overline{\eta}.$$

(This also works when $w_0 \leq 0$ or $w_1 < 0$.) Even though the reproducing kernel in this example lies in a space of dimension 16, only 5 terms in the standard basis have non-zero coefficients.

Actually, the method in the previous paragraph is the systematic way to arrive at a formula for the reproducing kernel for $PG_{l,q}$ in general. Everything comes down to showing the invertibility of the matrix G, where

$$G_{cdij} = \langle \eta^c \overline{\eta}^d, \eta^i \overline{\eta}^j \rangle_w \tag{13}$$

for $(c,d), (i,j) \in I_l \times I_l$ and then finding the inverse matrix. Again, we label the rows and columns of G by the elements in AW. As is well known, invertibility is a generic property of G (that is, true for an open, dense set of matrices G). So there are many, many examples of sesquilinear forms (including positive definite inner products) defined on the non-commutative algebra $PG_{l,q}$, making it into a reproducing kernel space. Then it becomes clear that it is straightforward to give any finite-dimensional algebra $\mathcal{A}$, commutative or not (but such that every element in $\mathcal{A}$ is in the image of some functional calculus), an inner product so that $\mathcal{A}$ has a reproducing kernel. The infinite-dimensional case will require more care due to the usual technical details.

But is the matrix G associated to the sesquilinear form (3) invertible?

Theorem 1. *Taking G to be the matrix (13) associated to the sesquilinear form defined by (3), we have that $\det G = \pm(w_{l-1})^{l^2} \neq 0$ for every $l \geq 2$.*

Proof. We will argue by induction on l. First, we will show the cases $l = 2$ and $l = 3$. Then we will use the induction hypothesis that the result holds for the case $l - 2$ for $l \geq 4$ in order to prove the result for the case l.

We have shown above that $\det G = -(w_1)^4 = -(w_1)^{2^2} \neq 0$ when $l = 2$. So we must establish this result for $l = 3$ as well. We claim in that case that $\det G = (w_2)^9 \neq 0$. First we calculate the matrix entries of G, which is a 9×9 matrix, and get

$$
G = \begin{pmatrix}
w_0 & 0 & 0 & w_1 & 0 & 0 & 0 & 0 & w_2 \\
0 & w_1 & 0 & 0 & 0 & 0 & 0 & w_2 & 0 \\
0 & 0 & w_1 & 0 & 0 & 0 & w_2 & 0 & 0 \\
w_1 & 0 & 0 & w_2 & 0 & 0 & 0 & 0 & 0 \\
0 & 0 & 0 & 0 & 0 & w_2 & 0 & 0 & 0 \\
0 & 0 & 0 & 0 & w_2 & 0 & 0 & 0 & 0 \\
0 & 0 & w_2 & 0 & 0 & 0 & 0 & 0 & 0 \\
0 & w_2 & 0 & 0 & 0 & 0 & 0 & 0 & 0 \\
w_2 & 0 & 0 & 0 & 0 & 0 & 0 & 0 & 0
\end{pmatrix}
$$

with respect to the ordered basis $\{1, \theta, \overline{\theta}, \theta\overline{\theta}, \theta^2, \overline{\theta}^2, \theta\overline{\theta}^2, \theta^2\overline{\theta}, \theta^2\overline{\theta}^2\}$. Now the last 5 columns have all entries equal to zero, except for one entry equal to w_2. Similarly, the last 5 rows have all entries equal to zero, except for one entry equal to w_2. We calculate the determinant by expanding successively along each of the 5 last columns, thereby obtaining 5 factors of w_2 and a sign (either plus or minus). With each expansion the corresponding row is also eliminated and this will eliminate the first 2 of the last 5 rows, but leave the remaining 3 rows. So we now expand along these remaining last 3 rows, getting 3 more factors of w_2 as well as a sign. We then have remaining a 1×1 matrix whose entry comes from the 4th row and 4th column of the above matrix. And that entry is again w_2. So the determinant of G in the case $l = 3$ is a sign times 9 factors of w_2. We leave it to the reader to check that the overall sign is positive and so we get $\det G = (w_2)^9 = (w_2)^{3^2}$ as claimed.

Now we assume $l \geq 4$ and prove this case by induction. The induction hypothesis that we will use is that in the case for $l - 2$ the matrix G (which is an $(l-2)^2 \times (l-2)^2$ matrix) has determinant $\pm(w_{l-3})^{(l-2)^2}$. This is why we started this argument by proving separately the cases $l = 2$ and $l = 3$.

We start by using the same argument of expansion of the determinant as used above when $l = 3$. The matrix G is an $l^2 \times l^2$ matrix. We consider this matrix in a basis made by ordering the basis AW in such a way that the last elements are all of the form $\theta^{l-1}\overline{\theta}^k$ or of the form $\theta^k\overline{\theta}^{l-1}$, where $k \in I_l$. Notice that the element $\theta^{l-1}\overline{\theta}^{l-1}$ is the only basis element in AW that has both of these forms. So, there are $2\operatorname{card}(I_l) - 1 = 2l - 1$ such elements. We claim that each of these elements has exactly one non-zero inner product with the elements in the ordered basis AW and that the value of that inner product is w_{l-1}. Starting with $\theta^{l-1}\overline{\theta}^k$ we note that

$$\langle \theta^i \overline{\theta}^j, \theta^{l-1}\overline{\theta}^k \rangle_w = \delta_{i+k,j+l-1} w_{j+l-1}\chi_l(j+l-1).$$

This is zero if $j > 0$ because of the χ_l factor. But for $j = 0$ we have $\langle \theta^i, \theta^{l-1}\overline{\theta}^k \rangle_w = \delta_{i+k,l-1} w_{l-1}$ which is only non-zero for $i = l-1-k$, in which case we have

$$\langle \theta^{l-1-k}, \theta^{l-1}\overline{\theta}^k \rangle_w = w_{l-1}.$$

A similar calculation shows that $\langle \overline{\theta}^{l-1-k}, \theta^k \overline{\theta}^{l-1} \rangle_w = w_{l-1}$, while all other elements in AW have zero inner product with $\theta^k \overline{\theta}^{l-1}$. So we expand the determinant of G along the last $2l-1$ columns, obtaining $2l-1$ factors of w_{l-1} and some sign, either plus or minus. The corresponding rows that are eliminated, according to the above, are labelled by *all* the powers of θ alone or the powers of $\overline{\theta}$ alone. Of these powers, only the two powers θ^{l-1} and $\overline{\theta}^{l-1}$ label one of the last $2l-1$ rows. So we proceed by expanding along the remaining $2l-3$ last rows, thereby obtaining $2l-3$ more factors of w_{l-1} and some sign. These $(2l-1)+(2l-3) = 4l-4$ factors of w_{l-1} as well as the sign multiply the determinant of a square matrix which has $l^2 - (4l-4) = (l-2)^2$ rows and the same number of columns.

Now, we claim that we can calculate the determinant of this remaining $(l-2)^2 \times (l-2)^2$ matrix, call it M, using the induction hypothesis. However, M is not the matrix G for the case $l-2$, but is related to it as we shall see. From the labeling of the rows and columns of G for the case l, the matrix M inherits a labeling, namely its rows and columns are labeled by the basis elements $\theta^i \overline{\theta}^j$ of AW for $1 \le i, j \le l-2$. This is clear by recalling the labeling of the columns and rows which were eliminated in the above expansions. But for any $a, b, c, d \in I_l$ we have that the entries of G are

$$\langle \theta^a \overline{\theta}^b, \theta^c \overline{\theta}^d \rangle_w = \delta_{a+d,b+c} w_{a+d}\chi_l(a+d).$$

In particular, this holds for $a, b, c, d \in I_l \setminus \{0, l-1\}$ in which case we can write

$$\langle \theta^a \overline{\theta}^b, \theta^c \overline{\theta}^d \rangle_w = \delta_{a+d-2,b+c-2} w_{a+d}\chi_{l-2}(a+d-2),$$

and so these are the entries in the matrix M. By changing to new variables $a', b', c', d' \in I_{l-2}$ where $a' = a-1, b' = b-1, c' = c-1, d' = d-1$ we see that the matrix entries of M are $\delta_{a'+d',b'+c'} w_{a'+d'+2}\chi_{l-2}(a'+d')$. But the entries for the matrix G in the case $l-2$ are $\langle \theta^{a'} \overline{\theta}^{b'}, \theta^{c'} \overline{\theta}^{d'} \rangle_w = \delta_{a'+d',b'+c'} w_{a'+d'}\chi_{l-2}(a'+d')$ for all $a', b', c', d' \in I_{l-2}$. So, except for a shift by $+2$ in the sub-indices of the weights, the entries in M correspond to the entries in G for the case $l-2$. Consequently, by the induction hypothesis as stated above, we have that $\det M = \pm(w_{(l-3)+2})^{(l-2)^2} = \pm(w_{l-1})^{(l-2)^2}$. Putting all this together, we have that

$$\det G = \pm(w_{l-1})^{(4l-4)} \det M = \pm(w_{l-1})^{(4l-4)}(w_{l-1})^{(l-2)^2} = \pm(w_{l-1})^{l^2}$$

which proves our result. $\qquad\qquad\qquad\qquad\qquad\qquad\qquad\qquad\qquad\qquad\square$

We leave it to the interested reader to track down the correct sign in the previous result. We also note that w_{l-1} is the coefficient of 1 in the definition of the weight in (1), according to our convention. Since $w_{l-1} \neq 0$, the main result of this section now follows immediately. Here it is:

Theorem 2. *For all $l \geq 2$ and for all $q \in \mathbb{C} \setminus \{0\}$ the paragrassmann space $PG_{l,q}$ has a unique reproducing kernel with respect to the inner product $\langle \cdot, \cdot \rangle_w$ defined in equation (3).*

Remark. $PG_{l,q}$ is a non-commutative algebra and so it is a quantum space in the broadest interpretation of that terminology, that is, it is not isomorphic to an algebra of functions defined on some set (the 'classical' space). Also it is not even a Hilbert space with respect to the sesquilinear form that we are using on it. Nonetheless, contrary to what one might expect from studying the theory of reproducing kernel Hilbert spaces, this space *does* have a reproducing kernel.

Also, we would like to comment that the existence of the reproducing kernel for $PG_{l,q}$ is a consequence of the definition of the sesquilinear form and is not dependent on the parameter q.

8. Concluding remarks

In this paper we have introduced ideas from the analysis of reproducing kernel Hilbert spaces of functions to the study of the non-commutative space $PG_{l,q}$ of paragrassmann variables. We are rather confident that other ideas from analysis will find application to $PG_{l,q}$, and this will be one direction for future research. Even more important could be the application of these ideas from analysis to other classes of non-commutative spaces. We expect that there could be many such applications.

Acknowledgment

This paper was inspired by a talk based on [1] given by Jean-Pierre Gazeau during my sabbatical stay at the Laboratoire APC, Université Paris Diderot (Paris 7) in the spring of 2011. Jean-Pierre was my academic host for that stay, and so I thank him not only for stirring my curiosity in this subject but also for his most kind hospitality which was, as the saying goes, above and beyond the call of duty. (I won't go into the details, but it really was. Way beyond, actually.) Merci beaucoup, Jean-Pierre! Also my thanks go to Rodrigo Fresneda for very useful comments as well as for being my most gracious host at the UFABC in São Paulo, Brazil in April, 2012 where work on this paper continued. Muito obrigado, Rodrigo!

References

[1] M. El Baz, R. Fresneda, J.-P. Gazeau and Y. Hassouni, *Coherent state quantization of paragrassmann algebras*, J. Phys. A: Math. Theor. **43** (2010) 385202 (15pp). Also see the Erratum for this article in arXiv:1004.4706v3.

[2] S.B. Sontz, *Paragrassmann Algebras as Quantum Spaces, Part II: Toeplitz Operators*, eprint, arXiv:1205.5493.

[3] J. Glimm and A. Jaffe, Quantum Physics, A Functional Integral Point of View, Springer, 1981.

[4] R. Fresneda, Private communication, 12 April, 2012.

[5] N. Aronszajn, *Theory of Reproducing Kernels*, Trans. Am. Math. Soc. **68** (1950) 337–404.

[6] S. Bergman, The kernel function and conformal mapping, Am. Math. Soc., Providence, 1950.

[7] B.C. Hall, *Holomorphic methods in analysis and mathematical physics*, in: Contemporary Mathematics, vol. 260, eds. S. Pérez-Esteva and C. Villegas-Blas, Am. Math. Soc., Providence, 2000.

[8] V. Bargmann, *On a Hilbert space of analytic functions and an associated integral transform, part I.* Commun. Pure Appl. Math. **14** (1961), 187–214.

[9] Jean-Pierre Gazeau, Coherent States in Quantum Physics, Wiley-VCH, 2009.

Stephen Bruce Sontz
Centro de Investigación en Matemáticas, A.C.
CIMAT
Guanajuato, Mexico
e-mail: sontz@cimat.mx

Part II

Groups, Algebras
and Symmetries

Geometric Methods in Physics. XXXI Workshop 2012
Trends in Mathematics, 67–86

Rolling of Coxeter Polyhedra Along Mirrors

Dmitri V. Alekseevski, Peter W. Michor and Yurii A. Neretin

Abstract. The topic of the paper are developments of n-dimensional Coxeter polyhedra. We show that the surface of such polyhedron admits a canonical cutting such that each piece can be covered by a Coxeter $(n-1)$-dimensional domain.

Mathematics Subject Classification (2010). Primary 51F15, 53C20, 20F55, 22E40.

Keywords. Reflection groups, Coxeter groups, Lobachevsky space, isometries, polyhedra, developments, trees.

1. Introduction. Coxeter groups

1.1. Coxeter groups in spaces of constant curvature

Consider a Riemannian space $\mathbb{M}^n$ of constant curvature, i.e., a Euclidean space $\mathbb{R}^n$, a sphere $\mathbb{S}^{n-1}$, or a Lobachevsky space $\mathbb{L}^n$ (on geometry of such spaces, see [1]).

Let $C \subset \mathbb{M}^n$ be an intersection of a finite or locally finite collection of half-spaces[1].

Consider reflections of C with respect to all $(n-1)$-dimensional faces. Next, consider "new polyhedra" and their reflections with respect to their faces, etc. The domain C is said to be a *Coxeter domain* if we get a tiling of the whole space in this way. The group of isometries generated by all such reflections is said to be a *reflection group* or a *Coxeter group* (in a narrow sense, see below). We say that a Coxeter group is *cocompact* if the initial domain C is compact. In this case, we say that C is a *Coxeter polyhedron*.

P.W.M. was supported by "Fonds zur Förderung der wissenschaftlichen Forschung, Projekt P 14195 MAT";

Yu.A.N was supported by Austrian "Fonds zur Förderung der wissenschaftlichen Forschung", projects 19064, 22122 and also by the Russian Agency on Nuclear Energy, the Dutch fund NWO, grant 047.017.015, and the Japan–Russian grant JSPS–RFBR 07-01-91209.
[1] A natural example with an infinite collection of half-spaces is given in Figure 9.

Evidently, if C is a Coxeter domain, then the dihedral angles between two neighboring faces of C are of the form $\frac{\pi}{m}$, where $m \geqslant 2$ is an integer. In particular, they are *acute*, i.e., $\leqslant 90°$.

Denote the faces of the polyhedron C by $F_1, \ldots, F_p$, denote by $s_1, \ldots, s_p$ the corresponding reflections. Denote by π/m_{ij} the angles between adjacent faces. Evidently,

$$s_j^2 = 1, \qquad (s_i s_j)^{m_{ij}} = 1. \tag{1}$$

1.2. More terminology

Consider a Coxeter tiling of $\mathbb{M}^n$. Below a "*chamber*" is any (n-dimensional) polyhedron of the tiling. A "*face*" or "*facet*" is an $(n-1)$-dimensional face of some chamber; a *hyperedge* is an $(n-2)$-dimensional edge; a *stratum* is an arbitrary stratum of codim $\geqslant 1$ of some chamber; a *vertex* is a vertex.

Also "*mirrors*" are hyperplanes of reflections. They divide the space $\mathbb{M}^n$ into chambers. The group G acts on the set of chambers simply transitively. We denote the reflection with respect to a mirror Y by s_Y.

Each facet is contained in a unique mirror.

1.3. General Coxeter groups

Take a symmetric $p \times p$ matrix M with positive integer elements, set $m_{jj} = 1$; we admit $m_{ij} = \infty$. An *abstract Coxeter group* is a group with generators $s_1, \ldots, s_n$ and relations (1).

For such a group we draw a graph (we use the term "*Coxeter scheme*") in the following way. Vertices of the graph correspond to generators. We connect i and jth vertices by $(m_{ij} - 2)$ edges. In fact, we draw a multiple edge if $k \leqslant 6$, otherwise we write a number k on the edge.

This rule also assign a graph to each Coxeter polyhedron.

1.4. Spherical Coxeter groups

By definition, a spherical Coxeter group, say Γ, acts by orthogonal transformations of the Euclidean space $\mathbb{R}^{n+1}$. A group Γ is said to be *reducible* if there exists a proper Γ-invariant subspace in $\mathbb{R}^{n+1}$. Evidently, the orthogonal complement to a Γ-invariant subspace is Γ-invariant.

The classification of irreducible Coxeter groups is well known[2], see Bourbaki [2]. The list consists of Weyl groups of semisimple Lie algebras (= Killing's list of root systems) + dihedral groups + groups of symmetries of the icosahedron and 4-dimensional hypericosahedron (the table is given Section 3).

This also gives a classification of reducible groups.

---

[2] Actually, these objects were known to Ludwig Schläfli and Wilhelm Killing in XIX century. In 1924, Hermann Weyl identified these groups as reflection groups, in 1934 Harold Coxeter gave a formal classification and also classified Euclidean groups.

1.5. Coxeter equipments

Next, consider an arbitrary Coxeter polyhedron in $\mathbb{R}^n$, $\mathbb{S}^n$, or $\mathbb{L}^n$. Consider a stratum H of codimension k, it is an intersection of k faces, $H = F_{i_1} \cap \cdots \cap F_{i_k}$. The reflections with respect to the faces $F_{i_1}, \ldots, F_{i_k}$ generate a Coxeter group, denote it by $\Gamma(H) = \Gamma(F_{i_1}, \ldots, F_{i_k})$.

This group is a spherical Coxeter group. Namely, for $x \in H$ consider the orthocomplement in the tangent space at x to the stratum H and the sphere in this orthocomplement. Then $\Gamma(H)$ is a reflection group of this Euclidean sphere.

If $H \subset H'$, then we have the tautological embedding

$$\iota_{H',H} : \Gamma(H') \to \Gamma(H).$$

If $H \subset H' \subset H''$, then

$$\iota_{H'',H} = \iota_{H',H}\iota_{H'',H'}.$$

Such a collection of groups and homomorphisms is said to be a *Coxeter equipment*.

1.6. Cocompact Euclidean Coxeter groups

Here classification is also simple and well known, see Bourbaki [2]. Any such group Γ contains a normal subgroup $\mathbb{Z}^n$ acting by translations and $\Gamma/\mathbb{Z}^n$ is a spherical Coxeter group.

1.7. Coxeter groups in Lobachevsky spaces

We report from Vinberg [3], Vinberg, Shvartsman, [4]. The situation differs drastically.

a) Coxeter polygons on Lobachevsky plane are arbitrary k-gons with angles of the form π/m_j. The sum of exterior angles must satisfy $\sum(\pi - \pi/m_j) > 2\pi$. If $k > 5$ this condition holds automatically. For $k = 4$ this excludes rectangles, also few triangles are forbidden (in fact, spherical and Euclidean triangles). A Coxeter k-gon with prescribed angles depends on $(k - 3)$ parameters.

b) In dimensions $n > 2$ Coxeter polyhedra are rigid. There are many Coxeter groups in spaces of small dimensions ($n = 3, 4, 5$), but for $n \geqslant 30$ there is no Coxeter group with compact fundamental polyhedron at all. For $n > 996$ there is no Coxeter group of finite covolume (Prokhorov, Khovanskii, 1986, see [5]); the maximal dimensions of known examples are: 8 for compact polyhedra (Bugaenko), and 21 for a polyhedron of finite volume (Borcherds). For $n = 3$ there is a nice Andreev description [6] of all Coxeter polyhedra, it is given in the following two subsections.

1.8. Acute angle polyhedra in $\mathbb{L}^3$

First, we recall the famous (and highly non-trivial) Steinitz Theorem (see, e.g., [7]) about possible combinatorial structure of convex polyhedra in $\mathbb{R}^3$.

Since the boundary of a polyhedron is a topological sphere S^2, edges form a connected graph on the sphere, it divides the sphere into polygonal domain (we use the term 'face' for such a domain).

There are the following evident properties of the graph:
- each edge is contained in 2 faces;
- each face has $\geqslant 3$ vertices;
- the intersection of any pair of faces can by the empty set, a vertex, or an edge.

Theorem (Ernst Steinitz). *Any graph on the sphere S^2 satisfying the above conditions can be realized as a graph of edges of a convex polyhedron.*

Our next question is the existence of a convex polyhedron in $\mathbb{L}^3$ of a given combinatorial structure where each dihedral (i.e., between two adjacent faces) angle is a given *acute* angle ('acute' or also 'non-obtuse' means $\leqslant \pi/2$) There are the following a priori properties of such polyhedra:

1) All spatial angles are simplicial, i.e., each vertex of the graph is contained in 3 edges. The angles φ_1, φ_2, φ_3 in a given vertex satisfy

$$\varphi_1 + \varphi_2 + \varphi_3 > 2\pi. \qquad (2)$$

2) At each vertex, the set of all dihedral angles determines all other angles in each face at this vertex (by the spherical cosine theorem). A face must be a Lobachevsky polygon, i.e., the sum of its exterior angles must be $\geqslant 2\pi$. Since all dihedral angles are acute, angles in each face are also acute. Therefore our conditions forbid only rectangles and some triangles.

3) The following restriction is non-obvious: We say that a *k-prismatic element* of a convex polyhedron C is a sequence

$$F_1, \quad F_2, \quad \ldots, \quad F_k, \quad F_{k+1} := F_1$$

of faces such that F_k and F_{k+1} have a common edge, and all triple intersections $F_i \cap F_j \cap F_k$ are empty.

Lemma (Andreev). *For any prismatic element in an acute angle polyhedron, the sum of exterior dihedral angles is $> 2\pi$.*

Theorem (Andreev). *Consider a Steinitz-admissible 3-valent spherical graph with > 4 vertices[3]. Prescribe a dihedral acute angle to each edge in such a way that:*
- *the inequality (2) in each vertex is satisfied;*
- *all 3- and 4-prismatic elements satisfy the previous lemma;*
- *we forbid the configuration given in Figure 1.*

Under these assumptions, there exists a unique convex polyhedron $\subset \mathbb{L}^3$ of the given combinatorial structure and with the given acute angles.

The uniqueness is a rigidity theorem of Cauchy type (see [8],[7]). The existence is a deep unusual fact; it is a special case of a theorem of Aleksandrov type [8] obtained by Rivin, see [9], [10].

For some applications of the Andreev and Rivin Theorems to elementary geometry, see Thurston [11], Rivin [12].

---

[3] Simplices are exceptions. However, their examination is simple, Lanner, 1950, see, e.g., [4].

FIGURE 1. The following configuration with dihedral angles $= \pi/2$ on *thick* edges is forbidden in the Andreev Theorem. In this case, we would get a quadrangle with right angles, but such quadrangles do not exist in Lobachevsky space.

1.9. Andreev polyhedra

Andreev's Theorem provides us a description of all Coxeter polyhedra in $\mathbb{L}^3$. Now all angles have the form π/m_{ij} with integer $m_{ij} > 1$. We simply write the labels m_{ij} on the corresponding edges.

Below the term *"Andreev polyhedron"* will mean a compact Coxeter polyhedron in $\mathbb{L}^3$.

All possible pictures at vertices of Andreev polyhedra are given in Figure 2.

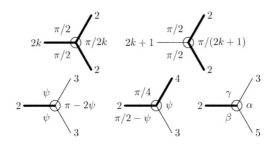

FIGURE 2. We draw all possible types of vertices of an Andreev polyhedron. We present the labels m_j on the edges and flat angles in faces. Here $\psi = \arctan \sqrt{2}$ and α, β, γ are explicit angles with $\alpha + \beta + \gamma = \pi/2$. Evaluations of all these angles are given in figures in Section 2.

We draw a *thick* line iff the label is even.

1.10. Results of the paper

Consider a convex polyhedron C in a space $\mathbb{M}^n$ of constant curvature. Following Alexandrov [8], we regard the boundary $\Xi = \partial C$ of C as an $(n-1)$-dimensional manifold of constant curvature with singularities. In the case $n = 3$, we get a two-dimensional surface with conic singularities of negative curvature (see, e.g., Figure 2, in all the cases the sum of angles at a singularity is $< 2\pi$).

Now, cut Ξ along hyperedges with *even* labels (i.e., hyperedges with dihedral angles $\pi/2k$). Let $\Omega_1, \Omega_2, \dots$ be the connected pieces of the cut surface.

Theorem 1. *The universal covering $\Omega_j^{\sim}$ of each Ω_j is a Coxeter domain in $\mathbb{M}^{n-1}$.*

Proof for Andreev polyhedra. We simply look at Figure 2. In all the cases, angles between *thick* edges are Coxeter. ☐

We also describe tiling of mirrors, groups of transformations of mirrors induced by the initial Coxeter group (Theorem 3) and the Coxeter equipments of $\Omega_j^\sim$ (Theorem 4).

The addendum to the paper contains two examples of 'calculation' of developments, for an Andreev prism $\subset \mathbb{L}^3$ and for a Coxeter simplex $\subset \mathbb{L}^4$. The proof of the Andreev Theorem is nonconstructive. In various explicit cases, our argumentation allows to construct an Andreev polyhedron from the a priori information about its development. Our example illustrates this phenomenon.

On the other hand, there arises a natural problem of elementary geometry:

– *Which Andreev polyhedra are partial developments of 4-dimensional Coxeter polyhedra? Is it possible to describe all 3-dimensional polyhedra that are faces of 4-dimensional Coxeter polyhedra?*

Our main argument (Rolling Lemma 1) is very simple, it is valid in a wider generality, we briefly discuss such possibilities in the next two subsections.

1.11. Polyhedral complexes and projective Coxeter polyhedra

Theorem (Tits). *Any Coxeter group can be realized as a group of transformations of an open convex subset of a real projective space $\mathbb{RP}^n$ which is generated by a collection of reflections s_1, ..., s_p with respect to hyperplanes[4] intersecting the subset. The closure of a chamber is a convex polyhedron.*

See also Vinberg [13].

1.12. A more general view

Nikolas Bourbaki[5] proposed a way to build topological spaces from Coxeter groups. M. Davis used this approach in numerous papers (see, e.g., [15], [16]) and the book [17]; in particular he constructed nice examples/counterexamples in topology.

Also it is possible to consider arbitrary Riemannian manifolds equipped with a discrete isometric action of a Coxeter group such that the set of fixed points of each generator is a (totally geodesic) hypersurface, and such that the generators act as reflections with respect to these submanifolds. In this context, a chamber itself can be a topologically non-trivial object, see [17], [18].

2. Rolling of chamber

In this section, $\mathbb{M}^n$ is a space $\mathbb{L}^n$, $\mathbb{S}^n$, $\mathbb{R}^n$ of constant curvature equipped with a Coxeter group Γ or, more generally, any space described in Subsection 1.11.

Fix a mirror $\mathbb{X}^{n-1}$ in $\mathbb{M}^n$. Consider intersections of $\mathbb{X}^{n-1}$ with other mirrors Y_α. The set $\mathbb{X}^{n-1} \setminus \bigcup Y_\alpha$ is a disjoint union of open facets. Thus, we get a tiling of $\mathbb{X}^{n-1}$ by facets.

Our aim is to describe this tiling in the terms of the geometry of a chamber.

---

[4]A reflection is determined by a fixed hyperplane and a reflected transversal line.

[5]Apparently, he used the work by Jacque Tits [14]; the latter text is inaccessible for the authors.

2.1. Rolling Lemma

Lemma 1. *Let $I \subset \mathbb{X}^{n-1}$ be an $(n-2)$-dimensional hyper-edge of our tiling. Let $F, H \subset \mathbb{X}^{n-1}$ be the facets adjacent to I.*

  a) *If the label m_I of I is even, then I is contained in a certain mirror Y_α orthogonal to X. In particular $s_{Y_\alpha} F = H$.*
  b) *Let the label m be odd. Let C be a chamber adjacent to the facet F. Let G be another facet of C adjacent to the same hyper-edge I. Then G is isometric to H. More precisely, there is $\gamma \in \Gamma$ fixing all the points of I such that $\gamma G = H$.*

The PROOF is given in Figure 3.

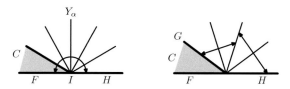

FIGURE 3. Even and odd labels. Proof of the Rolling Lemma.

2.2. Algorithm generating the tiling

Let C be a chamber adjacent to a facet $F \subset \mathbb{X}^{n-1}$. Consider an hyper-edge I of C lying in $\mathbb{X}^{n-1}$.

OPERATION 1. Let the hyper-edge I be odd. Consider a facet $G \neq F$ of C adjacent to I, consider the corresponding γ from Lemma 1 and draw γG on $\mathbb{X}^{n-1}$.

OPERATION 2. If the hyper-edge I is even, then we reflect F in $\mathbb{X}^{n-1}$ with respect to I.

We perform all the possible finite sequences of such operations. By the Rolling Lemma, we get the whole tiling of the mirror $\mathbb{X}^{n-1}$.

Remark. Let $\mathbb{M}^n = \mathbb{R}^3, \mathbb{S}^3, \mathbb{L}^3$ be a usual 3-dimensional space of constant curvature. Operation 1 corresponds to rolling of a polyhedron C along the hyperplane $\mathbb{X}^{n-1} \sim \mathbb{M}^2$ over the edge I. $\qquad \square$

2.3. The group preserving the mirror $\mathbb{X}^{n-1}$

For a mirror $\mathbb{X}^{n-1}$, consider the group $\Gamma_* = \Gamma_*(\mathbb{X}^{n-1})$ of all the isometries of $\mathbb{X}^{n-1}$ induced by elements of Γ preserving $\mathbb{X}^{n-1}$.

If $\gamma \in \Gamma$ preserves $\mathbb{X}^{n-1}$, then $s_{\mathbb{X}^{n-1}}\gamma$ also preserves $\mathbb{X}^{n-1}$ and agrees with γ on $\mathbb{X}^{n-1}$. Thus each element of Γ_* is induced by two different elements of Γ.

Observation 1. *Let $F_1, F_2 \subset \mathbb{X}^{n-1}$ be equivalent facets. There is a unique element $\mu \in \Gamma_*(\mathbb{X}^{n-1})$ such that $\mu F_1 = F_2$.*

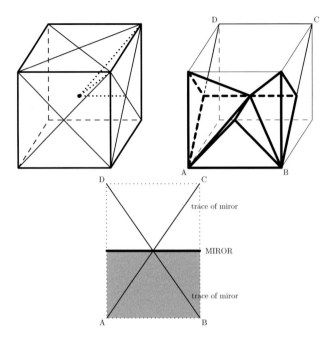

FIGURE 4. Example of rolling: the reflection group A_3 in $\mathbb{R}^3$. The mirrors are planes passing through opposite edges of the cube. There are 24 Weyl chambers, which are simplicial cones with dihedral angles $\pi/3$, $\pi/3$, $\pi/2$ (we draw them as simplices). Rolling of a Weyl chamber by the mirror $ABCD$ produces a half-plane.

We can also regard A_3 as a reflection group on the 2-dimensional sphere $\mathbb{S}^2$.

2.4. Reflections in mirrors and the new chamber

Consider all the mirrors $Z_\alpha \subset \mathbb{M}^n$ orthogonal to our mirror $\mathbb{X}^{n-1}$. The corresponding reflections s_{Z_α} generate a reflection group on $\mathbb{X}^{n-1}$; denote this group by $\Delta = \Delta(\mathbb{X}^{n-1})$.

Observation 2. Δ is a normal subgroup in $\Gamma_*(\mathbb{X}^{n-1})$.

Indeed, if s is a reflection, then $\gamma^{-1}s\gamma$ is a reflection. $\qquad\square$

Consider a chamber C of $\mathbb{M}^n$ lying on $\mathbb{X}^{n-1}$ (i.e., having a facet in $\mathbb{X}^{n-1}$) and consider all possible sequences of admissible rolling, i.e., we allow Operation 1 of Algorithm 2.2 and we forbid Operation 2. Denote by $B \subset \mathbb{X}^{n-1}$ the domain obtained by rolling, tiled by the traces of facets of C making contact with $\mathbb{X}^{n-1}$ during rolling.

Theorem 2. B is a chamber of the reflection group $\Delta(\mathbb{X}^{n-1})$.

Proof. We can not roll further if and only if we meet a "vertical" mirror. $\qquad\square$

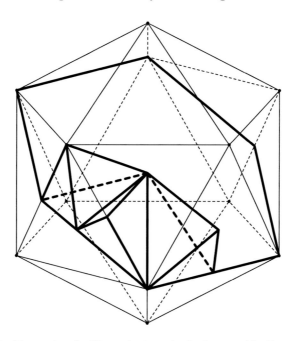

FIGURE 5. Example of rolling: the icosahedral group H_3. It is generated by reflections with respect to bisectors of segments connecting midpoints of opposite edges of the icosahedron. The bisectors separate $\mathbb{R}^3$ into 120 simplicial cones with dihedral angles $\pi/2$, $\pi/3$, $\pi/5$. In the figure the simplicial cones are cut by the surface of the icosahedron.

We show an admissible rolling of a Weyl chamber along a mirror. The final chamber in the mirror is a quadrant.

EXAMPLES OF ROLLING. Some examples of rolling corresponding to the usual spherical Coxeter groups

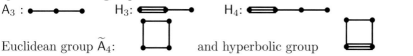

A_3 : H_3: H_4:

Euclidean group $\widetilde{A}_4$:   and hyperbolic group

are given in figures 4–9. In these figures, we also evaluate the new chamber B.

Lemma 2. *Each $(n-3)$-dimensional stratum of our tiling of $\mathbb{X}^{n-1}$ is contained in a mirror of the group $\Delta(\mathbb{X}^{n-1})$.*

Proof. This stratum is equipped with a finite 3-dimensional Coxeter group (i.e., A_3, BC_3, H_3, $A_1 \oplus G_2^m$, $A_1 \oplus A_1 \oplus A_1$, see the table below). For each mirror of such a group there exists an orthogonal mirror. □

2.5. Rolling scheme

Denote by $\Xi(C)$ the surface of the initial chamber C, let $\Xi'(C)$ be the surface with all even edges deleted.

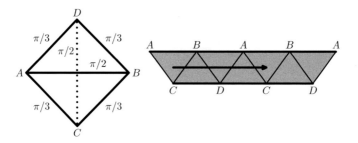

FIGURE 6. Example of rolling: the (affine) Euclidean reflection group $\widetilde{A}_4$ in $\mathbb{R}^3$. A chamber is the simplex $ABCD$. Rolling through AB and CD is forbidden. Deleting these edges from the surface of the simplex, we get a non-simply connected surface. Hence, the process of rolling is infinite. The arrow shows the deck transformation induced by the generator of the fundamental group.

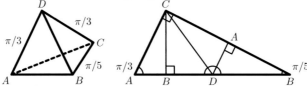

FIGURE 7. Example of rolling: the hypericosahedral group H_4 acting on the 3-dimensional sphere $\mathbb{S}^3$. The chamber is the spherical simplex drawn in the figure (we omit all labels $\pi/2$ on edges).

The angle $= \pi$ on the development at D was evaluated in Figure 4. The right angle at C was evaluated in figure 5.

The spherical triangle ABC is present on the circumscribed sphere in the next figure (in spite of the absence of the sphere itself).

Lemma 3. Ξ' *does not contain* $(n-3)$-*dimensional strata of* C.

This is rephrasing of Lemma 2. $\qquad\qquad\qquad\qquad\qquad\qquad\qquad$ □

Consider the graph, whose vertices are the facets of Ξ'; vertices are connected by an edge if the corresponding facets are neighbors in Ξ'. We call this graph the *Rolling scheme*. In fact, the Rolling scheme is the Coxeter scheme 1.3 with removed even (and infinite) edges.

Proposition 1. *The surface* Ξ' *is homotopically equivalent to the Rolling scheme.*

2.6. Proof of Proposition 1

Let U be a convex polyhedron in $\mathbb{R}^n$, denote by Ξ its surface. Choose a point A_j in interior of each $(n-1)$-dimensional face. Choose a point B_k in interior of each $(l-2)$-dimensional boundary stratum (hyperedge) of U.

Draw the segment $[A_j, B_k]$ iff the face contains the hyperedge. Thus we get a graph T on the surface of the polyhedron C whose vertices are enumerated by

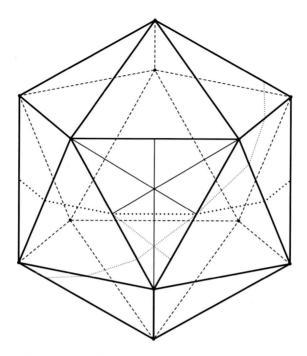

FIGURE 8. Example of rolling: the hypericosahedral group H_4 acting in $\mathbb{R}^4$. The figure presents the tiling of a mirror, i.e., of $\mathbb{R}^3$, by simplicial cones. We draw intersections of simplicial cones with the boundary of the icosahedron. Consider 3 types of 'axes' of the icosahedron:

    A) segments connecting midpoints of opposite edges;

    B) segments connecting central points of opposite faces;

    C) diagonals connecting opposite vertices.

    Consider bisectors of all such segments. Type A bisectors are mirrors. They divide $\mathbb{R}^3$ into 120 simplicial chambers. Six chambers are presented in the front face of the icosahedron.

    Adding bisectors of type B and C we obtain a partition of $\mathbb{R}^3$ into 480 simplicial cones. This is the desired tiling.

    In this figure, we present subdivisions of two chambers. A proof of this picture is contained in Figure 7

faces of U and edges are enumerated by hyperedges of U. Denote by Ξ^∇ the surface of the polyhedron S without boundary strata of dimension $(n-3)$.

Lemma 4. *The graph T is a deformation retract of Ξ^∇. Moreover, it is possible to choose a homotopy that preserves all faces and all hyperedges.*

Proof. See Figure 10.   □

    Proposition 1 follows from Lemma 4.   □

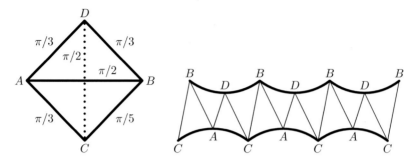

FIGURE 9. Example. A Coxeter simplex in $\mathbb{L}^3$. Its development is an infinite 'strip' $\subset \mathbb{L}^2$ bounded by two infinite polygonal curves, interior angles between segments of polygonal curves are $\pi/2$ and π.

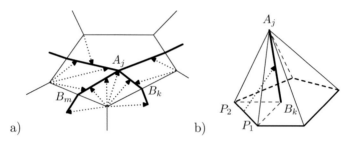

FIGURE 10. Proof of Lemma 4.

a) $n = 3$. Graph on a surface of a 3-dimensional polytop and a retraction. Recall that we have removed vertices.

b) $n = 4$. A piece of a 3-face of 4-dimensional polyhedron. Recall that 1-dimensional edges are removed. Inside a simplex $P_1 P_2 A_j B_k$ the retraction is the projection to $A_j B_k$ with center on the segment $P_1 P_2$. Note that all segments connecting $A_j B_k$ and $P_1 P_2$ are pairwise non-intersecting.

2.7. Action of the fundamental group on mirror

Let F be a facet in $\mathbb{X}^{n-1}$, let C be a chamber of $\mathbb{M}^n$ lying on F, and let $B \supset F$ be the chamber of the reflection group $\Delta(\mathbb{X}^{n-1})$ obtained by rolling C, as described in Subsection 2.4.

Let Ω be a connected component of Ξ' containing the facet F.

Let $F_1, \ldots, F_r$ be facets $\subset \Omega$. We can think that each facet has its own color; thus the mirror $\mathbb{X}^{n-1}$ is painted in r colors. Moreover, for each facet $H \in \mathbb{X}^{n-1}$ there is a canonical bijection ('parametrization') from the corresponding $F_i \subset \Omega$ to H. We say that a bijection $\mathbb{X}^{n-1} \to \mathbb{X}^{n-1}$ (or $B \to B$) is an isomorphism if it preserves the coloring and commutes with the parameterizations.

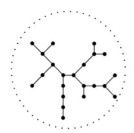

FIGURE 11. A graph of vicinity of $(n-1)$-dimensional facets in the new $(n-1)$-dimensional chamber B is a tree.

Proposition 2.
a) *The chamber $B \subset \mathbb{X}^{n-1}$ is the universal covering of Ω.*
b) *Any deck transformations of B is an isomorphism $B \to B$ and admits a unique extension to an isomorphism of the mirror $\mathbb{X}^{n-1}$.*
c) *Each isomorphism $\mu \in \Gamma_*(\mathbb{X}^{n-1})$ preserving B is induced by a deck transformation.*

Proof. a) Denote by $\Omega^\sim$ the universal covering of Ω. The chamber B was constructed as the image of $\Omega'^\vee$. Moreover, the map $\Omega^\sim \to C$ is locally bijective. On the other hand, a chamber on a simply connected manifold is simply connected see (see [18], 2.14); therefore $B \simeq \Omega^\sim$.

b) A deck transformation $B \to B$ is an isometry by the rolling rules. Let a deck transformation send a facet F to F'. Then the facets F, F' are Γ-equivalent, and the corresponding map in Γ is an isometry of $\mathbb{X}^{n-1}$.

c) Let $F \subset \mathbb{X}^{n-1}$ be a facet. We take the deck transformation sending F to F'. $\qquad \square$

2.8. Description of $\Gamma_*(\mathbb{X}^{n-1})$

Theorem 3. *The group $\Gamma_*(\mathbb{X}^{n-1})$ is a semidirect product $\mathrm{Deck}(B) \ltimes \Delta(\mathbb{X}^{n-1})$.*

Proof. Indeed, the group $\Delta(\mathbb{X}^{n-1})$ acts simply transitively on the set of chambers in $\mathbb{X}^{n-1}$; the group $\mathrm{Deck}(B)$ acts simply transitively on the set of facets of a given type in the chamber B. $\qquad \square$

3. Reduction of equipment

We keep the notation of the previous section. Our aim is to describe the Coxeter equipment of the new chamber B.

3.1. Combinatorial structure of the tiling of the chamber

Consider a graph $\mathfrak{F}$ whose vertices are enumerated by $(n-1)$-facets lying in B, two vertices are connected by an edge if they have a common $(n-2)$-dimensional stratum (a former hyperedge in $\mathbb{M}^n$).

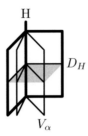

FIGURE 12. Subdivision of the cone normal to a stratum.

Table. Reduction of spherical Coxeter schemes

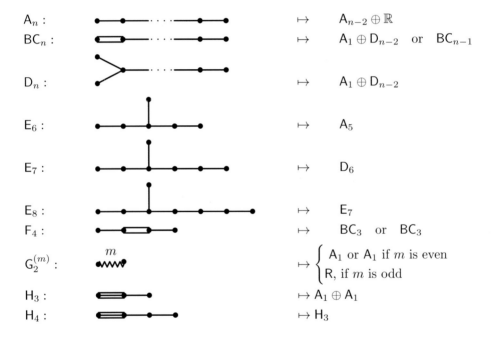

Observation 3. $\mathfrak{F}$ *is a tree.*

Proof. Indeed, the universal covering of a graph is a tree. □

If the initial rolling scheme is a tree, then we get the same tree. If the rolling scheme contains a cycle, then we get an infinite tree (examples: Figures 6, 9, the rolling schemes contain 1 cycle).

3.2. New equipment

All the strata of B of dimension $< (n-2)$ are contained in the boundary of B. These strata of B have their own equipments (in the sense of the Coxeter manifold $\mathbb{X}^{n-1}$).

For a boundary stratum H of B and some point $y \in H$, denote by $N_H \subset T_y \mathbb{X}^{n-1}$ the normal subspace to $H \subset \mathbb{X}^{n-1}$. The *normal cone* $D_H \subset N_H$ is the cone consisting of vectors looking inside B. Some of $(n-2)$-dimensional strata (former hyperedges) V_α contain H and thus we get the subdivision of the normal cone D_H by tangent spaces to $(n-2)$-dimensional strata, see Figure 12.

We wish to describe the equipment of $B \subset \mathbb{X}^{n-1}$ and the subdivisions of normal cones D_H.

3.3. Finite Coxeter groups

Let Γ be a finite Coxeter group acting in $\mathbb{R}^n$. Let $\mathbb{X}_j^{n-1}$ be the mirrors, let v_j be the vectors orthogonal to the corresponding mirrors. For a vector v_k denote by $R = R_k$ the set of all i such that v_i is orthogonal to v_k.

The reflection group $\Delta(\mathbb{X}_k^{n-1})$ is generated by reflections with respect to mirrors $\mathbb{X}_i^{n-1}$, where i ranges in R.

A. *Let the Coxeter group Γ be irreducible.* We come to the list given in the table. Some comments:

1) $\mathsf{G}_2^{(m)}$ denotes the group of symmetries of a regular plane m-gon, R denotes the one-element group acting in $\mathbb{R}^1$; all other notations are standard, see [2].
2) In some cases, there are two Γ-nonequivalent mirrors, then we write both possible variants.

The rolling scheme (see 2.5) is the Coxeter scheme without even edges.

EXAMPLE. a) For the Weyl chamber E_8, its complete development is the Weyl chamber E_7.

b) For the Weyl chamber BC_n, one of the facets is the Weyl chamber BC_{n-1}. All the remaining facets are connected by the rolling graph; the development is the Weyl chamber $\mathsf{A}_1 \oplus \mathsf{D}_{n-2}$.

PROOF OF THE TABLE is a case-by-case examination of root systems; for the groups H_3 and H_4 the proofs are given in Figures 5, 7, 8 (on the other hand the reader can find a nice coordinate description of the hypericosahedron in [2].

B. *If the Coxeter group Γ be reducible,*

$$\Gamma = \Gamma_1 \times \Gamma_2 \times \cdots$$

then its Weyl chamber is the product of the Weyl chambers for the corresponding chambers $C = C_1 \times C_2 \times \cdots$. The Coxeter scheme of Γ is the union of the Coxeter schemes of Γ_j, hence the rolling graph of Γ is the union of the rolling graphs for all the Γ_j. Now we reduce one of factors $C_j \mapsto B_j$ according to the rules given in the Table, and we get a Weyl chamber $B_j \times \prod_{i \neq j} C_i$.

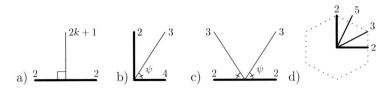

FIGURE 13. Subdivisions of a Coxeter polygon on the Lobachevsky plane (we also draw the labels on lines). There are only 4 possible variants of meetings between lines of a subdivision and the boundary. Here $\tan\psi = \sqrt{2}$; for case d) see Figure 5.

In cases c) and d), the corresponding trihedral angle of the Andreev polytope is covered by our bended polygon.

3.4. Reduction of equipment

Let H be an $(n-k)$-dimensional stratum of C $(k \geqslant 3)$, let $\Gamma_H(C)$ be the corresponding Coxeter group, and let $\mathfrak{N}_H(C)$ be its chamber in the normal cone. Denote by $\Gamma_H(B)$ the corresponding group of the equipment of B and by $\mathfrak{N}_H(B)$ the corresponding chamber in the normal cone.

Theorem 4. *The group $\Gamma_H(B)$ is obtained by reduction of the group $\Gamma_H(B)$ and the subdivision of $\mathfrak{N}_H(B)$ is a partial development of the Weyl chamber $\mathfrak{N}_H(C)$*

Proof. Is obvious. We consider rolling of C with fixed hyperedge H. The subdivision of the cone D_H is obtained by rolling with respect to the hyperedges containing H.   □

4. Addendum. Elementary geometry of Andreev polyhedra

4.1. Rolling of Andreev polyhedra and billiard trajectories in Coxeter polygons

Firstly, our construction gives some information about developments of Andreev polyhedra.

Let us roll an Andreev polyhedron $\subset \mathbb{L}^3$ along a mirror $\simeq \mathbb{L}^2$. In this case, the chamber B of a mirror is a convex plane Coxeter domain. By construction, B is subdivided into several convex polygons by a certain family of lines.

Proposition 3. *All the possible variants of meetings of lines of the subdivision and the boundary of B are presented in Figure 13.*

Proof. We watch all the possible variants of reduction of 3-dimensional finite Coxeter groups to a mirror. The parts a), b), c), d) of Figure 13 correspond to G_2^{2k+1}, BC_3, $A_3 = D_3$, H_3, respectively.   □

Observation 4. *The surface of an Andreev polyhedron is glued from several bended Coxeter polygons; the rules of bending and the rules of gluing are very rigid.*

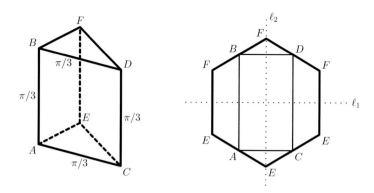

FIGURE 14. An example of an Andreev polyhedron in $\mathbb{L}^3$; we label the dihedral angles $\pi/3$, all other dihedral angles are $\pi/2$.

Its development is a (nonregular) 6-gon, whose angles are $\pi/2$. The lines ℓ_1, ℓ_2 are axes of symmetry. The polygonal curve $ABDCA$ is a billiard trajectory.

It is easy to reconstruct the lengths of edges of the prism from the combinatorial structure of the development and the billiard trajectory. Indeed, we know the angles of the triangle AEC and of the "trapezoids" $ABFE$, and the equiangular quadrangle $ABDC$.

EXAMPLES of rolling of a 3-dimensional Coxeter polyhedron are given in Figures 9 and 14.   ☐

4.2. Example: Rolling along Andreev polyhedra

Secondly, take a Coxeter polyhedron in $\mathbb{L}^4$. Rolling it along the 3-dimensional Lobachevsky space, we obtain a Coxeter polyhedron in $\mathbb{L}^3$ and also some strange subdivision of this polyhedron.

We present an example. Consider the simplex Σ in $\mathbb{L}^4$ defined by the Coxeter scheme

$$\underset{A}{\bullet}\!\!=\!\!\underset{B}{\bullet}\!\!-\!\!\underset{C}{\bullet}\!\!-\!\!\underset{D}{\bullet}\!\!=\!\!\underset{E}{\bullet}. \tag{3}$$

By $A, \ldots, E$ we denote the vertices of the simplex opposite to the corresponding faces. See Figure 15.

Comments to Figure 15. The development of Σ is a prism drawn in Figure 15. We write labels for the dihedral angles $\neq \pi/2$. Below a "stratum" means a stratum of the tiling; in particular, the vertical "edge" AB consists of two 1-dimensional *strata* BC and CA and three 0-dimensional ones, A, B, and C.

1. This is a development. Hence any two strata (segments, triangles) having the same notation are equal (for instance $CD = CD$, $CE = CE$, $\triangle CBE = \triangle CBE$, etc.).
2. Each stratum (a vertex, a segment) is equipped with a Coxeter group (this group is visible from its dihedral angles).

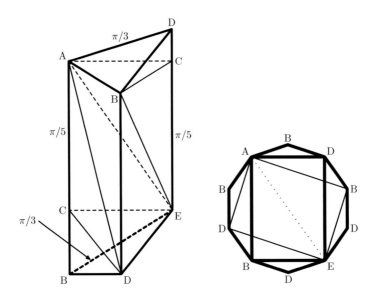

FIGURE 15. This prism in $\mathbb{L}^3$ is a complete development of the Coxeter simplex $ABCDE$ in $\mathbb{L}^4$ described in 4.2. It carries all 2-dimensional hyperedges of the initial simplex.

The development of the prism is the regular 10-gon with right angles (it also carries all 1-dimensional strata of the 4-dimensional simplex).

3. Subdivision of the normal cone D_H to a stratum H (a vertex, a segment) is determined by the reduction procedure from Subsection 3.3.

For instance, in the vertex A we have the subdivision of the spherical triangle H_3 drawn in Figure 7,

$$\overset{B\ \ C\ \ D\ \ E}{\bullet\!\!-\!\!\bullet\!\!-\!\!\bullet\!\!\Longrightarrow} \qquad \mapsto \qquad H_3.$$

In the normal cone to the edge-stratum DE of the prism, we have the icosahedral subdivision, see Figure 8,

$$\overset{A\ \ B\ \ C}{\Longrightarrow\!\!-\!\!\bullet} \qquad \mapsto \qquad A_1 \oplus A_1.$$

The normal cone to the segment AE is drawn in Figure 4; in particular, both angles of incidence are $\arctan\sqrt{2}$,

$$\overset{B\ \ \ C\ \ \ D}{\bullet\!\!-\!\!\bullet\!\!-\!\!\bullet} \qquad \mapsto \qquad A_1 \oplus \mathbb{R}.$$

The "front" face $ABDB$ is orthogonal to the sections CDE and ADE (since the lines CD and AD of intersection are equipped with the group $A_1 \oplus A_1$).

$$\overset{A\ \ B}{\Longrightarrow} \qquad\qquad \overset{E}{\bullet} \qquad \mapsto \qquad \mathbb{R} \oplus A_1$$

etc., etc.

4. The prism has two planes of symmetry. This is by chance, partially this is induced by a symmetry of the initial Coxeter scheme 3. The latter symmetry implies the equality of *strata*:

$$AB = DE, \quad AC = CE, \quad AD = BC, \quad BC = CD.$$

5. Our prism generates a reflection group in $\mathbb{L}^3$. The reader can easily imagine a neighborhood of our prism in $\mathbb{L}^3$. For instance, near the vertex A we have the picture drawn in Figure 8

6. The development of the prism is a regular 10-gon having right angles; the reflection of the "billiard trajectory" $ABEDA$ is of type d) in Figure 13. The regularity property follows by reduction from $\mathbb{L}^4$, but it is not self-obvious from the picture of the 3-dimensional prism. Obviously, diagonals[6] AB are orthogonal to diagonals DE at the points of intersection (see the left side of the figure; but this is not a self-obvious property of this regular 10-gon).

7. We observe the second copy of the polygonal line $ADEBA$ in the development. Bending the 10-gon by this line, we obtain a prism congruent to our prism.

In fact, our 10-gon is the picture on the intersection of two mirrors, denote them by Y_1, Y_2. We can roll the simplex Σ along each mirror Y_1, Y_2 and then we roll it again over the intersection $Y_1 \cap Y_2$. We obtain two different pictures on the 10-gon and both are presented in Figure 15.

References

[1] Alekseevskij, D.V.; Vinberg, E.B.; Solodovnikov, A.S. *Geometry of spaces of constant curvature.* Geometry, II, 1–138, Encyclopaedia Math. Sci., 29, Springer, Berlin, 1993.

[2] Bourbaki, N. *Eléments de mathématique. Fasc. XXXIV. Groupes et algèbres de Lie. Chapitre IV: Groupes de Coxeter et systhèmes de Tits. Chapitre V: Groupes engendrés par des réflexions. Chapitre VI: systèmes de racines.* Hermann, Paris 1968.

[3] Vinberg, E.B. *Hyperbolic reflection groups*, Russian Math. Surveys 40 (1985), 31–75.

[4] Vinberg, E.B.; Shvartsman, O.V. *Discrete groups of motions of spaces of constant curvature.* Geometry, II, 139–248, Encyclopaedia Math. Sci., 29, Springer, Berlin, 1993.

[5] Khovanskii, A. *Combinatorics of sections of polytopes and Coxeter groups in Lobachevsky spaces.* The Coxeter legacy, 129–157, Amer. Math. Soc., Providence, RI, 2006.

[6] Andreev, E.M. *Convex polyhedra in Lobachevsky spaces.* Mat. Sb. (N.S.) 81 (123) 1970 445–478. English transl.: Math. USSR Sb., 10(5), 1970, 413–440.

[7] Lyusternik, L.A. *Convex figures and polyhedra.* Translated and adapted from the first Russian edition (1956) by Donald L. Barnett, D.C. Heath and Co., Boston, Mass. 1966

---

[6]There are two diagonals AB.

[8] Aleksandrov, A.D. *Convex polyhedra.* Gosudarstv. Izdat. Tehn.-Teor. Lit., Moscow, 1950; German translation: Akademie-Verlag, Berlin, 1958; English translation: Springer, 2005

[9] Rivin, I., Hodgson, C.D. *A characterization of compact convex polyhedra in hyperbolic 3-space.* Invent. Math. 111 (1993), no. 1, 77–111.

[10] Hodgson, C.D. *Deduction of Andreev's theorem from Rivin's characterization of convex hyperbolic polyhedra.* Topology '90 (Columbus, OH, 1990), 185–193, Ohio State Univ. Math. Res. Inst. Publ., 1, de Gruyter, Berlin, 1992.

[11] Thurston, W. *The Geometry and Topology of Three-Manifolds.* The text is available via http://www.msri.org/communications/books/gt3m

[12] Rivin, I. *A characterization of ideal polyhedra in hyperbolic 3-space.* Ann. of Math. (2) 143 (1996), no. 1, 51–70.

[13] Vinberg, E.B. *Discrete linear groups that are generated by reflections.*Math. USSR Izvestia 5 (1971), 1083–1119.

[14] Tits, J. *Groupes et géometrie de Coxeter,* I.H.E.S., 1961, mimeographed notes.

[15] Davis, M.W. *Coxeter groups and aspherical manifolds.* Algebraic topology, Aarhus 1982 (Aarhus, 1982), 197–221, Lecture Notes in Math., 1051, Springer, Berlin, 1984

[16] Davis, M. *Groups generated by reflections and aspherical manifolds not covered by Euclidean space.* Ann. Math., 117 (1983), 293–324

[17] Davis, M. *Geometry and topology of Coxeter manifolds.* Princeton Univ. Press, 2007.

[18] Alexeevski D.V., Kriegl, A., Losik, M., Michor P.W. *Reflection groups on Riemannian manifolds*, Annali Mat. Pur. Appl. 186 (2007), no. 1, 25–58.

Dmitri V. Alekseevski
School of Mathematics, Edinburgh University
Edinburgh EH9 3JZ, United Kingdom
e-mail: d.aleksee@ed.ac.uk

Peter W. Michor
Fakultät für Mathematik, Universität Wien
Nordbergstrasse 15, A-1090 Wien, Austria
    and
Erwin Schrödinger Institute of Mathematical Physics
Boltzmanngasse 9, A-1090 Wien, Austria
e-mail: Peter.Michor@esi.ac.at

Yurii A. Neretin
Fakultät für Mathematik, Universität Wien
Nordbergstrasse 15, A-1090 Wien, Austria
    and
Group of Math.Physics, ITEP
B. Cheremushkinskaya, 25, Moscow, 117259, Russia
    and
MechMath Department, Moscow State University
Vorob'yovy Gory, Moscow, Russia
e-mail: neretin@mccme.ru

Geometric Methods in Physics. XXXI Workshop 2012
Trends in Mathematics, 87–97

Boundedness for Pseudo-differential Calculus on Nilpotent Lie Groups

Ingrid Beltiţă, Daniel Beltiţă and Mihai Pascu

Abstract. We survey a few results on the boundedness of operators arising from the Weyl–Pedersen calculus associated with irreducible representations of nilpotent Lie groups.

Mathematics Subject Classification (2010). Primary 47G30; Secondary 22E25, 47B10.

Keywords. Weyl calculus; Lie group; Calderón–Vaillancourt theorem.

1. Introduction

Convolution operators and global Weyl calculus on nilpotent Lie groups have been extensively studied in many papers, in connection with various problems in partial differential equations and representation theory (see [1–10]). It has been repeatedly remarked that the global Weyl calculus is an extension of the classical Weyl calculus on $\mathbb{R}^n$; however, to see this, some further identifications and results are needed (see [8]). This phenomenon has roots in the fact that the global Weyl calculus is not injective, and though it is associated to a given irreducible representation, the link with the corresponding coadjoint orbit is not clear. These issues were resolved by N.V. Pedersen in [11], who constructed a Weyl calculus – that we call Weyl–Pedersen calculus – associated to an irreducible representation of a nilpotent Lie group, which is a bijection between good spaces of symbols defined on the corresponding orbit and operators defined in the Hilbert space of the representation. In addition, this calculus directly extends the classical Weyl calculus.

The aim of this paper is to survey some boundedness results for the Weyl–Pedersen calculus in the case of flat orbits and to give further applications to some three-step nilpotent Lie groups that have non-flat generic orbits. The results

This research has been partially supported by the Grant of the Romanian National Authority for Scientific Research, CNCS-UEFISCDI, project number PN-II-ID-PCE-2011-3-0131.

are generalizations of the classical Calderón–Vaillancourt theorem [12] and of the
Beals characterization of the pseudo-differential operators [13] (see also [14]).

Main definitions are given in Section 2, along with an example illustrating
the non-injectivity of the global Weyl calculus. The boundedness results are given
in Section 3.

Finally, let us mention here that other extensions of the classical Weyl calculus have been constructed in terms of representations, for instance the magnetic
calculus on $\mathbb{R}^n$ [15], and nilpotent Lie groups [16], Weyl calculus on nilpotent
p-adic Lie groups [17, 18].

A good source for the background information on nilpotent Lie groups and
their representations is [19].

2. Weyl calculi for representations of nilpotent Lie groups

Preliminaries on nilpotent Lie groups

Throughout this paper the nilpotent Lie groups are supposed to be connected
and simply connected. Therefore there is no loss of generality in assuming that a
nilpotent Lie group is a pair $G = (\mathfrak{g}, \cdot)$, where $\mathfrak{g}$ is a nilpotent Lie algebra (over $\mathbb{R}$
unless otherwise mentioned) with the Lie bracket $[\cdot, \cdot]$, and the group multiplication
$\cdot$ is given by the Baker–Campbell–Hausdorff series:

$$(\forall x, y \in \mathfrak{g}) \quad x \cdot y = x + y + \frac{1}{2}[x, y] + \frac{1}{12}\big([x, [x, y]] + [y, [y, x]]\big) + \cdots .$$

If the Lie algebra $\mathfrak{g}$ is nilpotent of step n, the group multiplication $\mathfrak{g} \times \mathfrak{g} \to \mathfrak{g}$,
$(x, y) \mapsto x \cdot y$ is a polynomial mapping of degree n. With this identification the
exponential from the Lie algebra to the group is then the identity, while the inverse
of $x \in \mathfrak{g}$ is $-x$, and the unit element is $0 \in \mathfrak{g}$.

We recall that for every $\xi \in \mathfrak{g}^*$ the corresponding coadjoint orbit is

$$\mathcal{O}_\xi := \{(\mathrm{Ad}_G^* x)\xi \mid x \in \mathfrak{g}\} \simeq G/G_\xi$$

where $G_\xi := \{x \in \mathfrak{g} \mid (\mathrm{Ad}_G^* x)\xi = \xi\}$ is the coadjoint isotropy group (or the
stabilizer of ξ), and $\mathrm{Ad}_G^* \colon G \times \mathfrak{g}^* \to \mathfrak{g}^*$ stands for the coadjoint action. We use the
notation $\mathfrak{g}_\xi$ for the corresponding Lie algebra, called the radical of ξ.

We will always denote by $\mathrm{d}x$ a fixed Lebesgue measure on $\mathfrak{g}$ and we recall
that this is also a two-sided Haar measure on the group G. We let $\mathrm{d}\xi$ be a Lebesgue
measure on $\mathfrak{g}^*$ with the property that if we define the Fourier transform for every
$a \in L^1(\mathfrak{g}^*)$ by

$$(\mathcal{F}a)(x) = \int_{\mathfrak{g}^*} \mathrm{e}^{-\mathrm{i}\langle\xi, x\rangle} a(\xi)\mathrm{d}\xi,$$

then we get a unitary operator $\mathcal{F}\colon L^2(\mathfrak{g}^*, \mathrm{d}\xi) \to L^2(\mathfrak{g}, \mathrm{d}x)$. We also denote by
$\mathcal{S}(\mathfrak{g})$ ($\mathcal{S}(\mathfrak{g}^*)$) the Schwartz space on $\mathfrak{g}$ (respectively $\mathfrak{g}^*$), by $\mathcal{S}'(\mathfrak{g})$ ($\mathcal{S}'(\mathfrak{g}^*)$) its
topological dual consisting of the tempered distributions, and by $\langle\cdot, \cdot\rangle\colon \mathcal{S}'(\mathfrak{g}) \times$

$\mathcal{S}(\mathfrak{g}) \to \mathbb{C}$ the corresponding duality pairing. The Fourier transform extends to a linear topological isomorphism $\mathcal{F}\colon \mathcal{S}'(\mathfrak{g}^*) \to \mathcal{S}'(\mathfrak{g})$.

We recall that according to a theorem of Kirillov (see [19, Chap. 2]), there exists a natural one-to-one correspondence between the coadjoint orbits and the equivalence classes of unitary irreducible representations of a nilpotent, connected and simply connected Lie group G.

Example 1. For every integer $n \geq 1$ let $(\cdot \mid \cdot)$ denote the Euclidean scalar product on $\mathbb{R}^n$. The *Heisenberg algebra* is $\mathfrak{h}_{2n+1} = \mathbb{R}^n \times \mathbb{R}^n \times \mathbb{R}$ with the bracket $[(q,p,t),(q',p',t')] = [(0,0,(p \mid q') - (p' \mid q))]$. The *Heisenberg group* is $\mathbb{H}_{2n+1} = (\mathfrak{h}_{2n+1}, \cdot)$ with the multiplication $x \cdot y = x + y + \frac{1}{2}[x,y]$. If we identify the dual space $\mathfrak{h}_{2n+1}^*$ with $\mathbb{R}^n \times \mathbb{R}^n \times \mathbb{R}$ in the usual way, then every coadjoint orbit belongs to one of the following families:

   i) the affine hyperplanes $\mathcal{O}_\hbar = \mathbb{R}^n \times \mathbb{R}^n \times \{1/\hbar\}$ with $\hbar \in \mathbb{R} \setminus \{0\}$;
   ii) the singletons $\mathcal{O}_{a,b} = \{(a,b,0)\}$ with $a, b \in \mathbb{R}^n$.

For every $\hbar \in \mathbb{R} \setminus \{0\}$ there is a unitary irreducible representation on the Hilbert space $L^2(\mathbb{R}^n)$, namely $\pi_\hbar\colon \mathbb{H}_{2n+1} \to \mathcal{B}(L^2(\mathbb{R}^n))$ defined by

$$(\pi_\hbar(q,p,t)f)(x) = e^{i((p|x)+\frac{1}{2}(p|q)+\frac{t}{\hbar})}f(q+x) \text{ for a.e. } x \in \mathbb{R}^n$$

for arbitrary $f \in L^2(\mathbb{R}^n)$ and $(q,p,t) \in \mathbb{H}_{2n+1}$. This is the *Schrödinger representation* of the Heisenberg group $\mathbb{H}_{2n+1}$, and corresponds to the coadjoint orbit $\mathcal{O}_\hbar$ in the first family above. Moreover, for every $a, b \in \mathbb{R}^n$ there is a unitary irreducible representation on a 1-dimensional Hilbert space, namely $\pi_{(a,b)}\colon \mathbb{H}_{2n+1} \to U(1) := \{z \in \mathbb{C} \mid |z| = 1\}$ defined by

$$\pi_{(a,b)}(q,p,t) = e^{i((p|a)+(q|b))}$$

for all $(q,p,t) \in \mathbb{H}_{2n+1}$, and corresponds to the orbit $\mathcal{O}_{a,b}$ in the second family.

Weyl calculi on nilpotent Lie groups

Definition 1 ([20], [21]). Let $G = (\mathfrak{g}, \cdot)$ be a nilpotent Lie group. If $\pi\colon \mathfrak{g} \to \mathcal{B}(\mathcal{H})$ is a unitary representation of G, then we can use Bochner integrals for extending it to a linear mapping

$$\pi\colon L^1(\mathfrak{g}) \to \mathcal{B}(\mathcal{H}), \quad \pi(b)v = \int_\mathfrak{g} b(x)\pi(x)v dx \text{ if } b \in L^1(\mathfrak{g}) \text{ and } v \in \mathcal{H}.$$

Let $\mathcal{T}$ be a locally convex space which is continuously and densely embedded in $\mathcal{H}$, and has the property that for every $h, \chi \in \mathcal{T}$ we have $(\pi(\cdot)h \mid \chi) \in \mathcal{S}(\mathfrak{g})$. Then we can further extend π to a mapping

$$\pi\colon \mathcal{S}'(\mathfrak{g}) \to \mathcal{L}(\mathcal{T}, \mathcal{T}^*), \quad (\pi(u)h \mid \chi) = \langle u, (\pi(\cdot)h \mid \chi)\rangle$$
$$\text{if } u \in \mathcal{S}'(\mathfrak{g}), h \in \mathcal{T}, \text{ and } \chi \in \mathcal{T},$$

where $\mathcal{T}^*$ is the strong antidual of $\mathcal{T}$, and $(\cdot \mid \cdot)$ denotes the anti-duality between $\mathcal{T}$ and $\mathcal{T}^*$ that extends the scalar product of $\mathcal{H}$. Here we have used the notation

$\mathcal{L}(\mathcal{T}, \mathcal{T}^*)$ for the space of continuous linear operators between the above spaces (these operators are thought of as possibly unbounded linear operators in $\mathcal{H}$).

In the same setting, the *global Weyl calculus for the representation* π is the mapping

$$\text{OP}\colon \mathcal{F}^{-1}L^1(\mathfrak{g}) \to \mathcal{B}(\mathcal{H}), \quad \text{OP}(a) = \pi(\mathcal{F}a)$$

which further extends to

$$\text{OP}\colon \mathcal{S}'(\mathfrak{g}^*) \to \mathcal{L}(\mathcal{T}, \mathcal{T}^*), \quad \text{OP} := \pi \circ \mathcal{F}^{-1}.$$

Example 2. Consider the (left) regular representation

$$\lambda\colon \mathfrak{g} \to \mathcal{B}(L^2(\mathfrak{g})), \quad (\lambda(x)\phi)(y) = \phi((-x) \cdot y).$$

If we extend it as above to $\lambda\colon L^1(\mathfrak{g}) \to \mathcal{B}(L^2(\mathfrak{g}))$ then we obtain for $b \in L^1(\mathfrak{g})$ and $\psi \in L^2(\mathfrak{g})$

$$(\lambda(b)\psi)(y) = \int_{\mathfrak{g}} b(x)\psi((-x) \cdot y)\mathrm{d}x = (b * \psi)(y)$$

hence we recover the convolution product, which makes $L^1(\mathfrak{g})$ into a Banach algebra.

The above construction of the global Weyl calculus for the regular representation is usually considered with $\mathcal{T} = \mathcal{S}(\mathfrak{g})$, yielding

$$\text{OP}\colon \mathcal{S}'(\mathfrak{g}^*) \to \mathcal{L}(\mathcal{S}(\mathfrak{g}), \mathcal{S}'(\mathfrak{g}))$$

and the related mapping

$$\lambda\colon \mathcal{S}'(\mathfrak{g}) \to \mathcal{L}(\mathcal{S}(\mathfrak{g}), \mathcal{S}'(\mathfrak{g}))$$

whose values are the (possibly unbounded) convolution operators on the nilpotent Lie group G.

In the case of the global Weyl calculus the symbol of an operator in the representation space of an irreducible representation may not be uniquely determined on the corresponding coadjoint orbit, unlike in the case of pseudo-differential Weyl calculus on $\mathbb{R}^n$ or two step nilpotent groups. (See Example 4 and also [9].) The Kirillov character formula says that when $a \in \mathcal{S}(\mathfrak{g}^*)$, and $\pi\colon G \mapsto \mathcal{B}(\mathcal{H})$ is a irreducible representation, then $\text{OP}(a)$ is a trace class operator and there exists a constant that depends on the unitary class of equivalence of π only, such that

$$\text{Tr}\,(\text{OP}(a)) = C \int_{\mathcal{O}} a(\xi)\mathrm{d}\xi,$$

where $\mathcal{O}$ is the coadjoint orbit corresponding to π. This seems to suggest that $\text{OP}(a)$, when $a \in C^\infty(\mathfrak{g}) \cap \mathcal{S}'(\mathfrak{g})$, depends only on the restriction of a to the coadjoint orbit $\mathcal{O}$. This is not always the case, as it could be seen from the next example [6, App. Sect. I]; see also [22, Ex. N4N1, pp. 9–10].

Example 3. Let $\mathfrak{g}$ be the 4-dimensional threadlike (or filiform) Lie algebra. Equivalently, $\mathfrak{g}$ is 3-step nilpotent, 4-dimensional, and its center is 1-dimensional. Then $\mathfrak{g}$ has a Jordan–Hölder basis $\{X_1, X_2, X_3, X_4\}$ satisfying the commutation relations $[X_4, X_3] = [X_4, X_2] = X_1$ and $[X_j, X_k] = 0$ if $1 \leq k < j \leq 4$ with $(j, k) \notin \{(4, 3), (4, 2)\}$. Let $\mathcal{O}$ be the coadjoint orbit of the functional

$$\xi_0 \colon \mathfrak{g} \to \mathbb{R}, \quad \xi_0(t_1 X_1 + t_2 X_2 + t_3 X_3 + t_4 X_4) = t_1.$$

Then $\dim \mathcal{O} = 2$, and if we identify $\mathfrak{g}^*$ to $\mathbb{R}^4$ by using the basis dual to $\{X_1, X_2, X_3, X_4\}$, then we have $\mathcal{O} = \{(1, -t, \frac{t^2}{2}, s) \mid s, t \in \mathbb{R}\} \subset \mathbb{R}^4$. A unitary irreducible representation $\pi \colon \mathfrak{g} \to \mathcal{B}(L^2(\mathbb{R}))$ associated with the coadjoint orbit $\mathcal{O}$ can be defined by

$$\mathrm{d}\pi(X_1) = \mathrm{i}\mathbf{1}, \quad \mathrm{d}\pi(X_2) = -\mathrm{i}t, \quad \mathrm{d}\pi(X_3) = \frac{\mathrm{i}t^2}{2}, \quad \mathrm{d}\pi(X_3) = -\frac{\mathrm{d}}{\mathrm{d}t},$$

where we denote by t both the variable of the functions in $L^2(\mathbb{R})$ and the operator of multiplication by this variable in $L^2(\mathbb{R})$.

It is clear that the function $a \colon \mathbb{R}^4 \to \mathbb{R}$, $a(y_1, y_2, y_3, y_4) = y_4^2(2y_3 - y_2^2)$, vanishes on the coadjoint orbit $\mathcal{O}$, and on the other hand it was noted in [6, pag. 236] that $\mathrm{OP}^\pi(a) = -\frac{1}{6}\mathbf{1}$.

N.V. Pedersen introduced in [11] an orbital Weyl calculus that is specific to a given orbit, or equivalently, to a class of unitary irreducible representations and that, in addition, gives isomorphism between Schwartz symbols defined on the orbit and regularizing operators defined in the space of the representation. The calculus may depend on the choice of a Jordan–Hölder basis.

To describe this Weyl–Pedersen calculus we need first some notation. Let $G = (\mathfrak{g}, \cdot)$ be a nilpotent Lie group of dimension $m \geq 1$ and assume that $\{X_1, \ldots, X_m\}$ is a Jordan–Hölder basis in $\mathfrak{g}$; so for $j = 1, \ldots, m$ if we define $\mathfrak{g}_j := \mathrm{span}\{X_1, \ldots, X_j\}$ then $[\mathfrak{g}, \mathfrak{g}_j] \subseteq \mathfrak{g}_{j-1}$, where $\mathfrak{g}_0 := \{0\}$. Let $\pi \colon \mathfrak{g} \to \mathcal{B}(\mathcal{H})$ be a unitary representation of G associated with a coadjoint orbit $\mathcal{O} \subseteq \mathfrak{g}^*$. Pick $\xi_0 \in \mathcal{O}$, denote $e := \{j \mid X_j \notin \mathfrak{g}_{j-1} + \mathfrak{g}_{\xi_0}\}$, and then define $\mathfrak{g}_e := \mathrm{span}\{X_j \mid j \in e\}$. We have $\mathfrak{g} = \mathfrak{g}_{\xi_0} + \mathfrak{g}_e$ and the mapping $\mathcal{O} \to \mathfrak{g}_e^*$, $\xi \mapsto \xi|_{\mathfrak{g}_e}$, is a diffeomorphism. Hence we can define an orbital Fourier transform $\mathcal{S}'(\mathcal{O}) \to \mathcal{S}'(\mathfrak{g}_e)$, $a \mapsto \widehat{a}$ which is a linear topological isomorphism and such that for every $a \in \mathcal{S}(\mathcal{O})$ we have

$$(\forall X \in \mathfrak{g}_e) \quad \widehat{a}(X) = \int_{\mathcal{O}} \mathrm{e}^{-\mathrm{i}\langle \xi, X \rangle} a(\xi) \mathrm{d}\xi.$$

Here we have the Lebesgue measure $\mathrm{d}x$ on $\mathfrak{g}_e$ corresponding to the basis $\{X_j \mid j \in e\}$ and $\mathrm{d}\xi$ is the Borel measure on $\mathcal{O}$ such that the aforementioned diffeomorphism $\mathcal{O} \to \mathfrak{g}_e^*$ is a measure preserving mapping and the Fourier transform $L^2(\mathcal{O}) \to L^2(\mathfrak{g}_e)$ is unitary. The inverse of this orbital Fourier transform is denoted by $a \mapsto \breve{a}$.

Definition 2 ([11]). With the notation above, the *Weyl–Pedersen calculus* associated to the unitary irreducible representation π is the mapping

$$\mathrm{Op}_\pi : \mathcal{S}(\mathcal{O}) \to \mathcal{B}(\mathcal{H}), \quad \mathrm{Op}_\pi(a) = \int_{\mathfrak{g}_e} \widehat{a}(x)\pi(x)\mathrm{d}x.$$

The space of smooth vectors $\mathcal{H}_\infty := \{v \in \mathcal{H} \mid \pi(\cdot)v \in \mathcal{C}^\infty(\mathfrak{g}, \mathcal{H})\}$ is dense in $\mathcal{H}$ and has the natural topology of a nuclear Fréchet space with the space of the antilinear functionals denoted by $\mathcal{H}_{-\infty} := \mathcal{H}_\infty^*$ (with the strong dual topology). One can show that the Weyl–Pedersen calculus extends to a linear bijective mapping

$$\mathrm{Op}_\pi : \mathcal{S}'(\mathcal{O}) \to \mathcal{L}(\mathcal{H}_\infty, \mathcal{H}_{-\infty}), \quad (\mathrm{Op}_\pi(a)v \mid w) = \langle \widehat{a}, (\pi(\cdot)v \mid w)\rangle$$

for $a \in \mathcal{S}'(\mathcal{O})$, $v, w \in \mathcal{H}_\infty$, where in the left-hand side $(\cdot \mid \cdot)$ denotes the extension of the scalar product of $\mathcal{H}$ to the sesquilinear duality pairing between $\mathcal{H}_\infty$ and $\mathcal{H}_{-\infty}$.

Note that in fact Op_π is associated to the coadjoint orbit corresponding to the representation π. Indeed if π_1 and π are unitary equivalent representations, Op_π and Op_{π_1} are defined on the same space of symbols, there is a unitary operator U such that $\mathrm{Op}_{\pi_1}(a) = U^*\mathrm{Op}_\pi(a)U$ for every $a \in \mathcal{S}(\mathcal{O})$, and this equality extends naturally to $a \in \mathcal{S}'(\mathcal{O})$. Therefore, whenever the orbit $\mathcal{O}$ is fixed and no confusion may arise, we use the notation Op instead of Op_π, for any irreducible representation corresponding to $\mathcal{O}$.

Remark 1. The map $\mathcal{S}(\mathfrak{g}^*) \to \mathcal{L}(\mathcal{H}_{-\infty}, \mathcal{H}_\infty)$, $a \mapsto \mathrm{OP}(a)$ is surjective [5], while $\mathcal{S}(\mathcal{O}) \to \mathcal{L}(\mathcal{H}_{-\infty}, \mathcal{H}_\infty)$, $a \mapsto \mathrm{Op}_\pi(a)$ is bijective [11]. In fact, it follows by [11, Thm.2.2.7] that $\mathrm{OP}(a) = \mathrm{Op}_\pi(b)$, where

$$\widehat{b}(x) = C_{\mathcal{O},e}\mathrm{Tr}(\pi(-x)\mathrm{OP}(a)), \quad x \in \mathfrak{g}_e,$$

and $C_{\mathcal{O},e}$ is a constant that depends on the Jordan–Hölder basis and on the coadjoint orbit $\mathcal{O}$.

Example 4. The Weyl–Pedersen calculus for the Schrödinger representation of the Heisenberg group $\mathbb{H}_{2n+1}$ is just the usual Weyl calculus from the theory of partial differential equations on $\mathbb{R}^n$, as developed for instance in [23] and [24]. In this case we have $\mathcal{H} = L^2(\mathbb{R}^n)$ and $\mathcal{H}_\infty = \mathcal{S}(\mathbb{R}^n)$.

3. Boundedness for the orbital Weyl calculus

Let $G = (\mathfrak{g}, \cdot)$ be a nilpotent Lie group with a unitary irreducible representation $\pi : \mathfrak{g} \to \mathcal{B}(\mathcal{H})$ associated with the coadjoint orbit $\mathcal{O} \subseteq \mathfrak{g}^*$. One proved in [25] that if the symbols belong to suitable modulation spaces $M^{\infty,1} \hookrightarrow \mathcal{S}'(\mathcal{O})$, then the corresponding values of the Weyl–Pedersen calculus belong to $\mathcal{B}(\mathcal{H})$. This condition does not require smoothness properties of symbols.

We will now describe a result established in [26] which extends both the classical Calderón–Vaillancourt theorem and Beals' criterion on the Weyl calculus, by linking growth properties of derivatives of smooth symbols to boundedness properties of the corresponding pseudo-differential operators in the case when the coadjoint orbit $\mathcal{O}$ is *flat*, in the sense that for some $\xi_0 \in \mathcal{O}$ we have $\mathcal{O} = \{\xi \in \mathfrak{g}^* \mid \xi|_{\mathfrak{z}} = \xi_0|_{\mathfrak{z}}\}$, where $\mathfrak{z}$ is the center of $\mathfrak{g}$. This is equivalent to the condition $\dim \mathcal{O} = \dim \mathfrak{g} - \dim \mathfrak{z}$, and it is also equivalent to the fact that the representation π is square integrable modulo the center of G. Since this hypothesis on coadjoint orbits might look more restrictive than it really is, we recall that every nilpotent Lie group embeds into a nilpotent Lie group whose generic coadjoint orbits (that is, the ones of maximal dimension) are flat [19, Ex. 4.5.14].

If $\mathcal{O}$ is a generic flat coadjoint orbit, then the Weyl–Pedersen calculus

$$\mathrm{Op}: \mathcal{S}'(\mathcal{O}) \to \mathcal{L}(\mathcal{H}_\infty, \mathcal{H}_{-\infty})$$

is a linear topological isomorphism which is uniquely determined by the condition that for every $b \in \mathcal{S}(\mathfrak{g})$ we have

$$\mathrm{Op}((\mathcal{F}^{-1}b)|_{\mathcal{O}}) = \int_{\mathfrak{g}} \pi(x)b(x)\mathrm{d}x,$$

(see [11, Th. 4.2.1]).

We define $\mathrm{Diff}\,(\mathcal{O})$ as the space of all linear differential operators D on $\mathcal{O}$ which are *invariant* to the coadjoint action, in the sense that

$$(\forall x \in \mathfrak{g})(\forall a \in C^\infty(\mathcal{O})) \quad D(a \circ \mathrm{Ad}_G^*(x)|_{\mathcal{O}}) = (Da) \circ \mathrm{Ad}_G^*(x)|_{\mathcal{O}}.$$

Let us consider the Fréchet space of symbols

$$\mathcal{C}_b^\infty(\mathcal{O}) = \{a \in C^\infty(\mathcal{O}) \mid Da \in L^\infty(\mathcal{O}) \text{ for all } D \in \mathrm{Diff}\,(\mathcal{O})\},$$

with the topology given by the seminorms $\{a \mapsto \|Da\|_{L^\infty(\mathcal{O})}\}_{D \in \mathrm{Diff}\,(\mathcal{O})}$.

Theorem 1 ([26]). *Let G be a connected, simply connected, nilpotent Lie group whose generic coadjoint orbits are flat. Let $\mathcal{O}$ be such an orbit with a corresponding unitary irreducible representation $\pi: G \to \mathcal{B}(\mathcal{H})$. Then for $a \in C^\infty(\mathcal{O})$ we have*

$$a \in \mathcal{C}_b^\infty(\mathcal{O}) \iff (\forall D \in \mathrm{Diff}\,(\mathcal{O})) \quad \mathrm{Op}(Da) \in \mathcal{B}(\mathcal{H}).$$

Moreover the Weyl–Pedersen calculus defines a continuous linear map

$$\mathrm{Op}: \mathcal{C}_b^\infty(\mathcal{O}) \to \mathcal{B}(\mathcal{H}),$$

and the Fréchet topology of $C_b^\infty(\mathcal{O})$ is equivalent to that defined by the family of seminorms $\{a \mapsto \|\mathrm{Op}(Da)\|\}_{D \in \mathrm{Diff}\,(\mathcal{O})}$.

Let $\mathcal{C}_\infty^\infty(\mathcal{O})$ be the space of all $a \in C^\infty(\mathcal{O})$ such that the function Da vanishes at infinity on $\mathcal{O}$, for every $D \in \mathrm{Diff}\,(\mathcal{O})$. It easily follows by the above theorem that if $a \in C^\infty(\mathcal{O})$ then

$$a \in \mathcal{C}_\infty^\infty(\mathcal{O}) \iff (\forall D \in \mathrm{Diff}\,(\mathcal{O})) \quad \mathrm{Op}(Da) \text{ is a compact operator.}$$

If π is the Schrödinger representation of the $(2n+1)$-dimensional Heisenberg group, then Theorem 1 gives the characterization of the symbols of type $S_{0,0}^0$ for the pseudo-differential Weyl calculus $\mathrm{Op} \colon \mathcal{S}'(\mathbb{R}^{2n}) \to \mathcal{L}(\mathcal{S}(\mathbb{R}^n), \mathcal{S}'(\mathbb{R}^n))$. Namely, for any symbol $a \in C^\infty(\mathbb{R}^{2n})$ we have

$$(\forall \alpha \in \mathbb{N}^{2n}) \quad \partial^\alpha a \in L^\infty(\mathbb{R}^{2n}) \iff (\forall \alpha \in \mathbb{N}^{2n}) \quad \mathrm{Op}(\partial^\alpha a) \in \mathcal{B}(L^2(\mathbb{R}^n)),$$

where ∂^α stand as usually for the partial derivatives. The above statement unifies the Calderón–Vaillancourt theorem and the so-called Beals criterion for the Weyl calculus.

Application to 3-step nilpotent Lie groups

Proposition 1. *Let $G = (\mathfrak{g}, \cdot)$ be a nilpotent Lie group with an irreducible representation $\pi \colon \mathfrak{g} \to \mathcal{B}(\mathcal{H})$ associated with the coadjoint orbit $\mathcal{O} \subseteq \mathfrak{g}^*$. If $H = (\mathfrak{h}, \cdot)$ is a normal subgroup of G, then the following assertions are equivalent:*

1. *The restricted representation $\pi|_\mathfrak{h} \colon \mathfrak{h} \to \mathcal{B}(\mathcal{H})$ is irreducible.*
2. *The mapping $\mathfrak{g}^* \to \mathfrak{h}^*$, $\xi \mapsto \xi|_\mathfrak{h}$ gives a diffeomorphism of $\mathcal{O}$ onto a coadjoint orbit of H, which will be denoted by $\mathcal{O}|_\mathfrak{h}$.*

If this is the case, then the irreducible representation $\pi|_\mathfrak{h}$ is associated with the coadjoint orbit $\mathcal{O}|_\mathfrak{h}$ of H.

Proof. Pick any Jordan–Hölder sequence that contains $\mathfrak{h}$. Then the implication 1. $\Rightarrow$ 2. follows at once by iterating [19, Th. 2.5.1], while the converse implication follows by using Vergne polarizations. The details of the proof will be given elsewhere. $\qquad\square$

Remark 2. In the setting of Proposition 1, if $X_1, \ldots, X_m$ is a Jordan–Hölder basis in $\mathfrak{g}$ such that $X_1, \ldots, X_k$ is a Jordan–Hölder basis in $\mathfrak{h}$ for $k = \dim \mathfrak{h}$, then the following assertions hold true:

- The coadjoint G-orbit $\mathcal{O}$ and the coadjoint H-orbit $\mathcal{O}|_\mathfrak{h}$ have the same set of jump indices $e \subseteq \{1, \ldots, k\}$. In particular $\mathfrak{g}_e = \mathfrak{h}_e \subseteq \mathfrak{h}$ and $\mathcal{O}$ cannot be flat.
- We have the H-equivariant diffeomorphism $\Theta \colon \mathcal{O} \to \mathcal{O}|_\mathfrak{h}$, $\xi \mapsto \xi|_\mathfrak{h}$. It intertwines the orbital Fourier transforms

$$\mathcal{S}'(\mathcal{O}) \to \mathcal{S}'(\mathfrak{g}_e) = \mathcal{S}'(\mathfrak{h}_e) \quad \text{and} \quad \mathcal{S}'(\mathcal{O}|_\mathfrak{h}) \to \mathcal{S}'(\mathfrak{h}_e).$$

Therefore, by using also the previous remark, we obtain

$$(\forall a \in \mathcal{S}'(\mathcal{O})) \quad \mathrm{Op}_\pi(a) = \mathrm{Op}_{\pi|_H}(a \circ \Theta^{-1}). \tag{1}$$

- The H-equivariant diffeomorphism Θ also induces a unital injective homomorphism of associative algebras

$$\mathrm{Diff}\,(\mathcal{O}) \hookrightarrow \mathrm{Diff}\,(\mathcal{O}|_\mathfrak{h}), \quad D \mapsto D^\mathfrak{h},$$

such that $D(a \circ \Theta) = (D^\mathfrak{h} a) \circ \Theta$ for all $a \in \mathcal{C}^\infty(\mathcal{O}|_\mathfrak{h})$ and $D \in \mathrm{Diff}\,(\mathcal{O})$. In particular, it follows that for $p \in \{0\} \cup [1, \infty]$ we get a continuous injective linear map

$$\mathcal{C}^{\infty,p}(\mathcal{O}) \hookrightarrow \mathcal{C}^{\infty,p}(\mathcal{O}|_\mathfrak{h}), \quad a \mapsto a \circ \Theta^{-1} \tag{2}$$

where we use the Fréchet spaces $\mathcal{C}^{\infty,p}$ introduced in [26, Th. 4.4].

In particular, Proposition 1, (1) and (2), along with Theorem 1 prove the next corollary.

Corollary 1. *If one of the equivalent conditions of Proposition 1 holds true and if the orbit $\mathcal{O}|_{\mathfrak{h}}$ is flat, then for every $a \in \mathcal{C}_b^\infty(\mathcal{O})$ we have $\mathrm{Op}_\pi(a) \in \mathcal{B}(\mathcal{H})$, and moreover the Weyl–Pedersen calculus defines a continuous linear map $\mathrm{Op}_\pi \colon \mathcal{C}_b^\infty(\mathcal{O}) \to \mathcal{B}(\mathcal{H})$.*

Note that the coadjoint orbit of the representation π in the above Corollary 1 may not be flat, and yet we have proved an L^2-boundedness assertion just like the one of Theorem 1. We now provide a specific example of a 3-step nilpotent Lie group taken from [10], which illustrates this result.

Example 5. Recall from Example 1 the Heisenberg algebra $\mathfrak{h}_{2n+1} = \mathbb{R}^{2n} \times \mathbb{R}$ with the Lie bracket given by $[(x, t), (x', t')] = (0, \omega(x, x'))$ for $x, x' \in \mathbb{R}^{2n}$ and $t, t' \in \mathbb{R}$, where $\omega \colon \mathbb{R}^{2n} \times \mathbb{R}^{2n} \to \mathbb{R}$ is the symplectic form given by $\omega((q, p), (q', p')) = (p \mid q') - (p' \mid q)$ for $(q, p), (q', p') \in \mathbb{R}^n \times \mathbb{R}^n = \mathbb{R}^{2n}$. It easily follows that if we consider the symplectic group

$$\mathrm{Sp}(\mathbb{R}^{2n}, \omega) = \{T \in M_{2n}(\mathbb{R}) \mid (\forall x, x' \in \mathbb{R}^{2n}) \quad \omega(Tx, Tx') = \omega(x, x')\}$$

then every $T \in \mathrm{Sp}(\mathbb{R}^{2n}, \omega)$ gives rise to an automorphism $\alpha_T \in \mathrm{Aut}(\mathfrak{h}_{2n+1})$ by the formula $\alpha_T(x, t) = (Tx, t)$ for all $x \in \mathbb{R}^{2n}$ and $t \in \mathbb{R}$. Moreover, the correspondence $\alpha \colon \mathrm{Sp}(\mathbb{R}^{2n}, \omega) \to \mathrm{Aut}(\mathfrak{h}_{2n+1}) \simeq \mathrm{Aut}(\mathbb{H}_{2n+1})$, $T \mapsto \alpha_T$, is a group homomorphism and is injective. On the other hand, if we write the elements of $\mathrm{Sp}(\mathbb{R}^{2n}, \omega)$ as 2×2-block matrices with respect to the decomposition $\mathbb{R}^{2n} = \mathbb{R}^n \times \mathbb{R}^n$, then it is well known that $\mathrm{Sp}(\mathbb{R}^{2n}, \omega)$ is a Lie group whose Lie algebra is

$$\mathfrak{sp}(\mathbb{R}^{2n}, \omega) = \left\{ \begin{pmatrix} A & B \\ C & -A^\top \end{pmatrix} \mid A, B = B^\top, C = C^\top \in M_n(\mathbb{R}) \right\}$$

where $B^\top$ stands for the transpose of a matrix B. In particular, $\mathfrak{sp}(\mathbb{R}^{2n}, \omega)$ has the following abelian Lie subalgebra

$$\mathfrak{s}_n(\mathbb{R}) = \left\{ \begin{pmatrix} 0 & 0 \\ C & 0 \end{pmatrix} \mid C = C^\top \in M_n(\mathbb{R}) \right\}$$

and the corresponding Lie subgroup of $\mathrm{Sp}(\mathbb{R}^{2n}, \omega)$ is the abelian Lie group

$$S_n(\mathbb{R}) = \left\{ \begin{pmatrix} 1 & 0 \\ C & 1 \end{pmatrix} \mid C = C^\top \in M_n(\mathbb{R}) \right\}.$$

It is easily seen that the group $S_n(\mathbb{R})$ acts (via α) on $\mathfrak{h}_{2n+1}$ by unipotent automorphisms, and therefore the corresponding semidirect product $G = S_n(\mathbb{R}) \ltimes_\alpha \mathbb{H}_{2n+1}$ is a nilpotent Lie group. Its Lie algebra is $\mathfrak{g} = \mathfrak{s}_n(\mathbb{R}) \ltimes \mathfrak{h}_{2n+1}$. If we denote by $\mathrm{sym}_n(\mathbb{R})$ the set of all symmetric matrices in $M_n(\mathbb{R})$ viewed as an abelian Lie algebra, then we have an isomorphism of Lie algebras $\mathfrak{g} \simeq \mathrm{sym}_n(\mathbb{R}) \ltimes \mathfrak{h}_{2n+1}$ with the Lie bracket given by

$$[(C, q, p, t), (C', q', p', t')] = (0, 0, Cq' - C'q, (p \mid q') - (p' \mid q))$$

for $C, C' \in \text{sym}_n(\mathbb{R})$, $q, q', p, p' \in \mathbb{R}^n$ and $t, t' \in \mathbb{R}$. It easily follows by the above formula that $[\mathfrak{g}, [\mathfrak{g}, [\mathfrak{g}, \mathfrak{g}]]] = \{0\}$ hence $\mathfrak{g}$ is a 3-step nilpotent Lie algebra. Moreover, $\mathfrak{h}_{2n+1} \simeq \{0\} \times \mathfrak{h}_{2n+1}$ is an ideal of $\mathfrak{g}$ and it was proved in [10] that for every coadjoint G-orbit $\mathcal{O} \subseteq \mathfrak{g}^*$ of maximal dimension we have $\dim \mathcal{O} = n$ and moreover the mapping $\xi \mapsto \xi|_{\mathfrak{h}_{2n+1}}$ gives a diffeomorphism of $\mathcal{O}$ onto a coadjoint $\mathbb{H}_{2n+1}$-orbit $\mathcal{O}|_{\mathfrak{h}_{2n+1}}$. Thus Proposition 1 and Remark 2 apply for the Weyl–Pedersen calculus of a unitary irreducible representation $\pi\colon \mathfrak{g} \to \mathcal{B}(\mathcal{H})$ associated with the coadjoint orbit $\mathcal{O}$. Note that this result is similar to, and yet quite different from [10, Th. 8.6–8.7], inasmuch as we work here with an irreducible representation (see Definition 2) instead of the regular representation of G (see Example 2).

References

[1] R.R. Coifman, G. Weiss, *Operators associated with representations of amenable groups, singular integrals induced by ergodic flows, the rotation method and multipliers*. Studia Math. **47** (1973), 285–303.

[2] P. Głowacki, *A symbolic calculus and L^2-boundedness on nilpotent Lie groups*. J. Funct. Anal. **206** (2004), no. 1, 233–251.

[3] P. Głowacki, *The Melin calculus for general homogeneous groups*. Ark. Mat. **45** (2007), no. 1, 31–48.

[4] M. Gouleau, *Algèbre de Lie nilpotente graduée de rang 3 et inverse d'un opérateur différentiel*. J. Lie Theory **12** (2002), no. 2, 325–356.

[5] R. Howe, *A symbolic calculus for nilpotent groups*. In: Operator Algebras and Group Representations (Neptun, 1980). Vol. I. Monogr. Stud. Math., 17, Pitman, Boston, MA, 1984, pp. 254–277.

[6] D. Manchon, *Calcul symbolique sur les groupes de Lie nilpotents et applications*. J. Funct. Anal. **102** (1991), no. 1, 206–251.

[7] A. Melin, *Parametrix constructions for right invariant differential operators on nilpotent groups*. Ann. Global Anal. Geom. **1** (1983), no. 1, 79–130.

[8] K.G. Miller, *Invariant pseudodifferential operators on two-step nilpotent Lie groups*. Michigan Math. J. **29** (1982), no. 3, 315–328.

[9] K.G. Miller, *Invariant pseudodifferential operators on two-step nilpotent Lie groups. II*. Michigan Math. J. **33** (1986), no. 3, 395–401.

[10] G. Ratcliff, *Symbols and orbits for 3-step nilpotent Lie groups*. J. Funct. Anal. **62** (1985), no. 1, 38–64.

[11] N.V. Pedersen, *Matrix coefficients and a Weyl correspondence for nilpotent Lie groups*. Invent. Math. **118** (1994), no. 1, 1–36.

[12] A.-P. Calderón, R. Vaillancourt, *A class of bounded pseudo-differential operators*. Proc. Nat. Acad. Sci. U.S.A. **69** (1972), 1185–1187.

[13] R. Beals, *Characterization of pseudodifferential operators and applications*. Duke Math. J. **44** (1977), no. 1, 45–57.

[14] J.-M. Bony, *Caractérisations des opérateurs pseudo-différentiels*. In: Séminaire sur les Équations aux Dérivées Partielles, 1996–1997, Exp. No. XXIII, 17 pp., École Polytech., Palaiseau, 1997.

[15] V. Iftimie, M. Măntoiu, R. Purice, *Magnetic pseudodifferential operators*. Publ. Res. Inst. Math. Sci. **43** (2007), no. 3, 585–623.

[16] I. Beltiţă, D. Beltiţă, *Continuity of magnetic Weyl calculus*. J. Funct. Anal. **260** (2011), no. 7, 1944–1968.

[17] S. Haran, *Quantizations and symbolic calculus over the p-adic numbers*. Ann. Inst. Fourier (Grenoble) **43** (1993), no. 4, 997–1053.

[18] A. Bechata, *Calcul pseudodifférentiel p-adique*. Ann. Fac. Sci. Toulouse Math. (6) **13** (2004), no. 2, 179–240.

[19] L.J. Corwin, F.P. Greenleaf, *Representations of Nilpotent Lie Groups and Their Applications*. Part I. Basic theory and examples. Cambridge Studies in Advanced Mathematics, 18. Cambridge University Press, Cambridge, 1990.

[20] R.F.V. Anderson, *The Weyl functional calculus*. J. Functional Analysis 4(1969), 240–267.

[21] R.F.V. Anderson, *The multiplicative Weyl functional calculus*. J. Functional Analysis **9**(1972), 423–440.

[22] N.V. Pedersen, *Geometric quantization and nilpotent Lie groups: A collection of examples*. Preprint, University of Copenhagen, Denmark, 1988.

[23] L. Hörmander, *The Weyl calculus of pseudodifferential operators*. Comm. Pure Appl. Math. **32** (1979), no. 3, 360–444.

[24] L. Hörmander, *The Analysis of Linear Partial Differential Operators*. III. Pseudo-differential operators. Reprint of the 1994 edition. Classics in Mathematics. Springer, Berlin, 2007.

[25] I. Beltiţă, D. Beltiţă, *Modulation spaces of symbols for representations of nilpotent Lie groups*. J. Fourier Anal. Appl. **17** (2011), no. 2, 290–319.

[26] I. Beltiţă, D. Beltiţă, *Boundedness for Weyl–Pedersen calculus on flat coadjoint orbits*. Preprint arXiv:1203.0974v1 [math.AP], 2012.

Ingrid Beltiţă and Daniel Beltiţă
Institute of Mathematics "Simion Stoilow"
of the Romanian Academy, research unit 1
P.O. Box 1-764
Bucharest, Romania
e-mail: Ingrid.Beltita@imar.ro
       Daniel.Beltita@imar.ro

Mihai Pascu
University "Petrol-Gaze" of Ploieşti
    and
Institute of Mathematics "Simion Stoilow"
of the Romanian Academy, research unit 1
P.O. Box 1-764
Bucharest, Romania
e-mail: Mihai.Pascu@imar.ro

Geometric Methods in Physics. XXXI Workshop 2012
Trends in Mathematics, 99–108
© 2013 Springer Basel

A Useful Parametrization of Siegel–Jacobi Manifolds

Stefan Berceanu

Abstract. We determine the homogeneous Kähler diffeomorphism which expresses the Kähler two-form on the Siegel–Jacobi ball $\mathcal{D}_n^J = \mathbb{C}^n \times \mathcal{D}_n$ as the sum of the Kähler two-form on $\mathbb{C}^n$ and the one on the Siegel ball $\mathcal{D}_n$. Similar considerations are presented for the Siegel–Jacobi upper half-plane $\mathcal{X}_n^J = \mathbb{C}^n \times \mathcal{X}_n$, where $\mathcal{X}_n$ denotes the Siegel upper half-plane.

Mathematics Subject Classification (2010). 32Q15,32A25,30H20,81R30 81S10.

Keywords. Jacobi group, coherent and squeezed states, Siegel–Jacobi domains, fundamental conjecture for homogeneous Kähler manifolds.

1. Introduction

In this note by Jacobi group we mean the semidirect product $G_n^J = H_n \rtimes \mathrm{Sp}(n, \mathbb{R})_\mathbb{C}$, where H_n denotes the $(2n + 1)$-dimensional Heisenberg group [1, 2]. The Siegel–Jacobi ball (called in [3] Siegel–Jacobi disk) is the Kähler homogeneous domain $\mathcal{D}_n^J := H_n/\mathbb{R} \times \mathrm{Sp}(n, \mathbb{R})_\mathbb{C}/\mathrm{U}(n) = \mathbb{C}^n \times \mathcal{D}_n$ attached to the Jacobi group, where $\mathcal{D}_n$ is the Siegel ball. The real Jacobi group is defined as $G_n^J(\mathbb{R}) = \mathrm{Sp}(n, \mathbb{R}) \ltimes H_n$, where now H_n is the real Heisenberg group. The Siegel–Jacobi upper half-plane (called in [4] Siegel–Jacobi space) is $\mathcal{X}_n^J := \mathcal{X}_n \times \mathbb{R}^{2n}$, where $\mathcal{X}_n = \mathrm{Sp}(n, \mathbb{R})/\mathrm{U}(n)$ is the Siegel upper half-plane. There is an group isomorphism $\Theta : G_n^J(\mathbb{R}) \to G_n^J$ and the action of G_n^J on $\mathcal{D}_n^J$ is compatible with the action of G_n^J on $\mathcal{X}_n^J$ [5, 6]. The holomorphic irreducible unitary representations of the Jacobi groups based on Siegel–Jacobi domains $\mathcal{D}_n^J$ and $\mathcal{X}_n^J$ have been constructed by Berndt, Böcherer, Schmidt [2, 7] and Takase [8–10], see also [11]. The geometric properties of the Siegel–Jacobi domains have been investigated by Satake [12], Berndt and Schmidt [2], Kähler[13], and Yang [3–5].

The Jacobi group is relevant for several branches of physics, as quantum mechanics, quantum optics and in particular squeezed states, geometric quantization [14–23]. In [6, 24] we have attached to the Jacobi group coherent states [25] based on the Siegel–Jacobi ball. The case G_1^J was studied in [26]. Starting from the Jacobi algebra $\mathfrak{g}_n^J$, we have calculated the scalar product K of two coherent states based on $\mathcal{D}_n^J$ [6, 24]. Following a general procedure [27], we have calculated the homogeneous Kähler two-form ω_n on the Siegel–Jacobi domain $\mathcal{D}_n^J$ from the scalar product K by a formula of the type $\omega = \mathrm{i} \sum_{\alpha,\beta} \frac{\partial^2}{\partial z_\alpha \partial \bar{z}_\beta} f$ [28], where f is the Kähler potential $f = \log K$ and z are local coordinates on the homogeneous manifold M. ω_n was calculated also in other papers by different methods [3, 4, 13].

In this paper we continue the investigation of the geometric properties of Siegel–Jacobi domains. We are interested to find a homogeneous Kähler transform FC which put the Kähler two-form ω_n on the Siegel–Jacobi ball $\mathcal{D}_n^J = \mathbb{C}^n \times \mathcal{D}_n$ as the sum of independent Kähler two-forms on the Siegel ball $\mathcal{D}_n$ and $\mathbb{C}^n$. In [29] we have presented results for the construction of the FC transform in the case of $\mathcal{D}_1^J$ and $\mathcal{X}_1^J$ and in [30] we give the full details. In [29] also we have announced the construction of the FC transform for Siegel–Jacobi domains. Here we give an exact formulation of the results in the context of the fundamental conjecture of homogeneous Kähler manifolds of Gindikin and Vinberg [28, 31, 32]. More details and proofs are given in [33].

The paper is laid out as follows. §2 starts with the definition of the Jacobi algebra $\mathfrak{g}_n^J$, then just recalls the construction of Perelomov coherent states defined on the Siegel–Jacobi ball. Several geometric properties of the Siegel–Jacobi manifold are recalled. §3 treats the real Jacobi group and the partial Cayley transform from $\mathcal{D}_n^J$ to $\mathcal{X}_n^J$. After shortly recalling the contents of the fundamental conjecture, Proposition 4 contains our main result on the determination of the homogeneous Kähler transform FC transform for the Siegel–Jacobi domains. Only indications for the proof are presented.

Notation. We denote by $\mathbb{R}$, $\mathbb{C}$ the field of real numbers, the field of complex numbers, respectively. $M_{mn}(\mathbb{F}) \cong \mathbb{F}^{mn}$ denotes the set of all $m \times n$ matrices with entries in the field $\mathbb{F}$. $M_{n1}(\mathbb{F})$ is identified with $\mathbb{F}^n$. Set $M(n,\mathbb{F}) = M_{nn}(\mathbb{F})$. For any $A \in M(n,\mathbb{F})$, A^t denotes the transpose matrix of A. The identity matrix of degree n is denoted by $\mathbb{I}_n$. If A is a linear operator, we denote by $A^\dagger$ its adjoint. We consider a complex separable Hilbert space $\mathfrak{H}$ endowed with a scalar product which is antilinear in the first argument, $\langle \lambda x, y \rangle = \bar{\lambda}\langle x, y \rangle$, $x, y \in \mathfrak{H}$, $\lambda \in \mathbb{C} \setminus 0$. If $X \in \mathfrak{g}$, we denote $\boldsymbol{X} = \mathrm{d}\pi(X)$, where $\mathfrak{g}$ is the Lie algebra of the group G and $\mathrm{d}\pi$ is the derived representation of G. We recall that a complex analytic manifold is *Kählerian* if it caries a Hermitian metric whose imaginary part ω is closed [34]. A Kähler manifold M is *homogeneous Kählerian* if the group of *automorphisms* (i.e., invertible holomorphic maps which invariates ω) of M is transitive [28, 35]. By a *Kähler homogeneous diffeomorphism*, we mean a diffeomorphism $\varphi : M \to N$ of homogeneous Kähler manifolds such that $\varphi^* \omega_N = \omega_M$.

2. The Siegel–Jacobi ball $\mathcal{D}_n^J$

Let $a_i^\dagger$ (a_i) be the boson creation (respectively, annihilation) operators, verifying the canonical commutation relations $[a_i, a_j^\dagger] = \delta_{ij}$; $[a_i, a_j] = [a_i^\dagger, a_j^\dagger] = 0$, and let us denote by $\mathfrak{h}_n$ the $(2n + 1)$-dimensional Heisenberg algebra, isomorphic to the algebra

$$\mathfrak{h}_n = \left\langle is1 + \sum_{i=1}^n (x_i a_i^\dagger - \bar{x}_i a_i) \right\rangle_{s\in\mathbb{R}, x_i\in\mathbb{C}}.$$

If $K^{0,+,-}$ are the generators of the real symplectic algebra $\mathfrak{sp}(n, \mathbb{R})_\mathbb{C}$ [6], then the Jacobi algebra is the semi-direct sum $\mathfrak{g}_n^J := \mathfrak{h}_n \rtimes \mathfrak{sp}(n, \mathbb{R})_\mathbb{C}$, where the ideal $\mathfrak{h}_n$ in $\mathfrak{g}_n^J$ is determined by the commutation relations:

$$[a_k^\dagger, K_{ij}^+] = [a_k, K_{ij}^-] = 0, \qquad [K_{ij}^0, a_k^\dagger] = \frac{1}{2}\delta_{jk}a_i^\dagger, \quad [a_k, K_{ij}^0] = \frac{1}{2}\delta_{ik}a_j,$$

$$[a_i, K_{kj}^+] = \frac{1}{2}\delta_{ik}a_j^\dagger + \frac{1}{2}\delta_{ij}a_k^\dagger, \quad [K_{kj}^-, a_i^\dagger] = \frac{1}{2}\delta_{ik}a_j + \frac{1}{2}\delta_{ij}a_k.$$

Perelomov's coherent state vectors [25] associated to the Jacobi group G_n^J with the Lie algebra $\mathfrak{g}_n^J$, based on the complex N-dimensional ($N = \frac{n(n+3)}{2}$) manifold – the Siegel–Jacobi ball $\mathcal{D}_n^J := H_n/\mathbb{R} \times \mathrm{Sp}(n, \mathbb{R})_\mathbb{C}/\mathrm{U}(n) - \mathbb{C}^n \times \mathcal{D}_n$ – are defined as [6, 25]

$$e_{z,W} = \exp(\boldsymbol{X})e_0, \ \boldsymbol{X} := \sum_{i=1}^n z_i a_i^\dagger + \sum_{i,j=1}^n w_{ij}\boldsymbol{K}_{ij}^+, \ z \in \mathbb{C}^n; W \in \mathcal{D}_n. \qquad (1)$$

The non-compact Hermitian symmetric space $\mathrm{Sp}(n, \mathbb{R})_\mathbb{C}/\mathrm{U}(n)$ admits a matrix realization as a bounded homogeneous domain, the Siegel ball $\mathcal{D}_n$ [34]

$$\mathcal{D}_n := \{W \in M(n, \mathbb{C}) : W = W^t, \mathbb{I}_n - W\bar{W} > 0\}.$$

The vector e_0 appearing in (1) verifies the relations [6]

$$\boldsymbol{a}_i e_o = 0, \ i = 1, \ldots, n; \ \boldsymbol{K}_{ij}^+ e_0 \neq 0, \ \boldsymbol{K}_{ij}^- e_0 = 0,$$

$$\boldsymbol{K}_{ij}^0 e_0 = \frac{k_i}{4}\delta_{ij}e_0, \ i, j = 1, \ldots, n. \qquad (2)$$

In (2), $e_0 = e_0^H \otimes e_0^K$, where e_0^H is the minimum weight vector (vacuum) for the Heisenberg group H_n, while e_0^K is the extremal weight vector for $\mathrm{Sp}(n, \mathbb{R})_\mathbb{C}$ corresponding to the weight k in (2) with respect to a unitary representation, see details in [6].

If we identify $\mathbb{R}^{2n}$ with $\mathbb{C}^n$, $\mathbb{R}^n \times \mathbb{R}^n \ni (P, Q) \mapsto \alpha: \alpha = P + iQ$, then we have the correspondence

$$M = \begin{pmatrix} a & b \\ c & d \end{pmatrix} \in M(2n, \mathbb{R}) \leftrightarrow M_\mathbb{C} = \mathcal{C}^{-1}M\mathcal{C} = \begin{pmatrix} p & q \\ \bar{q} & \bar{p} \end{pmatrix},$$

$$\mathcal{C} = \begin{pmatrix} i\mathbb{I}_n & i\mathbb{I}_n \\ -\mathbb{I}_n & \mathbb{I}_n \end{pmatrix}. \qquad (3)$$

In particular, to $g \in \mathrm{Sp}(n, \mathbb{R})$, we associate $g \mapsto g_{\mathbb{C}} \in \mathrm{Sp}(n, \mathbb{R})_{\mathbb{C}} \equiv \mathrm{Sp}(n, \mathbb{C}) \cap \mathrm{U}(n, n)$

$$g_{\mathbb{C}} = \begin{pmatrix} p & q \\ \bar{q} & \bar{p} \end{pmatrix}, \quad pp^* - qq^* = \mathbb{I}_n, \ pq^t = qp^t. \tag{4}$$

The following proposition describes the holomorphic, transitive and effective action of the Jacobi group on the Siegel–Jacobi ball and some geometric properties of $\mathcal{D}_n^J$ (cf. [6]):

Proposition 1. *Let $g \in \mathrm{Sp}(n, \mathbb{R})_{\mathbb{C}}$ of the form (4) and $\alpha \in \mathbb{C}^n$. The action of the Jacobi group G_n^J on the Siegel–Jacobi ball $\mathcal{D}_n^J$ is expressed as*

$$W_1 = g \cdot W = (pW + q)(\bar{q}W + \bar{p})^{-1}, \quad z_1 = (Wq^* + p^*)^{-1}(z + \alpha - W\bar{\alpha}). \tag{5}$$

The composition law is

$$(g_1, \alpha_1, t_1) \circ (g_2, \alpha_2, t_2) = \big(g_1 \circ g_2, g_2^{-1} \cdot \alpha_1 + \alpha_2, t_1 + t_2 + \Im(g_2^{-1} \cdot \alpha_1 \bar{\alpha}_2)\big),$$

and if g is as in (4), then $g \cdot \alpha := \alpha_g$ is given by $\alpha_g = p\alpha + q\bar{\alpha}$, and $g^{-1} \cdot \alpha = p^*\alpha - q^t\bar{\alpha}$.

The scalar product $K : M \times \bar{M} \to \mathbb{C}$, $K(\bar{x}, \bar{V}; y, W) = (e_{x,V}, e_{y,W})_k$ is:

$$(e_{x,V}, e_{y,W})_k = \det(U)^{k/2} \exp F(\bar{x}, \bar{V}; y, W), \ U = (\mathbb{I}_n - W\bar{V})^{-1}; \\ 2F(\bar{x}, \bar{V}; y, W) = 2\langle x, Uy \rangle + \langle V\bar{y}, Uy \rangle + \langle x, UW\bar{x} \rangle. \tag{6}$$

In particular, the reproducing kernel $K = (e_{z,W}, e_{z,W})$ is

$$K = \det(M)^{\frac{k}{2}} \exp F, M = (\mathbb{I}_n - W\bar{W})^{-1}, \\ 2F = 2\bar{z}^t M z + z^t \bar{W} M z + \bar{z}^t M W \bar{z}.$$

The homogeneous Kähler manifold $\mathcal{D}_n^J$ has the Kähler potential $f = \log K$, and the Kähler two-form ω_n, G_n^J-invariant to the action (5) is

$$-\mathrm{i}\omega_n = \frac{k}{2}\mathrm{Tr}(B \wedge \bar{B}) + \mathrm{Tr}(A^t\bar{M} \wedge \bar{A}), A = \mathrm{d}z + \mathrm{d}W\bar{\eta}, \\ B = M\,\mathrm{d}W, \ \eta = M(z + W\bar{z}). \tag{7}$$

The Hilbert space of holomorphic functions $\mathfrak{F}_K$ is endowed with the scalar product

$$(\phi, \psi) = \Lambda_n \int_{z \in \mathbb{C}^n; W \in \mathcal{D}_n} \bar{f}_\phi(z, W) f_\psi(z, W) \rho_1 \,\mathrm{d}z\,\mathrm{d}W, \tag{8}$$

$$\rho_1 = \det(\mathbb{I}_n - W\bar{W})^p \exp(-F),$$

$$\Lambda_n = \frac{k-3}{2\pi^{n(n+3)/2}} \prod_{i=1}^{n-1} \frac{(\frac{k-3}{2} - n + i)\Gamma(k+i-2)}{\Gamma[k+2(i-n-1)]}, \ p = k/2 - n - 2,$$

$$\mathrm{d}z = \prod_{i=1}^{n} \mathrm{d}\Re z_i\,\mathrm{d}\Im z_i; \ \mathrm{d}W = \prod_{1 \le i \le j \le n} \mathrm{d}\Re w_{ij}\,\mathrm{d}\Im w_{ij}.$$

Remarks. The Jacobi groups are unimodular, **non**-reductive, algebraic groups of Harish–Chandra type [12]. The Siegel–Jacobi domains are reductive [36], **non**-symmetric manifolds associated to the Jacobi groups by the generalized Harish–Chandra embedding. The Siegel–Jacobi domains are **non**-Einstein manifolds. In the case $n = 1$ the equations of geodesics in the variables $(w, z) \in \mathcal{D}_1^J$ $((w, \eta) \in \mathcal{D}_1 \times \mathbb{C})$ have been calculated in [26] (respectively, in [30]), while the sectional curvature and the scalar curvature have been calculated (the last being $-\frac{3}{2k}$) in [37]. The orthonormal basis $(\varphi_0, \varphi_1, \dots)$ of the symmetric Fock space (8) in which the Bergman kernel K (6) can be developed has been calculated in [11]. This gives the possibility to explicitly construct the (Kobayashi [38]) Kählerian embedding $\iota : \mathcal{D}_n^J \hookrightarrow \mathbb{CP}^\infty$ as $\iota = [\varphi_0 : \varphi_1 : \dots]$ and the Kähler two-form (7) is the pullback of the Fubini-Study Kähler two-form on $\mathbb{CP}^\infty$, $\omega_n|_{\mathcal{D}_n^J} = \iota^* \omega_{FS}|_{\mathbb{CP}^\infty}$. The canonical projection $\xi : \mathfrak{H} \setminus 0 \to \mathbb{P}(\mathfrak{H})$ is $\xi(\boldsymbol{z}) = [\boldsymbol{z}]$ and the Hermitian metric on $\mathbb{CP}^\infty$ is the Fubini-Studi metric $\mathrm{d}\, s^2|_{FS}([\boldsymbol{z}]) = \frac{(\mathrm{d}\,\bar{\boldsymbol{z}}, \mathrm{d}\,\boldsymbol{z})(\boldsymbol{z}, \boldsymbol{z}) - (\mathrm{d}\,\bar{\boldsymbol{z}}, \boldsymbol{z})(\boldsymbol{z}, \mathrm{d}\,\boldsymbol{z})}{(\boldsymbol{z}, \boldsymbol{z})^2}$. This is a direct proof of the fact that *the Siegel–Jacobi ball* $\mathcal{D}_n^J$ *is a coherent state manifold*, i.e., it is a submanifold of a (infinite-dimensional) projective Hilbert space, and that *the Jacobi group* G_n^J *is a coherent-state group* in the meaning of Lisiecki and Neeb [39, 40].

3. The Siegel–Jacobi upper half-plane $\mathcal{X}_n^J$

The real Jacobi group is defined as $G_n^J(\mathbb{R}) = \mathrm{Sp}(n, \mathbb{R}) \ltimes H_n$, where H_n is now the real Heisenberg group of real dimension $(2n + 1)$. If $g = (M, X, k), g' = (M', X', k') \in G_n^J(\mathbb{R})$, $X = (\lambda, \mu) \in \mathbb{R}^{2n}$, $(X, k) \in H_n$, then the composition law in $G_n^J(\mathbb{R})$ is

$$gg' = (MM', XM' + X', k + k' + XM'JX'^t).$$

The restricted real Jacobi group $G_n^J(\mathbb{R})_0$ consists of elements of the form above, but $g = (M, X)$.

The Siegel–Jacobi upper half-plane is

$$\mathcal{X}_n^J := \mathcal{X}_n \times \mathbb{R}^{2n}, \quad \text{where} \quad \mathcal{X}_n = \mathrm{Sp}(n, \mathbb{R})/\mathrm{U}(n)$$

is the Siegel upper half-plane realized as complex symmetric matrices with positive imaginary part.

Let us consider an element $h = (g, l)$ in $G_n^J(\mathbb{R})_0$, i.e.,

$$g = \begin{pmatrix} a & b \\ c & d \end{pmatrix} \in \mathrm{Sp}(n, \mathbb{R}), \; v \in \mathcal{X}_n, \; u \in \mathbb{C}^n \equiv \mathbb{R}^{2n}, \; l = (n, m) \in \mathbb{R}^{2n}. \tag{9}$$

Now we consider the partial Cayley transform [6] $\Phi : \mathcal{X}_n^J \to \mathcal{D}_n^J$, $\Phi(v, u) = (W, z)$

$$W = (v - \mathrm{i}\mathbb{1}_n)(v + \mathrm{i}\mathbb{1}_n)^{-1}, \quad z = 2\mathrm{i}(v + \mathrm{i}\mathbb{1}_n)^{-1} u, \tag{10}$$

with the inverse partial Cayley transform $\Phi^{-1} : \mathcal{D}_n^J \to \mathcal{X}_n^J$, $\Phi^{-1}(W, z) = (v, u)$

$$v = \mathrm{i}(\mathbb{1}_n - W)^{-1}(\mathbb{1}_n + W), \quad u = (\mathbb{1}_n - W)^{-1} z. \tag{11}$$

Let us now define $\Theta : G_n^J(\mathbb{R})_0 \to G_n^J$, $\Theta(h) = h_*$, $h = (g, n, m)$, $h_* = (g_{\mathbb{C}}, \alpha)$. It can be proved that [33] (see also [5, 11])

Proposition 2. Θ *is an group isomorphism and the action of G_n^J on $\mathcal{D}_n^J$ is compatible with the action of $G_n^J(\mathbb{R})_0$ on $\mathcal{X}_n^J$ through the biholomorphic partial Cayley transform* (10): *if $\Theta(h) = h_*$, then $\Phi h = h_* \Phi$. The action of $G_n^J(\mathbb{R})_0$ on $\mathcal{X}_n^J$ is given by $(g, l) \times (v, u) \to (v_1, u_1) \in \mathcal{X}_n^J$, where*

$$v_1 = (av + b)(cv + d)^{-1}; u_1 = (vc^t + d^t)^{-1}(u + vn + m). \tag{12}$$

The matrices g in (9) *and $g_{\mathbb{C}}$ in* (4) *are related by* (3) *while $\alpha = m + \mathrm{i}n$, $m, n \in \mathbb{R}^n$.*

Proposition 3. *The partial Cayley transform is a homogeneous Kähler diffeomorphism, $\Phi^* \omega_n = \omega_n' = \omega_n \circ \Phi$, where the Kähler two-form ω_n'* (13) *on $\mathcal{X}_n^J$ is $G_n^J(\mathbb{R})_0$-invariant to the action* (12),

$$-\mathrm{i}\omega_n' = \frac{k}{2}\mathrm{Tr}(H \wedge \bar{H}) + \frac{2}{\mathrm{i}}\mathrm{Tr}(G^t D \wedge \bar{G}), \quad \text{where}$$
$$D = (\bar{v} - v)^{-1}, H = D\,\mathrm{d}\,v; \; G = \mathrm{d}\,u - \mathrm{d}\,vD(\bar{u} - u). \tag{13}$$

4. The fundamental conjecture for the Siegel–Jacobi domains

Let us remind the *fundamental conjecture for homogeneous Kähler manifolds* (Gindikin–Vinberg): *every homogeneous Kähler manifold is a holomorphic fiber bundle over a homogeneous bounded domain in which the fiber (with the induced Kähler structure) is the product of a locally flat homogeneous Kähler manifold and a compact simply connected homogeneous Kähler manifold.* The compact case was considered by Wang [41]; Borel [35] and Matsushima [42] have considered the case of a transitive reductive group of automorphisms, while Gindikin and Vinberg [31] considered a transitive splittable solvable automorphism group [28]. We mention also the essential contribution of Piatetski–Shapiro in this field [43]. The complex version, in the formulation of Dorfmeister and Nakajima [32], essentially asserts that: *every homogeneous Kähler manifold M, as a complex manifold, is isomorphic with the product of a compact simply connected homogeneous manifold (generalized flag manifold), a homogeneous bounded domain, and $\mathbb{C}^n/\Gamma$, where Γ denotes a discrete subgroup of translations of $\mathbb{C}^n$,*

$$M \;=\; (G^{\mathbb{C}}/P) \qquad \times D \qquad \times (\mathbb{C}^n/\Gamma)$$

flag manifold	homogeneous	Kähler
P-parabolic	bounded domain	flat

Proposition 4. *Under the homogeneous Kähler transform*

$$FC(\eta, W) = (z, W), \; z = \eta - W\bar{\eta}, \tag{14}$$

$$FC^{-1}(z, W) = (\eta, W), \; \eta = (\mathbb{I}_n - W\bar{W})^{-1}(z + W\bar{z}), \tag{15}$$

the Kähler two-form (7) *on* $\mathcal{D}_n^J$, G_n^J-*invariant to the action* (5), *becomes the Kähler two-form on* $\mathcal{D}_n \times \mathbb{C}^n$, $FC^*\omega_n = \omega_{n,0}$,

$$-\mathrm{i}\omega_{n,0} = \tfrac{k}{2}\mathrm{Tr}(B \wedge \bar{B}) + \mathrm{Tr}(\mathrm{d}\,\eta^t \wedge \mathrm{d}\,\bar{\eta}),$$

invariant to the G_n^J-*action on* $\mathcal{D}_n \times \mathbb{C}^n$, $(g,\alpha) \times (\eta, W) \to (\eta_1, W_1)$,

$$\eta_1 = p(\eta + \alpha) + q(\bar{\eta} + \bar{\alpha}).$$

Under the homogeneous Kähler transform

$$FC_1 : u = \frac{1}{2\mathrm{i}}[(v + \mathrm{i}\mathbb{I}_n)\eta - (v - \mathrm{i}\mathbb{I}_n)\bar{\eta}],$$

$$FC_1^{-1} : \eta = (\bar{v} - \mathrm{i}\mathbb{I}_n)(\bar{v} - v)^{-1}(v - \mathrm{i}\mathbb{I}_n)[(v - \mathrm{i}\mathbb{I}_n)^{-1}u - (\bar{v} - \mathrm{i}\mathbb{I}_n)^{-1}\bar{u}],$$

the Kähler two-form (13) *becomes a Kähler two-form on* $\mathcal{X}_n \times \mathbb{C}^n$, $FC_1^*\omega_n' = \omega_{n,0}'$,

$$-\mathrm{i}\omega_{n,0}' = \tfrac{k}{2}\mathrm{Tr}(H \wedge \bar{H}) + \mathrm{Tr}(\mathrm{d}\,\eta^t \wedge \mathrm{d}\,\bar{\eta}), \quad H = (\bar{v} - v)^{-1}\,\mathrm{d}\,v. \tag{16}$$

The Kähler two-form (16) *is invariant to the action* $G_n^J(\mathbb{R})_0$ *on* $\mathcal{X}_n \times \mathbb{C}^n$, $(g,\alpha) \times (v, \eta) \to (v_1, \eta_1)$, *where* g *has the form* (9), v_1 *is given by the first equation* (12), *while*

$$\eta_1 = \frac{1}{2}(\eta + \alpha)[a + d + \mathrm{i}(b - c)] + \frac{1}{2}(\bar{\eta} + \bar{\alpha})[a - d - \mathrm{i}(b + c)].$$

Proof – the first step. Let us introduce [13], [1] the variables $P, Q \in \mathbb{R}^n$ such that $u = vP + Q$, where $(u, v) \in \mathbb{C}^n \times \mathcal{X}_n$ are local coordinates on the Siegel–Jacobi upper half-plane $\mathcal{X}_n^J$. Using the second equation in (11), we have $u = (\mathbb{I}_n - W\bar{W})^{-1}z$ and we introduce for v the expression given by the first equation in (11). We get $z = \eta - W\bar{\eta}$, where $\eta = P + \mathrm{i}Q$ has appeared already in (7). For A in (7) we get $A = \mathrm{d}\,\eta - W\,\mathrm{d}\,\bar{\eta}$. In (7), we make the transform (14). Also, from (14) and the first equation in (10), we get (15).

Corollary 1. *Under the* FC-*change of coordinates* $x = \eta - V\bar{\eta}$, $y = \xi - W\bar{\xi}$, *the reproducing kernel* (6) *becomes* $\mathcal{K} = K \circ FC$,

$$\mathcal{K}(\bar{\eta}, \bar{V}; \xi, W) = (\det U)^{k/2} \exp \mathcal{F}, \quad \text{where}$$

$$2\mathcal{F} = \mathcal{F}_0 + \Delta\mathcal{F}; \ \mathcal{F}_0 = \bar{\xi}^t\xi + \bar{\eta}^t\eta - \bar{\xi}^t W\bar{\xi} - \eta^t\bar{V}\eta,$$

$$\Delta\mathcal{F} = (\bar{\zeta}^t - \zeta^t\bar{V})U(\xi - W\bar{\xi}) + (\bar{\eta}^t - \eta^t\bar{V})U(-\zeta + W\zeta); \zeta = \eta - \xi.$$

For $\xi = \eta, V = W$, *we have* $\Delta\mathcal{F} = 0$,

$$\mathcal{K} = \det(M)^{\frac{k}{2}} \exp(\mathcal{F}), \quad \text{where } \mathcal{F} = \bar{\eta}^t\eta - \frac{1}{2}\eta^t\bar{W}\eta - \frac{1}{2}\bar{\eta}^t W\bar{\eta}, \tag{17}$$

and the scalar product (8) *becomes*

$$(\phi, \psi) = \Lambda_n \int_{\eta \in \mathbb{C}^n; \mathbb{I}_n - W\bar{W} > 0} \bar{f}_\phi(\eta, W)f_\psi(\eta, W)\rho_2\,\mathrm{d}\,\eta\,\mathrm{d}\,W,$$

$$\rho_2 = \det(\mathbb{I}_n - W\bar{W})^q \exp(-\mathcal{F}), q = k/2 - n - 1,$$

with $\mathcal{F}$ *given by* (17).

Acknowledgment

I am indebted to the Organizing Committee of the XXXI Workshop on Geometric Methods in Physics for the opportunity to report results at the Białowieża meeting. I am grateful to Professor Pierre Bieliavsky for useful discussions. This investigation was supported by the ANCS project program PN 09 37 01 02/2009 and by the UEFISCDI-Romania program PN-II Contract No. 55/05.10.2011.

References

[1] M. Eichler and D. Zagier, *The theory of Jacobi forms*, Progress in Mathematics **55** Birkhäuser, Boston, MA, 1985.

[2] R. Berndt and R. Schmidt, *Elements of the representation theory of the Jacobi group*, Progress in Mathematics **163** Birkhäuser Verlag, Base, 1998.

[3] J.-H. Yang, *Invariant metrics and Laplacians on the Siegel–Jacobi disk*, Chin. Ann. Math. **31B** (2010) 85–100.

[4] J.-H. Yang, *Invariant metrics and Laplacians on the Siegel–Jacobi spaces*, J. Number Theory, **127** (2007) 83–102.

[5] J.-H. Yang, *A partial Cayley transform for Siegel–Jacobi disk*, J. Korean Math. Soc. **45** (2008) 781–794.

[6] S. Berceanu, A holomorphic representation of Jacobi algebra in several dimensions, in *Perspectives in Operator Algebra and Mathematical Physics*, F.-P. Boca, R. Purice and S. Stratila, eds., The Theta Foundation, Bucharest 1-25, 2008; arXiv: math.DG/060404381.

[7] R. Berndt and S. Böcherer, *Jacobi forms and discrete series representations of the Jacobi group*, Math. Z. **204** (1990) 13–44.

[8] K. Takase, *A note on automorphic forms*, J. Reine Angew. Math. **409** (1990) 138–171.

[9] K. Takase, *On unitary representations of Jacobi groups*, J. Reine Angew. Math. **430** (1992) 130–149.

[10] K. Takase, *On Siegel modular forms of half-integral weights and Jacobi forms*, Trans. Amer. Math. Soc. **351** (1999) 735–780.

[11] S. Berceanu and A. Gheorghe, *On the geometry of Siegel–Jacobi domains*, Int. J. Geom. Methods Mod. Phys. **8** (2011) 1783-1798; arXiv:1011.3317v1 (2010).

[12] I. Satake, *Algebraic structures of symmetric domains*, Publ. Math. Soc. Japan **14**, Princeton Univ. Press, 1980.

[13] *Erich Kähler: Mathematische Werke; Mathematical Works*, R. Berndt and O. Riemenschneider, editors, Walter de Gruyter, Berlin-New York, 2003.

[14] E.H. Kennard, *Zur Quantenmechanik einfacher Bewegungstypen*, Zeit. Phys. **44** (1927) 326–352.

[15] P. Stoler, *Equivalence classes of minimum uncertainty packets*, Phys. Rev. D **1** (1970) 3217–3219.

[16] H.P. Yuen, *Two-photon coherent states of the radiation field*, Phys. Rev. A **13** (1976) 2226–2243.

[17] P. Kramer and M. Saraceno, *Semicoherent states and the group* ISp(2, $\mathbb{R}$), Physics **114A** (1982) 448–453.

[18] C. Quesne, *Vector coherent state theory of the semidirect sum Lie algebras wsp(2N, ℝ)*, J. Phys. A: Gen. **23** (1990) 847–862.

[19] L. Mandel and E. Wolf, *Optical coherence and quantum optics*, Cambridge University Press, 1995.

[20] S.T. Ali, J.-P. Antoine and J.-P. Gazeau, *Coherent states, wavelets, and their generalizations*, Springer-Verlag, New York, 2000.

[21] K. Shuman, *Complete signal processing bases and the Jacobi group*, J. Math. Anal. Appl. **278** (2003), 203–213.

[22] P.D. Drummond and Z. Ficek, editors, *Quantum Squeezing*, Springer, Berlin, 2004.

[23] S. Berceanu and A. Gheorghe, *Applications of the Jacobi group to Quantum Mechanics*, Romanian J. Phys. **53** (2008) 1013-1021; arXiv: 0812.0448 (math.DG).

[24] S. Berceanu, *Coherent states associated to the Jacobi group – a variation on a theme by Erich Kähler*, J. Geom. Symmetry Physics **9** (2007) 1–8.

[25] A.M. Perelomov, *Generalized Coherent States and their Applications*, Springer, Berlin, 1986.

[26] S. Berceanu, *A holomorphic representation of the Jacobi algebra*, Rev. Math. Phys. **18** (2006) 163–199; *Errata* Rev. Math. Phys. **24** (2012) 1292001 (2 pages).

[27] S. Berceanu, Realization of coherent state algebras by differential operators, in *Advances in Operator Algebras and Mathematical Physics*, F. Boca, O. Bratteli, R. Longo and H. Siedentop, editors, The Theta Foundation, Bucharest 1–24, 2005.

[28] S.G. Ginikin, I.I. Pjatecckiĭ-Šapiro, E.B. Vinberg, Homogeneous Kähler manifolds, in *Geometry of homogenous bounded domains*, E. Vesentini (ed.), Springer-Verlag, Berlin Heidelberg 2011, Lectures given at the Summer School of the C.I.M.E. held at Urbino, Italy, July 3–13, 1967.

[29] S. Berceanu, Classical and quantum evolution on the Siegel–Jacobi manifolds, in *Proceedings of the XXX Workshop on Geometric Methods in Physics*, Trends in Mathematics, Springer Basel AG, 43–52, 2012.

[30] S. Berceanu, *Consequences of the fundamental conjecture for the motion on the Siegel–Jacobi disk*, Int. J. Geom. Methods Mod. Phys. **10** (2013) 1250076 (18 pages); arXiv: 1110.5469v2 [math.DG] 23 Apr. 2012.

[31] E.B. Vinberg and S.G. Gindikin, *Kählerian manifolds admitting a transitive solvable automorphism group*, Math. Sb. 74 (116) (1967) 333–351.

[32] J. Dorfmeiser and K. Nakajima, *The fundamental conjecture for homogeneous Kähler manifolds*, Acta Mathematica **161** (1988) 23–70.

[33] S. Berceanu, *A convenient coordinatization of Siegel–Jacobi domains*, Rev. Math. Phys. **24** (2012) 1250024 (38 pages); arXiv: 1204.5610v2 [math.DG] 12 Nov. 2012.

[34] S. Helgason, *Differential Geometry, Lie Groups and Symmetric Spaces*, Academic Press, New York, 1978.

[35] A. Borel, *Kählerian coset spaces of semi-simple Lie groups*, Proc. Nat. Acad. Sci. USA **40** (1954) 1147–1151.

[36] K. Nomizu, *Invariant affine connections on homogeneous spaces*, American. J. Math. **76**, 33–65 (1954).

[37] J.-H. Yang, Y.-H. Yong, S.-N. Huh, J.-H. Shin, H.-G. Min, *Sectional curvatures of the Siegel–Jacobi space*, to appear in Bull. Korean Math. Soc. (2013), http://lecture.math.inha.ac.kr/ jhyang/sub_02.html.

[38] S. Kobayashi, *Geometry of bounded domains*, Trans. Amer. Math. Soc **92** (1959) 267–290.

[39] W. Lisiecki, *Coherent state representations. A survey*, Rep. Math. Phys. **35** (1995) 327–358.

[40] K.-H. Neeb, Holomorphy and Convexity in Lie Theory, de Gruyter Expositions in Mathematics 28, Walter de Gruyter, Berlin–New York (2000).

[41] H.C. Wang, *Closed manifolds with complex structure*, Amer. J. Math. **76** (1954) 1–32.

[42] Y. Matsushima, *Sur les espaces homogènes Kähleriens d'un group de Lie reductif*, Nagoya Math. J. **11** (1957) 53–60.

[43] J. Cogdell, S. Gindikin and P. Sarnak, Editors, *Selected works of Ilya Piatetski-Shapiro*, American Mathematical Society, Providence, Rhode Island, 2000.

Stefan Berceanu
Horia Hulubei National Institute for Physics and Nuclear Engineering
Department of Theoretical Physics
P.O.B. MG-6
RO-077125 Magurele, Romania
e-mail: Berceanu@theory.nipne.ro

Geometric Methods in Physics. XXXI Workshop 2012
Trends in Mathematics, 109–117
© 2013 Springer Basel

An Application of the Reduction Method to Sutherland type Many-body Systems

L. Fehér

Abstract. We study Hamiltonian reductions of the free geodesic motion on a non-compact simple Lie group using as reduction group the direct product of a maximal compact subgroup and the fixed point subgroup of an arbitrary involution commuting with the Cartan involution. In general, we describe the reduced system that arises upon restriction to a dense open submanifold and interpret it as a spin Sutherland system. This dense open part yields the full reduced system in important special examples without spin degrees of freedom, which include the BC_n Sutherland system built on 3 arbitrary couplings for $m < n$ positively charged and $(n-m)$ negatively charged particles moving on the half-line.

Mathematics Subject Classification (2010). 70H06, 53D20.

Keywords. Integrable many-body systems, Hamiltonian reduction.

1. Introduction

One of the most popular approaches to integrable classical mechanical systems is to realize systems of interest as reductions of higher-dimensional "canonical free systems". The point is that the properties of the reduced systems can be understood in elegant geometric terms. This approach was pioneered by Olshanetsky and Perelomov [1] and by Kazhdan, Kostant and Sternberg [2] who interpreted the celebrated rational Calogero and hyperbolic/trigonometric Sutherland systems as Hamiltonian reductions of free particles moving on Riemannian symmetric spaces. As reviewed in [3, 4, 5], these integrable many-body systems possess important generalizations based on arbitrary root systems and elliptic interaction potentials. They also admit relativistic deformations, extensions by "spin" degrees of freedom and generalizations describing interactions of charged particles. The Hamiltonian reduction approach to many of these systems was successfully worked out in the past (see, e.g., [3, 6] and their references), but in some cases its discovery still poses us interesting open problems.

In a recent joint work with V. Ayadi [7], we enlarged the range of the reduction method to cover the BC_n Sutherland system of charged particles defined by the following Hamiltonian:

$$H = \frac{1}{2}\sum_{j=1}^{n} p_j^2 - \sum_{1 \le j \le m < k \le n} \left(\frac{\kappa^2}{\cosh^2(q_j - q_k)} + \frac{\kappa^2}{\cosh^2(q_j + q_k)} \right)$$

$$+ \sum_{1 \le j < k \le m} \left(\frac{\kappa^2}{\sinh^2(q_j - q_k)} + \frac{\kappa^2}{\sinh^2(q_j + q_k)} \right)$$

$$+ \sum_{m < j < k \le n} \left(\frac{\kappa^2}{\sinh^2(q_j - q_k)} + \frac{\kappa^2}{\sinh^2(q_j + q_k)} \right) \tag{1}$$

$$+ \frac{1}{2}\sum_{j=1}^{n} \frac{(x_0 - y_0)^2}{\sinh^2(2q_j)} + \frac{1}{2}\sum_{j=1}^{m} \frac{x_0 y_0}{\sinh^2(q_j)} - \frac{1}{2}\sum_{j=m+1}^{n} \frac{x_0 y_0}{\cosh^2(q_j)}.$$

Here m and n are positive integers subject to $m < n$, while κ, x_0 and y_0 are real coupling parameters satisfying the conditions $\kappa \ne 0$ and $(x_0^2 - y_0^2) \ne 0$, which permit to consistently restrict the dynamics to the domain where

$$q_1 > q_2 > \cdots > q_m > 0 \qquad \text{and} \qquad q_{m+1} > q_{m+2} > \cdots > q_n > 0. \tag{2}$$

If $x_0 y_0 > 0$, then we can interpret the Hamiltonian (1) in terms of attractive-repulsive interactions between m positively charged and $(n-m)$ negatively charged particles influenced also by their mirror images and a positive charge fixed at the origin.

The derivation [7] of the Hamiltonian (1) relied on reducing the free geodesic motion on the group $Y := SU(n, n)$ using as symmetry group $Y_+ \times Y^+$, where $Y_+ < Y$ is a maximal compact subgroup and $Y^+ < Y$ is the (non-compact) fixed point subgroup of an involution of Y that commutes with the Cartan involution fixing Y_+. This allowed us to cover the case of 3 arbitrary couplings, extending the previous derivation [8] of 2-parameter special cases of the system. The $m = 0$ special case was treated in [9] by applying the symmetry group $Y_+ \times Y_+$.

The emergence of system (1) as reduced system required to impose very special constraints on the free motion. Thus it is natural to enquire about the reduced systems that would arise under other moment map constraints. In fact, the main purpose of this contribution is to characterize the reduced systems in a general case, where Y will be taken to be an arbitrary non-compact simple Lie group, $Y_+ \times Y^+$ will have similar structure as mentioned above, and the moment map constraint will be chosen arbitrarily.

In Section 2, we study reductions of the geodesic system on Y restricting all considerations to a dense open submanifold consisting of regular elements. In general, we shall interpret the reduced system as a spin Sutherland type system. In exceptional cases, the initial restriction to regular elements is immaterial in the sense that the moment map constraint enforces the same restriction. This happens in the reduction that yields the spinless system (1), as will be sketched in Section 3. Finally, we shall present a short conclusion in Section 4.

2. Spin Sutherland type systems from reduction

We need to fix notations and recall an important group theoretic result before turning to the reduction of our interest.

2.1. Generalized Cartan decomposition

Let Y be a *non-compact* connected simple real Lie group with Lie algebra $\mathcal{Y}$. Equip $\mathcal{Y}$ with the scalar product $\langle\,,\,\rangle$ given by a positive multiple of the Killing form. Suppose that Θ is a Cartan involution of Y (whose fixed point set is a maximal compact subgroup) and Γ is an arbitrary involution commuting with Θ. The corresponding involutions of $\mathcal{Y}$, denoted by θ and γ, lead to the orthogonal decomposition

$$\mathcal{Y} = \mathcal{Y}_+^+ + \mathcal{Y}_+^- + \mathcal{Y}_-^+ + \mathcal{Y}_-^-, \tag{3}$$

where the subscripts $\pm$ refer to eigenvalues ± 1 of θ and the superscripts to the eigenvalues of γ. We may also use the associated projection operators

$$\pi_\pm^\pm : \mathcal{Y} \to \mathcal{Y}_\pm^\pm, \tag{4}$$

as well as $\pi_+ = \pi_+^+ + \pi_+^-$ and $\pi^+ = \pi_+^+ + \pi_-^+$. We choose a maximal Abelian subspace

$$\mathcal{A} \subset \mathcal{Y}_-^-,$$

and define

$$\mathcal{C} := \mathrm{Cent}_{\mathcal{Y}}(\mathcal{A}) = \{\eta \in \mathcal{Y} \mid [\eta, \alpha] = 0 \ \forall \alpha \in \mathcal{A}\}.$$

An element $\alpha \in \mathcal{A}$ is called *regular* if its centralizer inside $\mathcal{Y}$ is precisely $\mathcal{C}$. The connected subgroup $A < Y$ associated with $\mathcal{A}$ is diffeomorphic to $\mathcal{A}$ by the exponential map. For later use, we fix a connected component $\check{\mathcal{A}}$ of the set of regular elements of $\mathcal{A}$, and introduce also the open submanifold

$$\check{A} := \exp(\check{\mathcal{A}}) \subset A.$$

The restriction of the scalar product to $\mathcal{C}$ is non-degenerate and thus we obtain the orthogonal decomposition

$$\mathcal{Y} = \mathcal{C} + \mathcal{C}^\perp. \tag{5}$$

According to (5), any $X \in \mathcal{Y}$ can be written uniquely as $X = X_{\mathcal{C}} + X_{\mathcal{C}^\perp}$. Equation (3) induces also the decomposition

$$\mathcal{C} = \mathcal{C}_+^+ + \mathcal{C}_+^- + \mathcal{C}_-^+ + \mathcal{C}_-^-, \quad \mathcal{C}_-^- = \mathcal{A},$$

and similarly for $\mathcal{C}^\perp$.

Let Y_+ and Y^+ be the fixed point subgroups of Θ and Γ, respectively, possessing as their Lie algebras

$$\mathcal{Y}_+ = \mathcal{Y}_+^+ + \mathcal{Y}_+^- \quad \text{and} \quad \mathcal{Y}^+ = \mathcal{Y}_+^+ + \mathcal{Y}_-^+.$$

Consider the group

$$Y_+^+ := Y_+ \cap Y^+$$

and its subgroup

$$M := \mathrm{Cent}_{Y_+^+}(\mathcal{A}). \tag{6}$$

Pretending that we deal only with matrix Lie groups, the elements $m \in M$ have the defining property $m \alpha m^{-1} = \alpha$ for all $\alpha \in \mathcal{A}$. Note that $\mathcal{C}_+^+$ is the Lie algebra of M.

We shall study the reductions of a free particle moving on Y utilizing the symmetry group

$$G := Y_+ \times Y^+ < Y \times Y.$$

It is a well-known group theoretic result (see, e.g., [10]) that every element $y \in Y$ can be written in the form

$$y = y_l a y_r \quad \text{with} \quad y_l \in Y_+, \, y_r \in Y^+, \, a \in A. \tag{7}$$

This is symbolically expressed as the set-equality

$$Y = Y_+ A Y^+. \tag{8}$$

Furthermore, the subset of regular elements given by

$$\check{Y} := Y_+ \check{A} Y^+ \tag{9}$$

is open and dense in Y. The decomposition of $y \in \check{Y}$ in the form (7) is unique up to the replacement $(y_l, y_r) \to (y_l m, m^{-1} y_r)$ with any $m \in M$. The product decomposition (8) is usually referred to as a generalized Cartan decomposition since it reduces to the usual Cartan decomposition in the case $\gamma = \theta$. This decomposition will play crucial role in what follows.

2.2. Generic Hamiltonian reduction

We wish to reduce the Hamiltonian system of a free particle moving on Y along geodesics of the pseudo-Riemannian metric associated with the scalar product $\langle \, , \, \rangle$. To begin, we trivialize T^*Y by right-translations, identify $\mathcal{Y}$ with $\mathcal{Y}^*$ (and similarly for $\mathcal{Y}_+$ and $\mathcal{Y}^+$) by the scalar product, and choose an arbitrary coadjoint orbit

$$\mathcal{O} := \mathcal{O}^l \times \mathcal{O}^r$$

of the symmetry group $G = Y_+ \times Y^+$. We then consider the phase space

$$P := T^*Y \times \mathcal{O} \simeq Y \times \mathcal{Y} \times \mathcal{O}^l \times \mathcal{O}^r = \{(y, J, \xi^l, \xi^r)\}$$

endowed with its natural symplectic form ω and the free Hamiltonian $\mathcal{H}$,

$$\mathcal{H}(y, J, \xi^l, \xi^r) := \frac{1}{2}\langle J, J \rangle.$$

The form ω can be written symbolically as $\omega = d\langle J, (dy)y^{-1}\rangle + \Omega$, where Ω is the canonical symplectic form of the orbit $\mathcal{O}$.

The action of $(g_l, g_r) \in G$ on P is defined by

$$\Psi_{(g_l, g_r)} : (y, J, \xi^l, \xi^r) \mapsto (g_l y g_r^{-1}, g_l J g_l^{-1}, g_l \xi^l g_l^{-1}, g_r \xi^r g_r^{-1}).$$

This Hamiltonian action is generated by the moment map $\Phi = (\Phi_l, \Phi_r) : P \to \mathcal{Y}_+ \times \mathcal{Y}^+$ whose components are

$$\Phi_l(y, J, \xi^l, \xi^r) = \pi_+(J) + \xi^l, \qquad \Phi_r(y, J, \xi^l, \xi^r) = -\pi^+(y^{-1}Jy) + \xi^r.$$

We restrict our attention to the "big cell" $\check{P}_{\mathrm{red}}$ of the full reduced phase space

$$P_{\mathrm{red}} := \Phi^{-1}(0)/G \tag{10}$$

that arises as the symplectic reduction of the dense open submanifold

$$\check{P} := T^*\check{Y} \times \mathcal{O} \subset P.$$

In other words, we wish to describe the set of G-orbits,

$$\check{P}_{\mathrm{red}} := \check{P}_c/G, \tag{11}$$

in the constraint surface

$$\check{P}_c := \Phi^{-1}(0) \cap \check{P}. \tag{12}$$

An auxiliary symplectic reduction of the orbit $(\mathcal{O}, \Omega)$ by the group M (6) will appear in our final result. Notice that M acts naturally on $\mathcal{O}$ by its diagonal embedding into $Y_+ \times Y^+$, i.e., by the symplectomorphisms

$$\psi_m : (\xi^l, \xi^r) \mapsto (m\xi^l m^{-1}, m\xi^r m^{-1}), \qquad \forall m \in M. \tag{13}$$

This action has its own moment map $\phi : \mathcal{O} \to (\mathcal{C}_+^+)^* \simeq \mathcal{C}_+^+$ furnished by

$$\phi : (\xi^l, \xi^r) \mapsto \pi_+^+(\xi_{\mathcal{C}}^l + \xi_{\mathcal{C}}^r),$$

defined by means of equations (4) and (5). The reduced orbit

$$\mathcal{O}_{\mathrm{red}} := \phi^{-1}(0)/M \tag{14}$$

is a stratified symplectic space in general [11]. In particular, $\mathcal{O}_{\mathrm{red}}$ contains a dense open subset which is a symplectic manifold and its complement is the disjunct union of lower-dimensional symplectic manifolds. Accordingly, when talking about the reduced orbit $(\mathcal{O}_{\mathrm{red}}, \Omega_{\mathrm{red}})$, Ω_{red} actually denotes a collection of symplectic forms on the various strata of $\mathcal{O}_{\mathrm{red}}$.

The key result for the characterization of $\check{P}_{\mathrm{red}}$ (11) is encapsulated by the following proposition, whose formulation contains the functions

$$w(x) := \frac{1}{\sinh(x)} \quad \text{and} \quad \chi(x) := \frac{1}{\cosh(x)}. \tag{15}$$

Proposition 1. *Every G-orbit in the constraint surface $\check{P}_c$ (12) possesses representatives of the form (e^q, J, ξ^l, ξ^r), where $q \in \check{\mathcal{A}}$, $p \in \mathcal{A}$, $\phi(\xi^l, \xi^r) = 0$ and J is given by the formula*

$$J = p - \xi^l - w(\mathrm{ad}_q) \circ \pi_+^+(\xi_{\mathcal{C}\perp}^r) - \coth(\mathrm{ad}_q) \circ \pi_+^+(\xi_{\mathcal{C}\perp}^l)$$
$$+ \pi_-^+(\xi_{\mathcal{C}}^r) + \chi(\mathrm{ad}_q) \circ \pi_-^+(\xi_{\mathcal{C}\perp}^r) - \tanh(\mathrm{ad}_q) \circ \pi_+^-(\xi_{\mathcal{C}\perp}^l). \tag{16}$$

Every element (e^q, J, ξ^l, ξ^r) of the above-specified form belongs to $\check{P}_c$, and two such elements belong to the same G-orbit if and only if they are related by the action of the subgroup $M_{\mathrm{diag}} < Y_+ \times Y^+$, under which q and p are invariant and the pair (ξ^l, ξ^r) transforms by (13). Consequently, the space of orbits $\check{P}_{\mathrm{red}}$ can be identified as

$$\check{P}_{\mathrm{red}} \simeq (\check{\mathcal{A}} \times \mathcal{A}) \times \mathcal{O}_{\mathrm{red}}.$$

This yields the symplectic identification $\check{P}_{\text{red}} \simeq T^*\check{\mathcal{A}} \times \mathcal{O}_{\text{red}}$, i.e., the reduced (stratified) symplectic form ω_{red} of $\check{P}_{\text{red}}$ can be represented as

$$\omega_{\text{red}} = d\langle p, dq \rangle + \Omega_{\text{red}}. \tag{17}$$

Here, $T^*\check{\mathcal{A}}$ is identified with $\check{\mathcal{A}} \times \mathcal{A} = \{(q,p)\}$ and $(\mathcal{O}_{\text{red}}, \Omega_{\text{red}})$ is the reduced orbit (14).

Proposition 1 is easily proved by solving the moment map constraint after "diagonalizing" $y \in \check{Y}$ utilizing the generalized Cartan decomposition (9). The expression (17) of ω_{red} follows by evaluation of the original symplectic form ω on the "overcomplete set of representatives" $\{(e^q, J, \xi^l, \xi^r)\}$ of the G-orbits in $\check{P}_c$. The operator functions of ad_q that appear in (16) are well defined since $q \in \check{\mathcal{A}}$ is regular. Indeed, ad_q in (16) always acts on $\mathcal{C}^\perp$, where it is invertible[1].

Now the formula of the reduced "kinetic energy" $\mathcal{H} = \frac{1}{2}\langle J, J \rangle$ is readily calculated.

Proposition 2. *The reduction of the free Hamiltonian $\mathcal{H}$ is given by the following M-invariant function, $\mathcal{H}_{\text{red}}$, on $T^*\check{\mathcal{A}} \times \phi^{-1}(0)$:*

$$\begin{aligned}
2\mathcal{H}_{\text{red}}(q, p, \xi^l, \xi^r) &= \langle p, p \rangle + \langle \xi^l_{\mathcal{C}}, \xi^l_{\mathcal{C}} \rangle + \langle \pi_-(\xi^r_{\mathcal{C}}), \pi_-(\xi^r_{\mathcal{C}}) \rangle \\
&\quad - \langle w^2(\text{ad}_q) \circ \pi^+(\xi^l_{\mathcal{C}\perp}), \pi^+(\xi^l_{\mathcal{C}\perp}) \rangle - \langle w^2(\text{ad}_q) \circ \pi_+(\xi^r_{\mathcal{C}\perp}), \pi_+(\xi^r_{\mathcal{C}\perp}) \rangle \\
&\quad + \langle \chi^2(\text{ad}_q) \circ \pi^-(\xi^l_{\mathcal{C}\perp}), \pi^-(\xi^l_{\mathcal{C}\perp}) \rangle + \langle \chi^2(\text{ad}_q) \circ \pi_-(\xi^r_{\mathcal{C}\perp}), \pi_-(\xi^r_{\mathcal{C}\perp}) \rangle \\
&\quad - 2\langle (w^2\chi^{-1})(\text{ad}_q) \circ \pi^+(\xi^l_{\mathcal{C}\perp}), \pi_+(\xi^r_{\mathcal{C}\perp}) \rangle \\
&\quad + 2\langle (\chi^2 w^{-1})(\text{ad}_q) \circ \pi_-(\xi^r_{\mathcal{C}\perp}), \pi^-(\xi^l_{\mathcal{C}\perp}) \rangle, \tag{18}
\end{aligned}$$

where (15) and $\chi^{-1}(x) := \cosh(x)$, $w^{-1}(x) := \sinh(x)$ are applied.

In the special case $\gamma = 0$, studied in [9], the formulae simplify considerably. Indeed, in this case $\pi_+^- = \pi_-^+ = 0$, and thus the second line of equation (16) and all terms in the last three lines of (18) except the one containing $w^2\chi^{-1}$ disappear. (This term can be recast in a more friendly form by the identity $(w^2\chi^{-1})(x) = \frac{1}{2}w^2(\frac{x}{2}) - w^2(x)$.) Although such simplification does not occur in general, we can interpret $\mathcal{H}_{\text{red}}$ as a spin Sutherland type Hamiltonian. This means that we view the components of q as describing the positions of point particles moving on the line, whose interaction is governed by hyperbolic functions of q and "dynamical coupling parameters" encoded by the "spin" degrees of freedom represented by $\mathcal{O}_{\text{red}}$.

3. A spinless example

We now recall the special case [7] whereby the previously described general construction leads to the BC_n Sutherland system (1). We start by fixing positive

---

[1]For example, the action of $w(\text{ad}_q)$ in (16) is defined by expanding $w(x)$ as x^{-1} plus a power series in x, and then substituting $(\text{ad}_q|_{\mathcal{C}\perp})^{-1}$ for x^{-1}.

integers $1 \leq m < n$. We then prepare the matrices

$$Q_{n,n} := \begin{bmatrix} 0 & \mathbf{1}_n \\ \mathbf{1}_n & 0 \end{bmatrix} \in gl(2n, \mathbb{C}), \quad I_m := \operatorname{diag}(\mathbf{1}_m, -\mathbf{1}_{n-m}) \in gl(n, \mathbb{C}),$$

where $\mathbf{1}_n$ denotes the $n \times n$ unit matrix, and introduce also

$$D_m := \operatorname{diag}(I_m, I_m) = \operatorname{diag}(\mathbf{1}_m, -\mathbf{1}_{n-m}, \mathbf{1}_m, -\mathbf{1}_{n-m}) \in gl(2n, \mathbb{C}).$$

We realize the group $Y := SU(n, n)$ as

$$SU(n, n) = \{y \in SL(2n, \mathbb{C}) \mid y^\dagger Q_{n,n} y = Q_{n,n}\},$$

and define its involutions Θ and Γ by

$$\Theta(y) := (y^\dagger)^{-1}, \qquad \Gamma(y) := D_m \Theta(y) D_m, \qquad \forall y \in Y.$$

The fixed point subgroups Y_+ and Y^+ turn out to be isomorphic to $S(U(n) \times U(n))$ and $S(U(m, n - m) \times U(m, n - m))$, respectively. We choose the maximal Abelian subspace $\mathcal{A}$ as

$$\mathcal{A} := \left\{ q := \begin{bmatrix} \mathbf{q} & 0 \\ 0 & -\mathbf{q} \end{bmatrix} : \mathbf{q} = \operatorname{diag}(q_1, \dots, q_n), \ q_k \in \mathbb{R} \right\}. \tag{19}$$

Its centralizer is $\mathcal{C} = \mathcal{A} + \mathcal{M}$ with

$$\mathcal{M} \equiv \mathcal{C}_+^+ = \left\{ d := i \begin{bmatrix} \mathbf{d} & 0 \\ 0 & \mathbf{d} \end{bmatrix} : \mathbf{d} = \operatorname{diag}(d_1, \dots, d_n), \ d_k \in \mathbb{R}, \ \operatorname{tr}(d) = 0 \right\}.$$

In particular, now $\mathcal{C}_+^- = \mathcal{C}_-^+ = \{0\}$. The "Weyl chamber" $\check{\mathcal{A}}$ can be chosen as those elements $q \in \mathcal{A}$ (19) whose components satisfy Eq. (2).

It is important for us that both $\mathcal{Y}_+$ and $\mathcal{Y}^+$ possess one-dimensional centres, whose elements can be viewed also as non-trivial one-point coadjoint orbits of Y_+ and Y^+. The centre of $\mathcal{Y}_+$ is generated by $C^l := iQ_{n,n}$, and the centre of $\mathcal{Y}^+$ is spanned by

$$C^r := i \begin{bmatrix} 0 & I_m \\ I_m & 0 \end{bmatrix}.$$

These elements enjoy the property

$$C^\lambda \in (\mathcal{C}^\perp)_+^+ \quad \text{for} \quad \lambda = l, r.$$

Taking non-zero real constants κ and x_0, we choose the coadjoint orbit of Y_+ to be

$$\mathcal{O}^l \equiv \mathcal{O}_{\kappa, x_0} := \{x_0 C^l + \xi(u) \mid u \in \mathbb{C}^n, \ u^\dagger u = 2\kappa n\},$$

where

$$\xi(u) := \frac{1}{2} \begin{bmatrix} X(u) & X(u) \\ X(u) & X(u) \end{bmatrix} \quad \text{with} \quad X(u) := i \left(uu^\dagger - \frac{u^\dagger u}{n} \mathbf{1}_n \right). \tag{20}$$

It is not difficult to see that the elements $\xi(u)$ in (20) constitute a minimal coadjoint orbit of an $SU(n)$ block of $Y_+ \simeq S(U(n) \times U(n))$. The orbit $\mathcal{O}^r$ of Y^+ is chosen to be $\{y_0 C^r\}$ with some $y_0 \in \mathbb{R}$, imposing for technical reasons that $(x_0^2 - y_0^2) \neq 0$.

With the above data, we proved that the *full* reduced phase space P_{red} (10) is given by the cotangent bundle $T^* \check{\mathcal{A}}$, i.e., $\check{P}_{\text{red}} = P_{\text{red}}$. Moreover, the reduced free Hamiltonian turned out to yield precisely the BC_n Sutherland Hamiltonian (1). The details can be found in [7].

It is an important feature of our example that $\mathcal{O}^r$ is a one-point coadjoint orbit that belongs to $(\mathcal{C}^\perp)^+_+$. Several terms of (18), including the unpleasant last term, disappear for any such orbit. An even more special feature of the example is that $\mathcal{O}_{\mathrm{red}}$ contains a single element, which means that no spin degrees of freedom are present. This can be traced back to the well-known fact that the reductions of the minimal coadjoint orbits of $SU(n)$ by the maximal torus, at zero moment map value, yield one-point spaces. This fact underlies all derivations of spinless Sutherland type systems from free geodesic motion that we are aware of, starting from the classical paper [2].

4. Conclusion

In this contribution, we described a general class of Hamiltonian reductions of free motion on a non-compact simple Lie group. All spin Sutherland type systems that we obtained are expected to yield integrable systems after taking into account their complete phase spaces provided by P_{red} (10). It could be interesting to investigate the fine details of these reduced phase spaces and to also investigate their quantization. Because of their more immediate physical interpretation, the exceptional spinless members (like the system (1)) of the pertinent family of spin Sutherland type systems deserve closer attention, and this may motivate one to ask about the list of all spinless cases that can occur in the reduction framework.

Acknowledgment

This work was supported in part by the Hungarian Scientific Research Fund (OTKA) under the grant K 77400.

References

[1] M.A. Olshanetsky and A.M. Perelomov, *Explicit solutions of some completely integrable systems.* Lett. Nuovo Cim. **17** (1976), 97–101.

[2] D. Kazhdan, B. Kostant and S. Sternberg, *Hamiltonian group actions and dynamical systems of Calogero type.* Comm. Pure Appl. Math. **XXXI** (1978), 481–507.

[3] M.A. Olshanetsky and A.M. Perelomov, *Classical integrable finite-dimensional systems related to Lie algebras.* Phys. Rept. **71** (1981), 313–400.

[4] S.N.M. Ruijsenaars, *Systems of Calogero-Moser type.* pp. 251–352 in: Proc. of the 1994 CRM–Banff Summer School 'Particles and Fields', Springer, 1999.

[5] B. Sutherland, *Beautiful Models.* World Scientific, 2004.

[6] L. Fehér and C. Klimčík, *Poisson-Lie interpretation of trigonometric Ruijsenaars duality.* Commun. Math. Phys. **301** (2011), 55–104.

[7] V. Ayadi and L. Fehér, *An integrable BC_n Sutherland model with two types of particles.* Journ. Math. Phys **52** (2011), 103506.

[8] M. Hashizume, *Geometric approach to the completely integrable Hamiltonian systems attached to the root systems with signature.* Adv. Stud. Pure Math. **4** (1984), 291–330.

[9] L. Fehér and B.G. Pusztai, *A class of Calogero type reductions of free motion on a simple Lie group.* Lett. Math. Phys. **79** (2007), 263–277.

[10] H. Schlichtkrull, *Harmonic analysis on semisimple symmetric spaces.* pp. 91–225 in: G. Heckman and H. Schlichtkrull, Harmonic Analysis and Special Functions on Symmetric Spaces, Academic Press, 1994.

[11] J.-P. Ortega and T.S. Ratiu, *Momentum Maps and Hamiltonian Reduction.* Progress in Mathematics 222, Birkhäuser, 2004.

L. Fehér
Department of Theoretical Physics
WIGNER RCP, RMKI
P.O.B. 49
H-1525 Budapest, Hungary

and

Department of Theoretical Physics
University of Szeged
Tisza Lajos krt 84–86
H-6720 Szeged, Hungary
e-mail: lfeher@physx.u-szeged.hu

Geometric Methods in Physics. XXXI Workshop 2012
Trends in Mathematics, 119–126

Lagrange Anchor for Bargmann–Wigner Equations

D.S. Kaparulin, S.L. Lyakhovich and A.A. Sharapov

Abstract. A Poincaré invariant Lagrange anchor is found for the non-Lagrangian relativistic wave equations of Bargmann and Wigner describing free massless fields of spin $s > 1/2$ in four-dimensional Minkowski space. By making use of this Lagrange anchor, we assign a symmetry to each conservation law.

Mathematics Subject Classification (2010). Primary 70S10; Secondary 81T70.

Keywords. Symmetries; conservation laws; Bargmann–Wigner equations; Lagrange anchor.

Introduction

The notions of symmetry and conservation law are of paramount importance for classical and quantum field theory. For Lagrangian theories both these notions are tightly connected to each other due to Noether's first theorem. Beyond the scope of Lagrangian dynamics, this connection has remained unclear, though many particular results and generalizations are known (see [1] for a review). In our recent works [2, 3] a general method has been proposed for connecting symmetries and conservation laws in not necessarily Lagrangian field theories. The key ingredient of the method is the notion of a Lagrange anchor introduced earlier [4] in the context of quantization of (non-)Lagrangian dynamics. Geometrically, the Lagrange anchor defines a map from the vector bundle dual to the bundle of equations of motion to the tangent bundle of the configuration space of fields such that certain compatibility conditions are satisfied. The existence of the Lagrange anchor is much less restrictive for the equations than the requirement to be Lagrangian or admit an equivalent Lagrangian reformulation.

The work is done partially under the project 2.3684.2011 of Tomsk State University and FTP, contract 14.B37.21.0911 and the RFBR grant 13-02-00551. AAS appreciates the financial support from Dynasty Foundation, SLL acknowledges support from the RFBR grant 11-01-00830-a.

The theory of massless higher-spin fields is an area of particular interest for application of the Lagrange anchor construction. Here one can keep in mind Vasiliev's higher-spin equations in the form of unfolded representation [5–7]. The unfolded field equations are not Lagrangian even at the free level and their quantization by the conventional methods is impossible. Finding a Lagrange anchor for these equations can be considered as an important step towards the consistent quantum theory of higher-spin fields. In our recent paper [8], a general construction for the Lagrange anchor was proposed for unfolded equations that admit an equivalent Lagrangian formulation.

In this paper, the general concept of Lagrange anchor is exemplified by the Bargmann–Wigner equations for free massless fields of spin $s \geq 1/2$ in the four-dimensional Minkowski space [9]. The choice of the example is not accidental. First of all, it has long been known that the model admits infinite sets of symmetries and conservation laws. These have been a subject of intensive studies by many authors during decades, see, e.g., [10–17] and references therein. However, a complete classification has been obtained only recently, first for the conservation laws [18] and then for the symmetries [19]. As the field equations are non-Lagrangian for $s > 1/2$, there is no immediate Noether correspondence between symmetries and conservation laws. The rich structure of symmetries and conservation laws in the absence of a Lagrangian formulation makes this theory an appropriate area for testing the concept of Lagrange anchor.

1. The Lagrange anchor in field theory

In this section we give a brief exposition of the Lagrange anchor construction. A more detailed discussion can be found in [4].

Consider a collection of fields $\phi^i(x)$ whose dynamics are governed by a system of PDEs

$$T_a(x, \phi^i(x), \partial_\mu \phi^i(x), \ldots) = 0 . \tag{1}$$

Here x's denote local coordinates on a space-time manifold X and indices i and a numerate the components of fields and field equations. As we do not assume the field equations (1) to come from the least action principle, the indices i and a may run through different sets. In what follows we accept Einstein's convention on summation by repeated indices.

Instead of working with the set of PDEs (1) it is convenient for us to introduce a single linear functional

$$T[\xi] = \int_X dx \xi^a T_a$$

of the test functions $\xi^a = \xi^a(x)$ with compact support. Then $\phi^i(x)$ is a solution to (1) iff $T[\xi] = 0$ for all ξ's.

Consider now the linear space of the variational vector fields of the form

$$V[\xi] = \int_X dx V^i(\xi) \frac{\delta}{\delta \phi^i(x)} , \tag{2}$$

where $V^i(\xi) = \hat{V}^i_a \xi^a(x)$ and

$$\hat{V}^i_a = \sum_{q=0}^{p} V^{i,\mu_1,\dots,\mu_q}_a (x, \partial_\mu \phi(x), \dots) \partial_{\mu_1} \dots \partial_{\mu_q}$$

is a matrix differential operators with coefficients being smooth functions of space-time coordinates, fields and their partial derivatives up to some finite order. Action of the variational vector fields on local functionals of ϕ's is defined by the usual rules of variational calculus.

The variational vector field (2) is called the *Lagrange anchor* if for any ξ_1 and ξ_2 there exist a test function ξ_3 such that the following condition is satisfied:

$$V[\xi_1]T[\xi_2] - V[\xi_2]T[\xi_1] = T[\xi_3]. \tag{3}$$

Clearly, if exists, the function ξ_3 is given by a bilinear differential operator acting on ξ_1 and ξ_2:

$$\xi^a_3 = C^a(\xi_1, \xi_2). \tag{4}$$

The coefficients of the operator C may depend on space-time coordinates x, fields ϕ and their derivatives.

The defining condition (3) means that the left-hand side vanishes whenever ϕ's satisfy the field equations (1).

The Lagrangian equations $\delta S / \delta \phi^i(x) = 0$ admit an identical (or canonical) Lagrange anchor determined by the operator $\hat{V}^j_i = \delta^j_i$. The defining condition (3) reduces to commutativity of variational derivatives

$$\frac{\delta^2 S}{\delta \phi^i(x) \delta \phi^j(x')} = \frac{\delta^2 S}{\delta \phi^j(x') \delta \phi^i(x)}.$$

If the Lagrange anchor is invertible in the class of differential operators, then the operator $\hat{V}^{-1}$ has the sense of an integrating multiplier in the inverse problem of calculus of variations. In this case, one can define the local action functional $S[\phi]$ such that $\delta S / \delta \phi^i = \hat{V}^{-1}_i(T)$.

The classification of Lagrange anchors for the equations of evolutionary type was obtained in [20]. In particular, it was shown that all the stationary and strongly integrable (we explain the notion of integrability below) Lagrange anchors for determined systems of evolutionary equations are in one-to-one correspondence with the Poisson structures that are preserved by evolution. Let us illustrate this fact by the example of autonomous system of ODEs in normal form

$$\dot{y}^i = F^i(y).$$

Consider the following ansatz for the Lagrange anchor:

$$V[\xi] = \int dt V^{ij}(y(t)) \xi_j(t) \frac{\delta}{\delta y^i(t)}. \tag{5}$$

Here $V^{ij}(y)$ is a contravariant tensor on the space of y's. Verification of the defining condition (3) yields

$$V^{ij} + V^{ji} = 0, \qquad F^k \partial_k V^{ij} + V^{ik} \partial_k F^j - V^{jk} \partial_k F^i = 0,$$

that is, $V^{ij}(y)$ must be an F-invariant bivector field on the phase space of the system. The corresponding bidifferential operator (4) is given by

$$\xi_k^3 = \partial_k V^{ij}\xi_i^1\xi_j^2\,.$$

(In this particular case it does not involve derivatives of ξ_1 and ξ_2.)

One more important notion related to the Lagrange anchor is that of integrability. The Lagrange anchor is said to be *strongly integrable* if the following two conditions are satisfied:

$$[V[\xi_1], V[\xi_2]] = V[C(\xi_1,\xi_2)]\,,$$

$$C^a(\xi_1, C(\xi_2,\xi_3)) + V[\xi_1]C^a(\xi_2,\xi_3) + cycle(\xi_1,\xi_2,\xi_3) = 0\,. \tag{6}$$

The first condition means that the variational vector fields $V[\xi]$ form an integrable distribution in the configuration space of fields. If the Lagrange anchor is *injective*, that is, $V[\xi] = 0$ implies $\xi = 0$, then the second relation follows from the first one due to the Jacobi identity for the commutator of vector fields. Taken together relations (6) define what is known in mathematics as the Lie algebroid with anchor V and bracket C, see, e.g., [21].

The canonical Lagrange anchor is strongly integrable since $C = 0$ in this case. The integrability condition for (5) requires the bivector $V = V^{ij}(y)\partial_i \wedge \partial_j$ to satisfy the Jacobi identity

$$V^{in}\partial_n V^{jk} + cycle(i,j,k) = 0\,.$$

It should be noted, that the strong integrability condition is *not* a part of the definition of Lagrange anchor. In many cases it can be considerably relaxed or even omitted. So, in general, the concept of Lagrange can not be substituted by that of Lie algebroid. A lot of examples of non-canonical Lagrange anchors for non-Lagrangian and non-Hamiltonian theories can be found in [2, 4, 8, 22–25].

2. The generalization of Noether theorem for non-Lagrangian theories

A vector field $j^\mu(x,\phi^i,\partial_\mu\phi^i,\dots)$ on X is called a conserved current if its divergence is proportional to the equations of motion (1), i.e.,

$$\partial_\mu j^\mu = \sum_{q=0}^{p} \Psi^{a,\mu_1\cdots\mu_q}(x,\phi^i(x),\partial_\mu\phi^i(x),\dots)\partial_{\mu_1}\dots\partial_{\mu_q}T_a\,. \tag{7}$$

The right-hand side is defined by some differential operator Ψ called the characteristic of the conserved current j. Two conserved currents j and j' are considered to be equivalent if $j^\mu - j'^\mu = \partial_\nu i^{\nu\mu} \pmod{T_a}$ for some bivector $i^{\mu\nu} = -i^{\nu\mu}$. Similarly, two characteristics Ψ and Ψ' are said to be equivalent if they correspond to equivalent currents. These equivalences can be used to simplify the form of characteristics. Namely, one can see that in each equivalence class of j there is a representative with Ψ being the zero-order differential operator Ψ^a. For such a representative equation (7) can be written as

$$T[\Psi] = \int_X \partial_\mu j^\mu \,. \tag{8}$$

It can be shown that there is a one-to-one correspondence between equivalence classes of conserved currents and characteristics [2].

Given a Lagrange anchor, one can assign to any characteristic Ψ a variational vector field $V[\Psi]$. The main observation made in [2] was that $V[\Psi]$ generates a symmetry of the field equations (1):

$$\delta_\varepsilon \phi^i = \varepsilon V^i(\Psi)\,, \qquad \delta_\varepsilon T[\xi] = \varepsilon V[\Psi]T[\xi] = \varepsilon T[C(\Psi,\xi) - V[\xi]\Psi]\,, \tag{9}$$

with ε being an infinitesimal constant parameter. These relations follow immediately from the definitions of the Lagrange anchor (3) and characteristic (8) upon substitution $\xi_1 = \Psi$.

Recall that any characteristic Ψ of Lagrangian equations $\delta S/\delta \phi^i(x) = 0$ generates a symmetry $\delta_\varepsilon \phi^i = \varepsilon \Psi^i$ of the action functional and thus the equations of motion. This statement constitutes the content of Noether's first theorem [1] on correspondence between symmetries and conservations laws. One the other hand, this correspondence is a simple consequence of a more general relation (9) if one takes the canonical Lagrange anchor $V^i(\xi) = \xi^i$ for Lagrangian equations. From this perspective, the assignment

$$\Psi \mapsto V[\Psi] \tag{10}$$

can be regarded as a generalization of the first Noether theorem to the case of non-Lagrangian PDEs. In general, the map (10) from the space of characteristics (= conservation laws) to the space of symmetries is neither surjective nor injective. The symmetries from the image of this map are called *characteristic symmetries*.

In the particular case of strongly integrable Lagrange anchor the space of characteristics can be endowed with the structure of Lie algebra. The corresponding Lie bracket reads

$$\{\Psi_1, \Psi_2\}^a = V[\Psi_1]\Psi_2^a - V[\Psi_2]\Psi_1^a + C^a(\Psi_1, \Psi_2)\,. \tag{11}$$

Furthermore, the anchor map (10) defines a homomorphism from the Lie algebra of characteristics to the Lie algebra of symmetries

$$[V[\Psi_1], V[\Psi_2]] = V[\{\Psi_1, \Psi_2\}]\,.$$

The bracket (11) generalizes the Dickey bracket of conserved currents [26] known in Lagrangian dynamics.

3. The Lagrange anchor and characteristic symmetries for the Bargmann–Wigner equations

In this section we illustrate the general concept of Lagrange anchor by the example of Bargmann–Wigner's equations. These equations describe free massless fields of spin $s > 0$ on $d = 4$ Minkowski space. The equations read

$$T^{\dot{\alpha}}_{\alpha_1 \cdots \alpha_{2s-1}} := \partial^{\alpha\dot{\alpha}} \varphi_{\alpha\alpha_1 \ldots \alpha_{2s-1}} = 0 \,,$$

where $\varphi_{\alpha_1 \ldots \alpha_{2s}}(x)$ is a symmetric, complex-valued spin-tensor field on $\mathbb{R}^{3,1}$. We use the standard notation of the two-component spinor formalism [9], e.g., $\partial^{\alpha\dot{\alpha}} = (\sigma^\mu)^{\alpha\dot{\alpha}} \partial/\partial x^\mu$, $\mu = 0, 1, 2, 3$, $\alpha, \dot{\alpha} = 1, 2$, and the spinor indices are raised/lowered with $\varepsilon_{\alpha\beta}$, $\varepsilon_{\dot{\alpha}\dot{\beta}}$ and the inverse $\varepsilon^{\alpha\beta}$, $\varepsilon^{\dot{\alpha}\dot{\beta}}$.

To make contact with the general definitions of the previous section let us mention that the indices of equations and fields are given by the multi-indices $a = (\dot{\alpha}, \alpha_1, \ldots, \alpha_{2s-1})$ and $i = (\alpha_1, \ldots, \alpha_{2s})$. It is well known that the Bargmann–Wigner equations are non-Lagrangian unless $s = 1/2$.

In [25], it was shown that the Bargmann–Wigner equations admit the following Poincaré-invariant and strongly integrable Lagrange anchor:

$$V(\xi)_{\alpha_1 \cdots \alpha_{2s}} = i^{2s} \partial_{(\alpha_2 \dot{\alpha}_2} \cdots \partial_{\alpha_{2s} \dot{\alpha}_{2s}} \bar{\xi}^{\dot{\alpha}_2 \cdots \dot{\alpha}_{2s}}_{\alpha_1)} \,. \tag{12}$$

The round brackets mean symmetrization. This Lagrange anchor is unique (up to equivalence) if the requirements of (i) field-independence, (ii) Poincaré-invariance and (iii) locality are imposed. Being independent of fields, the Lagrange anchor is integrable with $C = 0$.

Let Ψ be a characteristic of a conserved current j such that

$$\partial^{\alpha\dot{\alpha}} j_{\alpha\dot{\alpha}} = \Psi_{\dot{\alpha}}{}^{\alpha_1 \ldots \alpha_{2s-1}} T^{\dot{\alpha}}_{\alpha_1 \ldots \alpha_{2s-1}} + c.c. \,.$$

Then the Lagrange anchor (6) takes this characteristic to the symmetry

$$\delta_\varepsilon \varphi_{\alpha_1 \ldots \alpha_{2s}} = \varepsilon V(\Psi)_{\alpha_1 \cdots \alpha_{2s}} \,, \tag{13}$$

where $V(\Psi)$ is defined by (12).

Applying (13) to the characteristics obtained and classified in [18], we get all the characteristic symmetries. Since the Lagrange anchor is strongly integrable, characteristic symmetries form an infinite-dimensional Lie subalgebra in the Lie algebra of all symmetries. This subalgebra was previously unknown. For low spins ($s = 1/2, 1$) the Lie algebra of characteristic symmetries contains a finite-dimensional subalgebra which is isomorphic to the Lie algebra of conformal group. The elements of this subalgebra correspond to conserved currents that are expressible in terms of the energy-momentum tensor.

Conclusion

We have presented a Poincaré invariant Lagrange anchor for the Bargmann–Wigner equations. By making use this Lagrange anchor we have established a systematic connection between the symmetries and conservation laws of the equations. The Lagrange anchor, being independent of fields, is strongly integrable. As a consequence the symmetries associated with the conservation laws (characteristic symmetries) form an infinite-dimensional subalgebra in the full Lie algebra of symmetries. The physical meaning of this subalgebra remains unclear for us at the moment.

The Lagrange anchor (12) may be used for quantization of the Bargmann–Wigner equations. At the free level the corresponding generalized Schwinger–Dyson equations and probability amplitude was found in [25]. It can also be a good starting point for constructing the Lagrange anchor for Vasiliev's equations and development of a quantum theory of higher-spin interactions.

Acknowledgment

We are thankful to G. Barnich, E.D. Skvortsov, and M.A. Vasiliev for discussions on various topics related to this work and for relevant references.

References

[1] Kosmann-Schwarzbach Y., *The Noether theorems: Invariance and conservation laws in the twentieth century.* Springer, New York, 2011.

[2] Kaparulin D.S., Lyakhovich S.L., Sharapov A.A., *Rigid symmetries and conservation laws in non-Lagrangian field theory.* J. Math. Phys. **51** (2010), 082902.

[3] Kaparulin D.S., Lyakhovich S.L. and Sharapov A.A., *Local BRST cohomology in (non-)Lagrangian field theory.* JHEP 1109:006.

[4] P.O. Kazinski, S.L. Lyakhovich, and A.A. Sharapov, *Lagrange structure and quantization.* JHEP 0507:076.

[5] M.A. Vasiliev, *Actions, charges and off-shell fields in the unfolded dynamics approach.* Int. J. Geom. Meth. Mod. Phys. **3** (2006), 37–80.

[6] M.A. Vasiliev, *Higher spin theories in various dimensions.* Fortsch. Phys. **52** (2004), 702–717.

[7] X. Bekaert, S. Cnockaert, C. Iazeolla and M.A. Vasiliev, *Nonlinear higher spin theories in various dimensions.* Proc. of the First Solvay Workshop on Higher-Spin Gauge Theories (Brussels, 2004), 132–197.

[8] D.S. Kaparulin, S.L. Lyakhovich, A.A. Sharapov, *On Lagrange structure of unfolded field theory.* Int. J. Mod. Phys. **A26** (2011), 1347–1362.

[9] Penrose R., Rindler W., Spinors and space-time. Vol. I and II, Cambridge University Press, Cambridge, 1987.

[10] Lipkin D., *Existence of a new conservation law in electromagnetic theory.* J. Math. Phys. **5** (1964), 696–700.

[11] Morgan T., *Two classes of new conservation laws for the electromagnetic field and for other massless fields.* J. Math. Phys. **5** (1964), 1659–1660.

[12] Kibble T.W.B., *Conservation laws for free fields.* J. Math. Phys. **6** (1965), 1022–1026.

[13] Fairlie D.B., *Conservation laws and invariance principles.* Nuovo Cimento **37** (1965), 897–904.

[14] Fushchich W.I., Nikitin A.G., *Symmetries of equations of quantum mechanics.* Allerton Press Inc., New York, 1994.

[15] Konstein S.E., Vasiliev M.A., Zaikin V.N., *Conformal higher spin currents in any dimension and AdS/CFT correspondence.* JHEP 0012:018.

[16] Anco S., Pohjanpelto J., *Classification of local conservation laws of Maxwell's equations.* Acta Appl. Math. **69** (2001), 285–327.

[17] Vasiliev M.A., Gelfond O.A., Skvortsov E.D., *Conformal currents of fields of higher spins in Minkowski space.* Theor. Math. Phys. **154** (2008), 294–302.

[18] Anco S., Pohjanpelto J., *Conserved currents of massless fields of spin $s > 0$.* R. Soc. Lond. Proc. Ser. A Math. Phys. Eng. Sci.**459** (2003), 1215–1239.

[19] Anco S., Pohjanpelto J., *Generalized symmetries of massless free fields on Minkowski space.* SIGMA **4** (2008), 004.

[20] D.S. Kaparulin, S.L. Lyakhovich, and A.A. Sharapov, *BRST analysis of general mechanical systems.* arXiv:1207.0594.

[21] K. Mackenzie, *Lie Groupoids and Lie Algebroids in Differential Geometry.* London Math. Soc. Lecture Notes Series **124**, Cambridge Univ. Press, 1987.

[22] S.L. Lyakhovich and A.A. Sharapov, *Schwinger–Dyson equation for non-Lagrangian field theory.* JHEP 0602:007.

[23] S.L. Lyakhovich and A.A. Sharapov, *Quantization of Donaldson–Uhlenbeck–Yau theory.* Phys. Lett. **B656** (2007), 265–271.

[24] S.L. Lyakhovich and A.A. Sharapov, *Quantizing non-Lagrangian gauge theories: An augmentation method.* JHEP 0701:047.

[25] D.S. Kaparulin, S.L. Lyakhovich, and A.A. Sharapov, *Lagrange Anchor and Characteristic Symmetries of Free Massless Fields.* SIGMA **8** (2012), 021.

[26] Dickey L.A. Soliton equations and Hamiltonian systems, Advanced Series in Mathematical Physics, Vol. 12. Singapore, World Scientific, 1991.

D.S. Kaparulin
Department of Quantum Field Theory
Tomsk State University
Lenin ave. 36
Tomsk 634050, Russia

   and

Department of Higher Mathematics and Mathematical Physics
Tomsk Polytechnic University
Lenin ave. 30
Tomsk 634050, Russia
e-mail: dsc@phys.tsu.ru

S.L. Lyakhovich and A.A. Sharapov
Department of Quantum Field Theory
Tomsk State University
Lenin ave. 36
Tomsk 634050, Russia
e-mail: sll@phys.tsu.ru
       sharapov@phys.tsu.ru

Geometric Methods in Physics. XXXI Workshop 2012
Trends in Mathematics, 127–142
© 2013 Springer Basel

The Laplace Transform, Mirror Symmetry, and the Topological Recursion of Eynard–Orantin

Motohico Mulase

Abstract. This paper is based on the author's talk at the 2012 Workshop on Geometric Methods in Physics held in Białowieża, Poland. The aim of the talk is to introduce the audience to the Eynard–Orantin topological recursion. The formalism is originated in random matrix theory. It has been predicted, and in some cases it has been proven, that the theory provides an effective mechanism to calculate certain quantum invariants and a solution to enumerative geometry problems, such as open Gromov–Witten invariants of toric Calabi–Yau threefolds, single and double Hurwitz numbers, the number of lattice points on the moduli space of smooth algebraic curves, and quantum knot invariants. In this paper we use the Laplace transform of generalized Catalan numbers of an arbitrary genus as an example, and present the Eynard–Orantin recursion. We examine various aspects of the theory, such as its relations to mirror symmetry, Gromov–Witten invariants, integrable hierarchies such as the KP equations, and the Schrödinger equations.

Mathematics Subject Classification (2010). Primary: 14H15, 14N35, 05C30, 11P21; Secondary: 81T30.

Keywords. Mirror symmetry, Laplace transform, higher-genus Catalan numbers, topological recursion.

1. Introduction

The purpose of this paper is to give an introduction to the Eynard–Orantin topological recursion [1], by going through a simple mathematical example. Our example is constructed from the Catalan numbers, their higher-genus analogues, and the mirror symmetry of these numbers.

There have been exciting new developments around the Eynard–Orantin theory in the last few years that involve various quantum topological invariants, such as single and double Hurwitz numbers, open Gromov–Witten invariants, and quantum knot polynomials. A big picture is being proposed, from which, for example, we can understand the relation between the A-polynomial [2] of a knot and its colored Jones polynomials as the same as the **mirror symmetry** in string theory.

From the rigorous mathematical point of view, the predictions on this subject coming from physics are conjectural. In mathematics we need a simple example, for which we can prove all the predicted properties, and from which we can see what is going on in a more general context. The aim of this paper is to present such an example of the Eynard–Orantin theory.

The formalism of our interest is originated in the large N asymptotic analysis of the correlation functions of resolvents of a random matrix of size $N \times N$ [3, 4]. The motivation of Eynard and Orantin [1] is to find applications of the computational mechanism beyond random matrix theory. Their formula takes the shape of an integral recursion equation on a given Riemann surface Σ called the **spectral curve** of the theory. At that time already Mariño was developing the idea of **remodeled B-model** of topological string theory on a Riemann surface Σ in [5]. He noticed the geometric significance of [1], and formulated a precise theory of remodeling B-model with Bouchard, Klemm, and Pasquetti in [7]. This work immediately attracted the attention of the mathematics community. The currently accepted picture is that the remodeled B-model defines symmetric differential forms on Σ via the Eynard–Orantin recursion, and that these differentials forms are the **Laplace transform** of the quantum topological invariants that appear on the A-model side of the story. In this context *the Laplace transform plays the role of the mirror symmetry.*

This picture tells us that once we identify the spectral curve Σ, we can calculate the quantum topological invariants in terms of complex analysis on Σ. The effectiveness of this mechanism has been mathematically proven for single Hurwitz numbers [8, 9], orbifold (or double) Hurwitz numbers [10], enumeration of the lattice points of $\mathcal{M}_{g,n}$ [11–13], the Poincaré polynomials of $\mathcal{M}_{g,n}$ [14], the Weil–Petersson volume and its higher analogues of $\overline{\mathcal{M}}_{g,n}$ [15–19], and the higher-genus Catalan numbers [20]. A spectacular conjecture of [7] states that the Laplace transform of the open Gromov–Witten invariants of an arbitrary toric Calabi–Yau threefold satisfies the Eynard–Orantin topological recursion. A significant progress toward this conjecture has been made in [21].

Furthermore, an unexpected application of the Eynard–Orantin theory has been proposed in knot theory [22–27]. A key ingredient there is the **quantum curve** that characterizes quantum knot invariants.

The word *quantum* means many different things in modern mathematics. For example, a quantum curve is a holonomic system of linear differential equations whose Lagrangian is an algebraic curve embedded in the cotangent bundle of a base curve. Quantum knot invariants, on the other hand, are invariants of knots defined by representation theory of quantum algebras, and quantum algebras are deformations of usual algebras. In such a diverse usage, the only common feature is the aspect of non-commutative deformations. Therefore, when two completely different quantum objects turn out to be the same, we expect a deep mathematical theory behind the scene. In this vein, within the last two years, mathematicians and physicists have discovered a new, miraculous mathematical procedure, although still conjectural, that directly relates quantum curves and quantum knot invariants.

The notion of quantum curves appeared in Aganagic, Dijkgraaf, Klemm, Mariño, and Vafa [28], and later in Dijkgraaf, Hollands, Sułkowski, and Vafa [29, 30]. When the A-model we start with has a vanishing obstruction class in algebraic K-theory, then it is expected that a quantum curve exists, and it is a differential operator. Let us call it P. A quantum knot invariant is a function. Call it Z. Then the conjectural relation is simply the **Schrödinger equation** $PZ = 0$. For this equation to make sense, in addition to the very existence of P, we need to identify the variables appearing in P and Z. The key observation is that both P and Z are defined on the same Riemann surface, and that it is exactly the spectral curve of the Eynard–Orantin topological recursion, being realized as a Lagrangian immersion. Moreover, the total symbol of the operator P defines the Lagrangian immersion.

What is the significance of this Schrödinger equation $PZ = 0$? Recently Gukov and Sułkowski [27], based on [25], provided the crucial insight that when the underlying spectral curve is defined by the A-polynomial of a knot, the algebraic K-theory obstruction vanishes, and the equation $PZ = 0$ becomes the same as the AJ-conjecture of Garoufalidis [31]. This means that the Eynard–Orantin theory conjecturally computes colored Jones polynomial as the partition function Z of the theory, starting from a given A-polynomial.

In what follows, we present a simple example of the story. Although our example is not related to knot theory, it exhibits all key ingredients of the theory, such as the Schrödinger equation, relations to quantum topological invariants, the Eynard–Orantin recursion, the KP equations, and mirror symmetry.

At the Białowieża Workshop in summer 2012, Professor L.D. Faddeev gave a beautiful talk on the quantum dilogarithm, Bloch groups, and algebraic K-theory [32]. Our example of this paper does not illustrate the fundamental connection to these important subjects, because our spectral curve (4) has genus 0, and the K-theoretic obstruction to quantization, similar to the idea of K_2-Lagrangian of Kontsevich, vanishes. Further developments are expected in this direction.

2. Mirror dual of the Catalan numbers and their higher genus extensions

The **Catalan numbers** appear in many different places of mathematics and physics, often quite unexpectedly. The *Wikipedia* lists some of the mathematical interpretations. The appearance in string theory [33] is surprising. Here let us use the following definition:

$$C_m = \left\{ \begin{array}{l} \text{the number of ways to place } 2m \text{ pairs of} \\ \text{parentheses in a legal manner.} \end{array} \right\} \tag{1}$$

A *legal* manner means the usual way we stack them together. If we have one pair, then $C_1 = 1$, because ( ) is legal, while)(is not. For $m = 2$, we have (( )) and ( )( ), hence $C_2 = 2$. Similarly, $C_3 = 5$ because there are five legal combinations:

$$((( ))), (( ))( ), (( )( )), ( )(( )), ( )( )( ).$$

This way of exhaustive listing becomes harder and harder as m grows. We need a better mechanism to find the value, and also a general *closed* formula, if at all possible. Indeed, we have the *Catalan recursion equation*

$$C_m = \sum_{a+b=m-1} C_a C_b, \tag{2}$$

and a closed formula

$$C_m = \frac{1}{m+1} \binom{2m}{m}. \tag{3}$$

Although our definition (1) does not make sense for $m = 0$, the closed formula (3) tells us that $C_0 = 1$, and the recursion (2) works only if we define $C_0 = 1$. We will give a proof of these formulas later.

Being a ubiquitous object, the Catalan numbers have many different generalizations. What we are interested here is not those kind of generalized Catalan numbers. We want to define *higher-genus* Catalan numbers. They are necessary if we ask the following question:

Question 1. *What is the mirror symmetric dual object of the Catalan numbers?*

The mirror symmetry was conceived in modern theoretical physics as a duality between two different Calabi–Yau spaces of three complex dimensions. According to this idea, the universe consists of the visible 3-dimensional spatial component, 1-dimensional time component, and an invisible 6-dimensional component. The invisible component of the universe is considered as a complex 3-dimensional Calabi–Yau space, and the quantum nature of the universe, manifested in quantum interactions of elementary particles and black holes, is believed to be hidden in the geometric structure of this invisible manifold. The surprising discovery is that the same physical properties can be obtained from two different settings: a Calabi–Yau space X with its Kähler structure, or another Calabi–Yau space Y with its complex structure. The duality between these two sets of data is the mirror symmetry.

The phrase, "having the same quantum nature of the universe," does not give a mathematical definition. The idea of Kontsevich [34], the **Homological Mirror Symmetry**, is to define the mirror symmetry as the equivalence of *derived* categories. Since categories do not necessarily require underlying spaces, we can talk about mirror symmetries among more general objects. For instance, we can ask the above question.

What I'd like to explain in this paper is that the answer to the question is a simple *function*

$$x = z + \frac{1}{z}. \tag{4}$$

It is quite radical: the mirror symmetry holds between the Catalan numbers and a function like (4)!

If we naively understand the homological mirror symmetry as the derived equivalence between symplectic geometry (the A-model side) and holomorphic complex geometry (the B-model side), then it is easy to guess that (4) should define a B-model. According to Ballard [35], the mirror symmetric partner to this function is the projective line $\mathbb{P}^1$, together with its standard Kähler structure. The higher-genus Catalan numbers we are going to define below are associated with the Kähler geometry of $\mathbb{P}^1$. Their mirror symmetric partners are the symmetric differential forms that the Eynard–Orantin theory defines on the Riemann surface of the function $x = z + \frac{1}{z}$.

It is more convenient to give a different definition of the Catalan numbers that makes the higher-genus extension more straightforward. Consider a graph Γ drawn on a sphere S^2 that has only one vertex. Since every edge coming out from this vertex has to come back, the vertex has an even degree, say $2m$. This means $2m$ **half-edges** are incident to the unique vertex. Let us place an outgoing arrow to one of the half-edges near at the vertex (see Figure 1). Since Γ is drawn on S^2, the large loop of the left of Figure 1 can be placed as in the right graph. These are the same graph on the sphere.

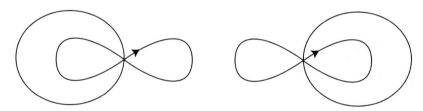

FIGURE 1. Two ways of representing the same arrowed graph on S^2 with one vertex. This graph corresponds to $((( )))$.

Lemma 1. *The number of arrowed graphs on S^2 with one vertex of degree $2m$ is equal to the Catalan number C_m.*

Proof. We assign to each edge forming a loop a pair of parentheses. Their placement is nested according to the graph. The starting parenthesis '(' corresponds to the unique arrowed half-edge. We then examine all half-edges by the counter clock-wise order. When a new loop is started, we open a parenthesis '('. When it is closed to form a loop, we complete a pair of parentheses by placing a ')'. In this way we have a bijective correspondence between graphs on S^2 with one vertex of degree $2m$ and the nested pairs of $2m$ parentheses. $\square$

Now a higher-genus generalization is easy. A **cellular graph** of type (g,n) is the one-skeleton of a cell-decomposition of a connected, closed, oriented surface of genus g with n 0-cells labeled by the index set $[n] = \{1, 2, \ldots, n\}$. Two cellular graphs are identified if an orientation-preserving homeomorphism of a surface into another surface maps one cellular graph to another, honoring the labeling of each vertex. Let $D_{g,n}(\mu_1, \ldots, \mu_n)$ denote the number of connected cellular graphs Γ of type (g,n) with n labeled vertices of degrees $(\mu_1, \ldots, \mu_n)$, counted with the weight $1/|\mathrm{Aut}(\Gamma)|$. It is generally a rational number. The orientation of the surface induces a cyclic order of incident half-edges at each vertex of a cellular graph Γ. Since $\mathrm{Aut}(\Gamma)$ fixes each vertex, it is a subgroup of the Abelian group $\prod_{i=1}^{n} \mathbb{Z}/\mu_i\mathbb{Z}$ that rotates each vertex and the incident half-edges. Therefore,

$$C_{g,n}(\mu_1, \ldots, \mu_n) = \mu_1 \cdots \mu_n D_{g,n}(\mu_1, \ldots, \mu_n) \tag{5}$$

is always an integer. The cellular graphs counted by (5) are connected graphs of genus g with n vertices of degrees $(\mu_1, \ldots, \mu_n)$, and at the jth vertex for every $j = 1, \ldots, n$, an arrow is placed on one of the incident μ_j half-edges (see Figure 2). The placement of n arrows corresponds to the factors $\mu_1 \cdots \mu_n$ on the right-hand side. We call this integer the **Catalan number** of type (g,n). The reason for this naming comes from the fact that $C_{0,1}(2m) = C_m$, and the following theorem.

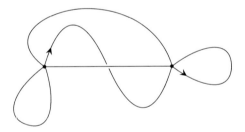

FIGURE 2. A cellular graph of type $(1,2)$.

Theorem 1. *The generalized Catalan numbers of (5) satisfy the following equation.*

$$C_{g,n}(\mu_1, \ldots, \mu_n) = \sum_{j=2}^{n} \mu_j C_{g,n-1}(\mu_1 + \mu_j - 2, \mu_2, \ldots, \widehat{\mu_j}, \ldots, \mu_n)$$

$$+ \sum_{\alpha+\beta=\mu_1-2} \Bigg[C_{g-1,n+1}(\alpha, \beta, \mu_2, \cdots, \mu_n)$$

$$+ \sum_{\substack{g_1+g_2=g \\ I \sqcup J = \{2,\ldots,n\}}} C_{g_1,|I|+1}(\alpha, \mu_I) C_{g_2,|J|+1}(\beta, \mu_J) \Bigg], \tag{6}$$

where $\mu_I = (\mu_i)_{i \in I}$ for an index set $I \subset [n]$, $|I|$ denotes the cardinality of I, and the third sum in the formula is for all partitions of g and set partitions of $\{2, \ldots, n\}$.

Proof. Consider an arrowed cellular graph Γ counted by the left-hand side of (6), and let $\{p_1, \ldots, p_n\}$ denote the set of labeled vertices of Γ. We look at the half-edge incident to p_1 that carries an arrow.

Case 1. The arrowed half-edge extends to an edge E that connects p_1 and p_j for some $j > 1$.

We shrink the edge E and join the two vertices p_1 and p_j together. By this process we create a new vertex of degree $\mu_1 + \mu_j - 2$. To make the counting bijective, we need to be able to go back from the shrunken graph to the original, provided that we know μ_1 and μ_j. Thus we place an arrow to the half-edge next to E around p_1 with respect to the counter-clockwise cyclic order that comes from the orientation of the surface. In this process we have μ_j different arrowed graphs that produce the same result, because we must remove the arrow placed around the vertex p_j in the original graph. This gives the right-hand side of the first line of (6). See Figure 3.

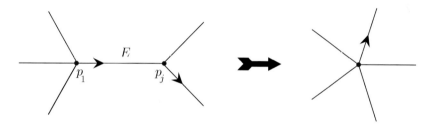

FIGURE 3. The process of shrinking the arrowed edge E that connects vertices p_1 and p_j, $j > 1$.

Case 2. The arrowed half-edge at p_1 is actually a loop E that goes out and comes back to p_1.

The process we apply is again shrinking the loop E. The loop E separates all other half-edges into two groups, one consisting of α of them placed on one side of the loop, and the other consisting of β half-edges placed on the other side. It can happen that $\alpha = 0$ or $\beta = 0$. Shrinking a loop on a surface causes pinching. Instead of creating a pinched (i.e., singular) surface, we separate the double point into two new vertices of degrees α and β. Here again we need to remember the position of the loop E. Thus we place an arrow to the half-edge next to the loop in each group. See Figure 4.

After the pinching and separating the double point, the original surface of genus g with n vertices $\{p_1, \ldots, p_n\}$ may change its topology. It may have genus $g - 1$, or it splits into two pieces of genus g_1 and g_2. The second line of (6) records all such possibilities. This completes the proof. $\qquad\square$

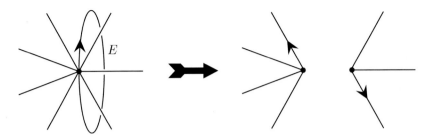

FIGURE 4. The process of shrinking the arrowed loop E that is attached to p_1.

Remark 1. For $(g, n) = (0, 1)$, the above formula reduces to

$$C_{0,1}(\mu_1) = \sum_{\alpha+\beta=\mu_1-2} C_{0,1}(\alpha)C_{0,1}(\beta), \tag{7}$$

which proves (2) since $C_{0,1}(2m) = C_m$.

Note that we *define* $C_{0,1}(0) = 1$. Only for the $(g, n) = (0, 1)$ case this irregularity of non-zero value happens for $\mu_1 = 0$. This is because a degree 0 single vertex is *connected,* and gives a cell-decomposition of S^2. We can imagine that a single vertex on S^2 has an infinite cyclic group as its automorphism, so that $C_{0,1}(0) = 1$ is consistent. In all other cases, if one of the vertices has degree 0, then the Catalan number $C_{g,n}$ is simply 0 because of the definition (5).

Following Kodama–Pierce [36], we introduce the generating function of the Catalan numbers by

$$z = z(x) = \sum_{m=0}^{\infty} C_m \frac{1}{x^{2m+1}}. \tag{8}$$

Then by the quadratic recursion (7), we find that the inverse function of $z(x)$ that vanishes at $x = \infty$ is given by

$$x = z + \frac{1}{z},$$

which is exactly (4). We remark that solving the above equation as a quadratic equation for z yields

$$z = \frac{x - \sqrt{x^2 - 4}}{2} = \frac{x}{2}\left(1 - \sqrt{1 - \left(\frac{2}{x}\right)^2}\right) = \frac{x}{2}\sum_{m=1}^{\infty}(-1)^{m-1}\binom{\frac{1}{2}}{m}\left(\frac{2}{x}\right)^{2m},$$

from which the closed formula (3) follows.

3. The Laplace transform of the generalized Catalan numbers

Let us compute the Laplace transform of the generalized Catalan numbers. Why are we interested in the Laplace transform? The answer becomes clear only after we examine the result of computation.

So we define the discrete Laplace transform

$$F_{g,n}^{C}(t_1,\ldots,t_n) = \sum_{(\mu_1,\ldots,\mu_n)\in\mathbb{Z}_+^n} D_{g,n}(\mu_1,\ldots,\mu_n)\, e^{-\langle w,\mu\rangle}$$

for (g,n) subject to $2g-2+n>0$, where the Laplace dual coordinates $w = (w_1,\ldots,w_n)$ of $(\mu_1,\ldots,\mu_n)$ is related to the function coordinate $t = (t_1,\ldots,t_n)$ by

$$e^{w_i} = x_i = z_i + \frac{1}{z_i} = \frac{t_i+1}{t_i-1} + \frac{t_i-1}{t_i+1}, \qquad i=1,2,\ldots,n,$$

and $\langle w,\mu\rangle = w_1\mu_1 + \cdots + w_n\mu_n$. The **Eynard–Orantin differential form** of type (g,n) is given by

$$W_{g,n}^{C}(t_1,\ldots,t_n) = d_1\cdots d_n F_{g,n}^{C}(t_1,\ldots,t_n)$$

$$= (-1)^n \sum_{(\mu_1,\ldots,\mu_n)\in\mathbb{Z}_+^n} C_{g,n}(\mu_1,\ldots,\mu_n)\, e^{-\langle w,\mu\rangle} dw_1 \cdots dw_n.$$

Due to the irregularity that a single point is a connected cellular graph of type $(0,1)$, we *define*

$$W_{0,1}^{C}(t) = -\sum_{\mu=0}^{\infty} C_{0,1}(\mu)\frac{1}{x^\mu}\cdot\frac{dx}{x} = -z(x)dx,$$

including the $\mu=0$ term. Since $dF_{0,1}^{C} = W_{0,1}^{C}$, we find

$$F_{0,1}^{C}(t) = -\frac{1}{2}z^2 + \log z + \text{const.} \tag{9}$$

Using the value of Kodama and Pierce [36] for $D_{0,2}(\mu_1,\mu_2)$, we calculate (see [20])

$$F_{0,2}^{C}(t_1,t_2) = -\log(1 - z_1z_2), \tag{10}$$

and hence

$$W_{0,2}^{C}(t_1,t_2) = \frac{dt_1\cdot dt_2}{(t_1-t_2)^2} - \frac{dx_1\cdot dx_2}{(x_1-x_2)^2} = \frac{dt_1\cdot dt_2}{(t_1+t_2)^2}.$$

The 2-form $\frac{dx_1\cdot dx_2}{(x_1-x_2)^2}$ is the local expression of the symmetric second derivative of the logarithm of Riemann's **prime form** on a Riemann surface. Thus $W_{0,2}^{C}$ is the difference of this quantity between the Riemann surface of $x = z + \frac{1}{z}$ and the x-coordinate plane. This relation is true for all known examples, and hence $W_{0,2}$ is *defined* as the second log derivative of the prime form of the spectral curve in [1]. It is important to note that in our definition, $W_{0,2}^{C}(t_1,t_2)$ is regular at the diagonal $t_1 = t_2$.

Note that the function $z(x)$ is absolutely convergent for $|x| > 2$. Since its inverse function is a rational function given by (4), the *Riemann surface* of the inverse function, i.e., the maximal domain of holomorphy of $x(z)$, is $\mathbb{P}^1 \setminus \{0, \infty\}$. At $z = \pm 1$ the function $x = z + \frac{1}{z}$ is branched, and this is why $z(x)$ has the radius of convergence 2, measured from ∞. The coordinate change

$$z = \frac{t+1}{t-1}$$

brings the branch points to 0 and ∞.

Theorem 2 ([37]). *The Laplace transform $F_{g,n}^C(t_{[n]})$ satisfies the following differential recursion equation for every (g,n) subject to $2g - 2 + n > 0$.*

$$\frac{\partial}{\partial t_1} F_{g,n}^C(t_{[n]})$$

$$= -\frac{1}{16} \sum_{j=2}^{n} \left[\frac{t_j}{t_1^2 - t_j^2} \left(\frac{(t_1^2-1)^3}{t_1^2} \frac{\partial}{\partial t_1} F_{g,n-1}^C(t_{[\hat{j}]}) - \frac{(t_j^2-1)^3}{t_j^2} \frac{\partial}{\partial t_j} F_{g,n-1}^C(t_{[\hat{1}]}) \right) \right]$$

$$- \frac{1}{16} \sum_{j=2}^{n} \frac{(t_1^2-1)^2}{t_1^2} \frac{\partial}{\partial t_1} F_{g,n-1}^C(t_{[\hat{j}]})$$

$$- \frac{1}{32} \frac{(t_1^2-1)^3}{t_1^2} \left[\frac{\partial^2}{\partial u_1 \partial u_2} F_{g-1,n+1}^C(u_1, u_2, t_2, t_3, \ldots, t_n) \right] \Bigg|_{u_1 = u_2 = t_1}$$

$$- \frac{1}{32} \frac{(t_1^2-1)^3}{t_1^2} \sum_{\substack{g_1+g_2=g \\ I \sqcup J = \{2,3,\ldots,n\}}}^{\text{stable}} \frac{\partial}{\partial t_1} F_{g_1, |I|+1}^C(t_1, t_I) \frac{\partial}{\partial t_1} F_{g_2, |J|+1}^C(t_1, t_J). \qquad (11)$$

Here we use the index convention

$$[n] = \{1, 2, \ldots, n\} \text{ and } [\hat{j}] = \{1, 2, \ldots, \hat{j}, \ldots, n\}.$$

The final sum is for partitions subject to the stability condition $2g_1 - 1 + |I| > 0$ and $2g_2 - 1 + |J| > 0$.

The proof follows from the Laplace transform of (6). Since the formula for the generalized Catalan numbers contain unstable geometries $(g,n) = (0,1)$ and $(0,2)$, we need to substitute the values (9) and (10) in the computation to derive the recursion in the form of (11).

Since the form of the equation (11) is identical to [14, Theorem 5.1], and since the initial values $F_{1,1}^C$ and $F_{0,3}^C$ of [37] agree with that of [14, (6.1), (6,2)], the same conclusion of [14] holds. Therefore,

Theorem 3. *The Laplace transform $F_{g,n}^C(t_1, \ldots, t_n)$ in the stable range $2g-2+n > 0$ satisfies the following properties.*

- *The reciprocity: $F_{g,n}^C(1/t_1, \ldots, 1/t_n) = F_{g,n}^C(t_1, \ldots, t_n)$.*
- *The polynomiality: $F_{g,n}^C(t_1, \ldots, t_n)$ is a Laurent polynomial of degree $3(2g - 2 + n)$.*

- *The highest degree asymptotics as the Virasoro condition: The leading terms of $F_{g,n}^C(t_1,\ldots,t_n)$ form a homogeneous polynomial defined by*

$$F_{g,n}^{C\text{-top}}(t_1,\ldots,t_n)$$
$$= \frac{(-1)^n}{2^{2g-2+n}} \sum_{\substack{d_1+\cdots+d_n \\ =3g-3+n}} \langle \tau_{d_1} \cdots \tau_{d_n} \rangle_{g,n} \prod_{i=1}^n \left[(2d_i - 1)!! \left(\frac{t_i}{2} \right)^{2d_i+1} \right],$$

  where $\langle \tau_{d_1} \cdots \tau_{d_n} \rangle_{g,n}$ is the ψ-class intersection numbers of the Deligne–Mumford moduli stack $\overline{\mathcal{M}}_{g,n}$. The recursion Theorem 2 restricts to the highest degree terms and produces the DVV formulation [38] of the Witten–Kontsevich theorem [39, 40], which is equivalent to the Virasoro constraint condition for the intersection numbers on $\overline{\mathcal{M}}_{g,n}$.
- *The Poincaré polynomial: The principal specialization $F_{g,n}^C(t,t,\ldots,t)$ is a polynomial in*

$$s = \frac{(t+1)^2}{4t}, \tag{12}$$

  and coincides with the virtual Poincaré polynomial of $\mathcal{M}_{g,n} \times \mathbb{R}_+^n$.
- *The Euler characteristic: In particular, we have*

$$F_{g,n}^C(1,1\ldots,1) = (-1)^n \chi(\mathcal{M}_{g,n}).$$

Remark 2. The above theorem explains why the Laplace transform of the generalized Catalan numbers is important. The function $F_{g,n}^C(t_1,\ldots,t_n)$ knows a lot of topological information of both $\mathcal{M}_{g,n}$ and $\overline{\mathcal{M}}_{g,n}$.

Taking the n-fold differentiation of (11), we obtain a residue form of the recursion. The formula given in (14) is an example of the Eynard–Orantin topological recursion.

Theorem 4 ([20]). *The Laplace transform of the Catalan numbers of type (g,n) defined as a symmetric differential form*

$$W_{g,n}^C(t_1,\ldots,t_n) = (-1)^n \sum_{(\mu_1,\ldots,\mu_n) \in \mathbb{Z}_+^n} C_{g,n}(\mu_1,\ldots,\mu_n) \, e^{-\langle w,\mu \rangle} dw_1 \cdots dw_n$$

satisfies the Eynard–Orantin recursion with respect to the Lagrangian immersion

$$\Sigma = \mathbb{C} \ni z \longmapsto (x(z), y(z)) \in T^*\mathbb{C}, \quad \begin{cases} x(z) = z + \frac{1}{z} \\ y(z) = -z \end{cases}. \tag{13}$$

The recursion formula is given by a residue transformation equation

$$W_{g,n}^C(t_1, \ldots, t_n)$$

$$= \frac{1}{2\pi i} \int_\gamma K^C(t, t_1) \left[\sum_{j=2}^n \left(W_{0,2}^C(t, t_j) W_{g,n-1}^C(-t, t_2, \ldots, \widehat{t_j}, \ldots, t_n) \right. \right.$$

$$\left. + W_{0,2}^D(-t, t_j) W_{g,n-1}^C(t, t_2, \ldots, \widehat{t_j}, \ldots, t_n) \right) + W_{g-1,n+1}^C(t, -t, t_2, \ldots, t_n)$$

$$+ \sum_{\substack{g_1+g_2=g \\ I \sqcup J = \{2,3,\ldots,n\}}}^{stable} W_{g_1, |I|+1}^C(t, t_I) W_{g_2, |J|+1}^C(-t, t_J) \right]. \tag{14}$$

The kernel function is defined to be

$$K^C(t, t_1) = \frac{1}{2} \frac{\int_t^{-t} W_{0,2}(\,\cdot\,, t_1)}{W_{0,1}(-t) - W_{0,1}(t)} = -\frac{1}{64} \left(\frac{1}{t+t_1} + \frac{1}{t-t_1} \right) \frac{(t^2-1)^3}{t^2} \cdot \frac{1}{dt} \cdot dt_1,$$

which is an algebraic operator contracting dt, while multiplying dt_1. The contour integration is taken with respect to t on the curve defined in Figure 5.

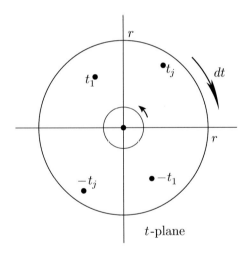

FIGURE 5. The integration contour γ. This contour encloses an annulus bounded by two concentric circles centered at the origin. The outer one has a large radius $r > \max_{j \in N} |t_j|$ and the negative orientation, and the inner one has an infinitesimally small radius with the positive orientation.

Remark 3. The recursion (14) is a universal formula compared to (11), because the only input is the spectral curve Σ that is realized as a Lagrangian immersion, which determines $W_{0,1}$, and $W_{0,2}$ can be defined by taking the difference of the log of prime forms of Σ and $\mathbb{C}$.

4. The partition function for the generalized Catalan numbers and the Schrödinger equation

Let us now consider the exponential generating function of the Poincaré polynomial $F_{g,n}^C(t, \ldots, t)$. This function is called the **partition function** for the generalized Catalan numbers:

$$Z^C(t, \hbar) = \exp\left(\sum_{g=0}^{\infty} \sum_{n=1}^{\infty} \frac{1}{n!} \hbar^{2g-2+n} F_{g,n}^C(t, t, \ldots, t)\right). \tag{15}$$

The constant ambiguity in (9) makes the partition function well defined up an overall non-zero constant factor.

Theorem 5 ([41]). *The partition function satisfies the following Schrödinger equation*

$$\left(\hbar^2 \frac{d^2}{dx^2} + \hbar x \frac{d}{dx} + 1\right) Z^C(t, \hbar) = 0, \tag{16}$$

where t is considered as a function in x by

$$t = t(x) = \frac{z(x) + 1}{z(x) - 1}$$

and (8). Moreover, the partition function has a matrix integral expression

$$Z^C(z, \hbar) = \int_{\mathcal{H}_{N \times N}} \det(1 - \sqrt{s}X)^N e^{-\frac{N}{2} \operatorname{trace}(X^2)} dX$$

with the identification (12) and $\hbar = 1/N$. Here dX is the normalized Lebesgue measure on the space of $N \times N$ Hermitian matrices $\mathcal{H}_{N \times N}$. It is a well-known fact that this matrix integral is the principal specialization of a KP τ-function [42].

The currently emerging picture [23, 25, 27] is the following. If we start with the A-polynomial of a knot K and consider the Lagrangian immersion it defines, like the one in (13), then the partition function Z of the Eynard–Orantin recursion, defined in a much similar way as in (15) but with a theta function correction factor of [23], *is* the colored Jones polynomial of K, and the corresponding Schrödinger equation like (16) is equivalent to the AJ-conjecture of [31].

Our example comes from an elementary enumeration problem, yet as Theorem 3 suggests, the geometric information contained in this example is quite non-trivial.

Acknowledgment

The paper is based on the author's talk at the *XXXI Workshop on the Geometric Methods in Physics* held in Białowieża, Poland, in June 2012. He thanks the organizers of the workshop for their hospitality and exceptional organization of the successful workshop. The author also thanks Gaëtan Borot, Vincent Bouchard, Bertrand Eynard, Marcos Mariño, Paul Norbury, Yongbin Ruan, Sergey Shadrin, Piotr Sułkowski, and Don Zagier for their tireless and patient explanations of their work to the author, and for stimulating discussions. The author's research was supported by NSF grants DMS-1104734 and DMS-1104751.

References

[1] B. Eynard and N. Orantin, *Invariants of algebraic curves and topological expansion*, Communications in Number Theory and Physics **1**, 347–452 (2007).

[2] D. Cooper, D.M. Culler, H. Gillet, D. Long, and P. Shalen, *Plane curves associated to character varieties of 3D manifolds*, Invent. Math. **118**, 47–84 (1994).

[3] A. Alexandrov, A. Mironov and A. Morozov, *Unified description of correlators in non-Gaussian phases of Hermitean matrix model*, arXiv:hep-th/0412099 (2004).

[4] B. Eynard, *Topological expansion for the 1-hermitian matrix model correlation functions*, arXiv:0407261 [hep-th] (2004).

[5] M. Mariño, *Open string amplitudes and large order behavior in topological string theory*, J. High Energy Physics **0803-060**, 1–33 (2008).

[6] M. Mariño, *Chern-Simons theory, matrix models, and topological strings*, Oxford University Press, 2005.

[7] V. Bouchard, A. Klemm, M. Mariño, and S. Pasquetti, *Remodeling the B-model*, Commun. Math. Phys. **287**, 117–178 (2008).

[8] B. Eynard, M. Mulase and B. Safnuk, *The Laplace transform of the cut-and-join equation and the Bouchard-Mariño conjecture on Hurwitz numbers*, Publications of the Research Institute for Mathematical Sciences **47**, 629–670 (2011).

[9] M. Mulase and N. Zhang, *Polynomial recursion formula for linear Hodge integrals*, Communications in Number Theory and Physics **4**, 267–294 (2010).

[10] V. Bouchard, D. Hernández Serrano, X. Liu, and M. Mulase, *Mirror symmetry of orbifold Hurwitz numbers*, preprint 2012.

[11] K. Chapman, M. Mulase, and B. Safnuk, *Topological recursion and the Kontsevich constants for the volume of the moduli of curves*, Communications in Number theory and Physics **5**, 643–698 (2011).

[12] P. Norbury, *Counting lattice points in the moduli space of curves*, arXiv:0801.4590 (2008).

[13] P. Norbury, *String and dilation equations for counting lattice points in the moduli space of curves*, arXiv:0905.4141 (2009).

[14] M. Mulase and M. Penkava, *Topological recursion for the Poincaré polynomial of the combinatorial moduli space of curves*, Advances in Mathematics **230**, 1322–1339 (2012).

[15] B. Eynard and N. Orantin, *Weil–Petersson volume of moduli spaces, Mirzakhani's recursion and matrix models*, arXiv:0705.3600 [math-ph] (2007).

[16] K. Liu and H. Xu, *Recursion formulae of higher Weil–Petersson volumes*, Int. Math. Res. Notices **5**, 835–859 (2009).

[17] M. Mirzakhani, *Simple geodesics and Weil–Petersson volumes of moduli spaces of bordered Riemann surfaces*, Invent. Math. **167**, 179–222 (2007).

[18] M. Mirzakhani, *Weil–Petersson volumes and intersection theory on the moduli space of curves*, J. Amer. Math. Soc. **20**, 1–23 (2007).

[19] M. Mulase and B. Safnuk, *Mirzakhani's Recursion Relations, Virasoro Constraints and the KdV Hierarchy*, Indian Journal of Mathematics **50**, 189–228 (2008).

[20] O. Dumitsrescu, M. Mulase, A. Sorkin and B. Safnuk, *The spectral curve of the Eynard–Orantin recursion via the Laplace transform*, arXiv:1202.1159 [math.AG] (2012).

[21] B. Eynard and N. Orantin, *Computation of open Gromov–Witten invariants for toric Calabi–Yau 3-folds by topological recursion, a proof of the BKMP conjecture*, arXiv:1205.1103v1 [math-ph] (2012).

[22] M. Aganagic and C. Vafa, *Large N duality, mirror symmetry, and a Q-deformed A-polynomial for K knots*, arXiv:1204.4709v4 [physics.hep-th] (2012).

[23] G. Borot and B. Eynard, *All-order asymptotics of hyperbolic knot invariants from non-perturbative topological recursion of A-polynomials*, arXiv:1205.2261v1 [math-ph] (2012).

[24] A. Brini, B. Eynard, and M. Mariño, *Torus knots and mirror symmetry* arXiv: 1105.2012 (2011).

[25] R. Dijkgraaf, H. Fuji, and M. Manabe, *The volume conjecture, perturbative knot invariants, and recursion relations for topological strings*, arXiv:1010.4542 [hep-th] (2010).

[26] H. Fuji, S. Gukov, and P. Sułkowski, *Volume conjecture: refined and categorified*, arXiv:1203.2182v1 [hep-th] (2012).

[27] S. Gukov and P. Sułkowski, *A-polynomial, B-model, and quantization*, arXiv: 1108.0002v1 [hep-th] (2011).

[28] M. Aganagic, R. Dijkgraaf, A. Klemm, M. Mariño, and C. Vafa, *Topological Strings and Integrable Hierarchies*, [arXiv:hep-th/0312085], Commun. Math. Phys. **261**, 451–516 (2006).

[29] R. Dijkgraaf, L. Hollands, and P. Sułkowski, *Quantum curves and D-modules*, Journal of High Energy Physics **0810.4157**, 1–58 (2009).

[30] R. Dijkgraaf, L. Hollands P. Sułkowski, and C. Vafa, *Supersymmetric gauge theories, intersecting branes and free Fermions*, Journal of High Energy Physics **0802.106**, (2008).

[31] S. Garoufalidis, *On the characteristic and deformation varieties of a knot*, Geometry & Topology Monographs **7**, 291–309 (2004).

[32] L.D. Faddeev, *Volkov's pentagon for the modular quantum dilogarithm*, arXiv: 1201.6464 [math.QA] (2012).

[33] H. Ooguri, A. Strominger, and C. Vafa, *Black Hole Attractors and the Topological String*, Phys. Rev. **D70:106007**, (2004).

[34] M. Kontsevich, *Homological algebra of mirror symmetry*, arXiv:alg-geom/9411018 (1994).

[35] M. Ballard, *Meet homological mirror symmetry*, in "Modular forms and string duality," Fields Inst. Commun. **54**, 191–224 (2008).

[36] Y. Kodama and V.U. Pierce, *Combinatorics of dispersionless integrable systems and universality in random matrix theory*, arXiv:0811.0351 (2008).

[37] M. Mulase and M. Zhou, *The Laplace transform and the Eynard–Orantin topological recursion*, preprint 2012.

[38] R. Dijkgraaf, E. Verlinde, and H. Verlinde, *Loop equations and Virasoro constraints in non-perturbative two-dimensional quantum gravity*, Nucl. Phys. **B348**, 435–456 (1991).

[39] M. Kontsevich, *Intersection theory on the moduli space of curves and the matrix Airy function*, Communications in Mathematical Physics **147**, 1–23 (1992).

[40] E. Witten, *Two-dimensional gravity and intersection theory on moduli space*, Surveys in Differential Geometry **1**, 243–310 (1991).

[41] M. Mulase and P. Sułkowski, *Spectral curves and the Schrödinger equations for the Eynard–Orantin recursion*, preprint 2012.

[42] M. Mulase, *Algebraic theory of the KP equations*, in "Perspectives in Mathematical Physics," R. Penner and S.-T. Yau, editors, International Press Company, 157–223 (1994).

Motohico Mulase
Department of Mathematics
University of California
Davis, CA 95616–8633, USA
e-mail: `mulase@math.ucdavis.edu`

Geometric Methods in Physics. XXXI Workshop 2012
Trends in Mathematics, 143–153

An Elementary Proof of the Formal Rigidity of the Witt and Virasoro Algebra

Martin Schlichenmaier

Abstract. A sketch of a proof that the Witt and the Virasoro algebra are infinitesimally and formally rigid is given. This is done by elementary and direct calculations showing that the 2nd Lie algebra cohomology of these algebras with values in the adjoint module is vanishing. The relation between deformations and Lie algebra cohomology is explained.

Mathematics Subject Classification (2010). Primary: 17B56; Secondary: 17B68, 17B65, 17B66, 14D15, 81R10, 81T40.

Keywords. Witt algebra; Virasoro algebra; Lie algebra cohomology; deformations of algebras; rigidity; conformal field theory.

1. Introduction

The simplest nontrivial infinite-dimensional Lie algebras are the Witt algebra and its central extension the Virasoro algebra. The Witt algebra is related to the Lie algebra of the group of diffeomorphisms of the unit circle. In the process of quantizing a classical system or regularizing a field theory one is typically forced to consider projective actions and hence central extensions come into play. We will introduce these algebras below.

Deformations are of fundamental importance in mathematics and physics. Infinitesimal and formal deformations of Lie algebras are classified in terms of the second Lie algebra cohomology with values in the adjoint module. In particular, if this cohomology space vanishes then the algebra will be infinitesimally and formally rigid. This means that all deformations over an infinitesimal base, resp. formal base will be equivalent to the trivial family.

This contribution reports on work which was finalized during a visit to the Institute Mittag-Leffler (Djursholm, Sweden) in the frame of the program "Complex Analysis and Integrable Systems", Fall 2011. Partial support by the ESF networking programme HCAA, and the Internal Research Project GEOMQ11, University of Luxembourg, is acknowledged.

The result which we discuss here is the fact that *both the Witt algebra $\mathcal{W}$ and the Virasoro algebra $\mathcal{V}$ are formally and infinitesimally rigid.* This theorem was stated by Fialowski in 1990 [1] but without giving a proof. Nevertheless, the result should be clearly attributed to her. In 2003, the author together with Fialowski gave a sketch of a proof [2], using density arguments and very deep results due to Tsujishita [3], Reshetnikov [4], and Goncharova [5]. This proof would require that we work in continuous cohomology, but here we need algebraic cohomology. Recently, I found a way [6] how to show in an elementary way that for the 2nd cohomology spaces with values in the adjoint module we have

$$\mathrm{H}^2(\mathcal{W};\mathcal{W}) = \mathrm{H}^2(\mathcal{V};\mathcal{V}) = \{0\}.$$

In this contribution I will give a sketch of the proof. For the convenience of the reader, I will explain the relevant Lie algebra cohomology, what deformations are and how both concepts are related.

I like to mention that Fialowski (based on her older calculations [1]) presented in the meantime a (different) elementary proof [7].

It has to be pointed out that, contrary to the finite-dimensional Lie algebra case, a vanishing of the cohomology space only shows infinitesimal and formal rigidity, but not rigidity with respect to algebraic-geometric or analytic families. Jointly with Fialowski we showed that there exist examples of locally non-trivial families of Lie algebras given by Krichever–Novikov type algebras containing the Witt resp. Virasoro algebra as special elements [2, 8, 9]. I will give such an example further down.

2. The Witt and the Virasoro algebra

As our proofs are completely algebraic we allow arbitrary base fields $\mathbb{K}$ of $\mathrm{char}(\mathbb{K}) = 0$. The *Witt algebra* $\mathcal{W}$ is the Lie algebra generated as vector space over $\mathbb{K}$ by the basis elements $\{e_n \mid n \in \mathbb{Z}\}$ with Lie structure

$$[e_n, e_m] = (m - n)e_{n+m}, \quad n, m \in \mathbb{Z}.$$

It is easy to verify that the Jacobi identity is fulfilled.

A geometric realization for $\mathbb{K} = \mathbb{C}$ is obtained by considering inside the Lie algebra $Vect(S^1)$ of vector fields on the circle S^1, the subalgebra $Vect_{pol}(S^1)$ of polynomial vector fields, i.e., those vector fields which are given as sum of finitely many Fourier modes. After complexifying $Vect_{pol}(S^1)$ the Witt algebra over $\mathbb{C}$ is obtained. In this realization the generators are given by $e_n = \exp(i\, n\, \varphi)\frac{d}{d\varphi}$, where φ is the angle variable along S^1. The Lie product is the usual bracket of vector fields.

These vector fields can be holomorphically extended to the punctured complex plane, and we obtain another realization of the Witt algebra, now as the algebra of meromorphic vector fields on the Riemann sphere $\mathbb{P}^1(\mathbb{C})$ which are holomorphic outside $\{0\}$ and $\{\infty\}$. In this realization $e_n = z^{n+1}\frac{d}{dz}$, where z is the quasi-global complex coordinate.

The Witt algebra could also be described as the Lie algebra of derivation of the associative algebra of Laurent polynomials $\mathbb{K}[z^{-1}, z]$ over $\mathbb{K}$.

The Witt algebra is a $\mathbb{Z}$-graded Lie algebra. The degree is given by $\deg(e_n) := n$. Obviously, the Lie product between elements of degree n and of degree m is of degree $n + m$ (if nonzero). The homogeneous spaces $\mathcal{W}_n$ of degree n are one-dimensional with basis e_n. From

$$[e_0, e_n] = n\, e_n = \deg(e_n)\, e_n \qquad (1)$$

it follows that the eigenspace decomposition of the element e_0, acting via the adjoint action on $\mathcal{W}$, coincides with the decomposition into homogeneous subspaces. Hence, e_0 is also called a grading element.

Furthermore, $\mathcal{W}$ is a perfect Lie algebra, i.e., $[\mathcal{W}, \mathcal{W}] = \mathcal{W}$.

The *Virasoro algebra* $\mathcal{V}$ is the universal (one-dimensional) central extension of $\mathcal{W}$. As vector space it is the direct sum $\mathcal{V} = \mathbb{K} \oplus \mathcal{W}$. If we set for $x \in \mathcal{W}$, $\hat{x} := (0, x)$, and $t := (1, 0)$ then its basis elements are $\hat{e}_n$, $n \in \mathbb{Z}$ and t with the Lie product

$$[\hat{e}_n, \hat{e}_m] = (m - n)\hat{e}_{n+m} \quad \frac{1}{12}(n^3 - n)\delta_n^{-m} t, \qquad [\hat{e}_n, t] = [t, t] = 0,$$

for all $n, m \in \mathbb{Z}$ [1]. If we set $\deg(\hat{e}_n) := \deg(e_n) = n$ and $\deg(t) := 0$ then $\mathcal{V}$ becomes also a graded algebra. Let ν be the Lie homomorphism mapping the central element t to 0 and $\hat{x}$ to x. We have the short exact sequence of Lie algebras

$$0 \longrightarrow \mathbb{K} \longrightarrow \mathcal{V} \overset{\nu}{\longrightarrow} \mathcal{W} \longrightarrow 0 . \qquad (2)$$

This sequence does not split, i.e., it is a non-trivial central extension.

In some abuse of notation we identify the element $\hat{x} \in \mathcal{V}$ with $x \in \mathcal{W}$ and after identification we have $\mathcal{V}_n = \mathcal{W}_n$ for $n \neq 0$ and $\mathcal{V}_0 = \langle e_0, t \rangle_{\mathbb{K}}$. The relation (1), inducing the eigenspace decomposition for the grading element $\hat{e}_0 = e_0$, remains true.

3. From deformations to cohomology

Let W be an arbitrary Lie algebra over $\mathbb{K}$. The Lie algebra W with its bracket $[.,.]$ might also be written with the help of an anti-symmetric bilinear map

$$\mu_0 : W \times W \to W, \qquad \mu_0(x, y) = [x, y],$$

fulfilling certain additional conditions corresponding to the Jacobi identity. We consider on the same vector space W is modeled on, a family of Lie structures

$$\mu_t = \mu_0 + t \cdot \psi_1 + t^2 \cdot \psi_2 + \cdots , \qquad (3)$$

---

[1] Here δ_k^l is the Kronecker delta which is equal to 1 if $k = l$, otherwise zero.

with bilinear maps $\psi_i : W \times W \to W$ such that $W_t := (W, \mu_t)$ is a Lie algebra and W_0 is the Lie algebra we started with. The family $\{W_t\}$ is a *deformation* of W_0.

For the deformation "parameter" t we have different possibilities.

1. The parameter t might be a variable which allows to plug in numbers $\alpha \in \mathbb{K}$. In this case W_α is a Lie algebra for every α for which the expression (3) is defined. The family can be considered as deformation over (a subset of) the affine line $\mathbb{K}[t]$ or over the convergent power series $\mathbb{K}\{\{t\}\}$. The deformation is called an *algebraic-geometric* or an *analytic deformation* respectively.

2. The parameter t might be a formal variable and we allow infinitely many terms in (3), independent of any convergency requirement. It might be the case that μ_t does not exist if we plug in for t any value different from 0. In this way we obtain deformations over the ring of formal power series $\mathbb{K}[[t]]$. The corresponding deformation is a *formal deformation*.

3. The parameter t is considered as an infinitesimal variable, i.e., we set $t^2 = 0$. We obtain *infinitesimal deformations* defined over the quotient $\mathbb{K}[X]/(X^2) = \mathbb{K}[[X]]/(X^2)$.

Even more general situations for the parameter space can be considered. See [2, 8, 9] for a general mathematical treatment.

There is always the trivially deformed family given by $\mu_t = \mu_0$ for all values of t. Two families μ_t and μ'_t deforming the same μ_0 are *equivalent* if there exists a linear automorphism (with the same vagueness about the meaning of t)

$$\psi_t = id + t \cdot \alpha_1 + t^2 \cdot \alpha_2 + \cdots$$

with $\alpha_i : W \to W$ linear maps such that

$$\mu'_t(x, y) = \psi_t^{-1}(\mu_t(\psi_t(x), \psi_t(y))).$$

A Lie algebra (W, μ_0) is called *rigid* if every deformation μ_t of μ_0 is locally equivalent to the trivial family. Intuitively, this says that W cannot be deformed.

Clearly, a question of fundamental interest is to decide whether a given Lie algebra is rigid. Moreover, the question of rigidity will depend on the category we consider. Depending on the set-up we will have to consider infinitesimal, formal, algebraic- geometric, and analytic rigidity. This question is directly related to cohomology. If we write down the Jacobi identity for the μ_t given by (3) then we obtain

$$\mu_t(\mu_t(x, y), z) + \text{cycl.perm.} = 0$$

We have to consider this to all orders in t. For t^0 it is just the Jacobi identity for W. The expression for t^1 writes as

$$\psi_1([x, y], z) + \text{cycl.perm.} + [\psi_1(x, y), z] + \text{cycl.perm.} = 0.$$

This says ψ_1 is a Lie algebra 2-cocycle – with values in the adjoint module. More precisely, the first non-vanishing ψ_i has to be a 2-cocycle. Furthermore, if μ_t and μ'_t are equivalent then the corresponding ψ_i and ψ'_i are cohomologous, i.e., their difference is a coboundary.

As it is crucial for the following we will recall the definition of the second *Lie algebra cohomology* of a Lie algebra W with values in the adjoint module W. A 2-cochain is an alternating bilinear map $\psi : W \times W \to W$. Such a 2-cochain ψ is called a 2-*cocycle* if it lies in the kernel of the (2-)coboundary operator δ_2 defined by

$$\delta_2\psi(x, y, z) := \psi([x, y], z) + \psi([y, z], x) + \psi([z, x], y)$$
$$- [x, \psi(y, z)] + [y, \psi(x, z)] - [z, \psi(x, y)].$$

The vector space of 2-cochains is denoted by $C^2(W; W)$. The 1-cochains $C^1(W; W)$ are simply linear maps $W \to W$.

A 2-cochain ψ is called a 2-*coboundary* if it lies in the image of the (1-)coboundary operator, i.e., there exists a linear map $\phi : W \to W$ such that

$$\psi(x, y) = (\delta_1\phi)(x, y) := \phi([x, y]) - [\phi(x), y] - [x, \phi(y)].$$

Two cocycles whose difference is a coboundary are called *cohomologous*. As $\delta_2 \circ \delta_1 = 0$ the quotient space of 2-cocycles modulo 2-coboundaries consisting of the cohomology classes is well defined and denoted by $\mathrm{H}^2(W; W)$.

In a completely similar way higher cochains $C^k(W; W)$, cocycles, etc. are defined. We will not need them here.

The following results are well known:

1. $\mathrm{H}^2(W; W)$ classifies infinitesimal deformations of W (Gerstenhaber [10]).
2. If $\dim \mathrm{H}^2(W; W) < \infty$, then all formal deformations up to equivalence can be realized in this vector space (Fialowski [11], Fuks and Fialowski [12]).
3. If $\mathrm{H}^2(W; W) = 0$, then W is infinitesimally and formally rigid (this follows directly from (1) and (2)).
4. If $\dim W < \infty$, then $\mathrm{H}^2(W; W) = 0$ implies that W is also rigid in the algebraic-geometric and analytic sense (Gerstenhaber [10], Nijenhus and Richardson [13])

4. The main theorem

In the case of infinite-dimensional Lie algebras quite often one assumes certain additional properties for the cocycles, e.g., they should be continuous with respect to a certain topology. This has to be taken into account for the interpretation of the results, see, e.g., [14, 15]. Here we deal always with *algebraic cocycles*. In other words, there are no additional conditions.

Theorem 1. *Both the second cohomology of the Witt algebra $\mathcal{W}$ and of the Virasoro algebra $\mathcal{V}$ (over a field $\mathbb{K}$ with $\mathrm{char}(\mathbb{K}) = 0$) with values in the adjoint module vanishes, i.e.,*

$$\mathrm{H}^2(\mathcal{W}; \mathcal{W}) = \{0\}, \qquad \mathrm{H}^2(\mathcal{V}; \mathcal{V}) = \{0\}.$$

From the discussion above follows immediately the

Corollary 1. *Both $\mathcal{W}$ and $\mathcal{V}$ are formally and infinitesimally rigid.*

For the history of the theorem, see the introduction, in particular see the work of Fialowski mentioned there. In the remaining sections I will give a sketch of my recent elementary proof [6].

Remark. In the case that the Lie algebra is finite dimensional formal rigidity also implies analytic rigidity. This means that locally all families are trivial. This is not true anymore in the infinite-dimensional case. Together with Alice Fialowski we showed that there exist examples of locally non-trivial families of Lie algebras given by Krichever–Novikov type algebras containing the Witt resp. Virasoro algebra as special elements [2, 8].

As an example take the family of Lie algebras parameterized by $(e_1, e_2) \in \mathbb{C}^2$ generated by $V_n, n \in \mathbb{Z}$ over $\mathbb{C}$ with Lie structure

$$[V_n, V_m] = \begin{cases} (m-n)V_{n+m}, & n, m \text{ odd}, \\ (m-n)\big(V_{n+m} + 3e_1 V_{n+m-2} & n, m \text{ even} \\ \quad + (e_1 - e_2)(e_1 - e_3)V_{n+m-4}\big), & \\ (m-n)V_{n+m} + (m-n-1)3e_1 V_{n+m-2} & n \text{ odd}, m \text{ even} \\ \quad + (m-n-2)(e_1 - e_2)(e_1 - e_3)V_{n+m-4}. & \end{cases}$$

They are constructed as families of Krichever–Novikov type algebras for tori with parameters e_1, e_2, e_3 given by the roots of the corresponding Weierstraß polynomials. (Using the relation $e_3 = -(e_1 + e_2)$.) Poles are allowed at $z = 0$ and $z = 1/2$ (modulo the lattice of the torus) [16]. For every pair (e_1, e_2), even for the degenerate cases when at least two of the e_i coincide, this gives a Lie algebra $\mathcal{L}^{(e_1, e_2)}$. For $(e_1, e_2) \neq (0, 0)$ the algebras $\mathcal{L}^{(e_1, e_2)}$ are not isomorphic to the Witt algebra $\mathcal{W}$, but $\mathcal{L}^{(0,0)} \cong \mathcal{W}$.

5. Some steps of the proof

For a complete proof I have to refer to [6].

5.1. Reduction to degree zero

Let W be an arbitrary $\mathbb{Z}$-graded Lie algebra, i.e., $W = \bigoplus_{n \in \mathbb{Z}} W_n$. A k-cochain ψ is homogeneous of degree d if there exists a $d \in \mathbb{Z}$ such that for all $i_1, i_2, \ldots, i_k \in \mathbb{Z}$ and homogeneous elements $x_{i_l} \in W$, of $\deg(x_{i_l}) = i_l$, (for $l = 1, \ldots, k$) such that

$$\psi(x_{i_1}, x_{i_2}, \ldots, x_{i_k}) \in W_n, \quad \text{with} \quad n = \sum_{l=1}^{k} i_l + d.$$

We denote the corresponding subspace of degree d homogeneous k-cochains by $C_{(d)}^k(W; W)$.

Every k-cochain can be written as a formal infinite sum

$$\psi = \sum_{d \in \mathbb{Z}} \psi_{(d)}, \quad \psi_{(d)} \in C_{(d)}^k(W; W). \tag{4}$$

Note that for a fixed k-tuple of elements only a finite number of the summands will produce values different from zero.

The coboundary operators δ_k are operators of degree zero, i.e., applied to a k-cocycle of degree d they will produce a $(k+1)$-cocycle also of degree d.

For our situation only $k = 2$ and $k = 1$ are needed. The cochain ψ will be a 2-cocycle if and only if all degree d components $\psi_{(d)}$ in (4) will be individually 2-cocycles. If $\psi_{(d)}$ is 2-coboundary, i.e., $\psi_{(d)} = \delta_1 \phi$ with a 1-cochain ϕ, then not necessarily ϕ will be a degree d cochain, but we can find another 1-cochain ϕ' of degree d such that $\psi_{(d)} = \delta_1 \phi'$.

This shows that every cohomology class $\alpha \in \mathrm{H}^2(W; W)$ can be decomposed as formal sum

$$\alpha = \sum_{d \in \mathbb{Z}} \alpha_{(d)}, \qquad \alpha_{(d)} \in \mathrm{H}^2_{(d)}(W; W).$$

The latter space consists of classes of cocycles of degree d modulo coboundaries of degree d.

For the rest let W be either $\mathcal{W}$ or $\mathcal{V}$ and assume first that the degree $d \neq 0$.

Theorem 2. *The following hold:*
  (a) $\mathrm{H}^2_{(d)}(\mathcal{W}; \mathcal{W}) = \mathrm{H}^2_{(d)}(\mathcal{V}; \mathcal{V}) = \{0\}, \quad for\ d \neq 0.$
  (b) $\mathrm{H}^2(\mathcal{W}; \mathcal{W}) = \mathrm{H}^2_{(0)}(\mathcal{W}; \mathcal{W}), \qquad \mathrm{H}^2(\mathcal{V}; \mathcal{V}) = \mathrm{H}^2_{(0)}(\mathcal{V}; \mathcal{V}).$

Proof. We start with a cocycle of degree $d \neq 0$ and make a cohomological change $\psi' = \psi - \delta_1 \phi$ where ϕ is the linear map

$$\phi : W \to W, \quad x \mapsto \phi(x) = \frac{\psi(x, e_0)}{d}.$$

Recall e_0 is the element of either $\mathcal{W}$ or $\mathcal{V}$ which gives the degree decomposition. This implies (note that $\phi(e_0) = 0$)

$$\psi'(x, e_0) = \psi(x, e_0) - \phi([x, e_0]) + [\phi(x), e_0]$$
$$= d\,\phi(x) + \deg(x)\phi(x) - (\deg(x) + d)\phi(x) = 0.$$

Now we evaluate the 2-cocycle condition for the cocycle ψ' on the triple (x, y, e_0) (we leave out the cocycle values which vanish due to $\psi'(x, e_0) = 0$)

$$0 = \psi'([y, e_0], x) + \psi'([e_0, x], y) - [e_0, \psi'(x, y)]$$
$$= (\deg(y) + \deg(x) - (\deg(x) + \deg(y) + d))\psi'(x, y) = -d\psi'(x, y).$$

As $d \neq 0$ we obtain $\psi'(x, y) = 0$ for all $x, y \in W$. Hence $\psi = \delta_1 \phi$ is a coboundary. $\qquad\square$

5.2. The degree zero part for the Witt algebra

Let ψ be a degree zero 2-cocycle for $\mathcal{W}$. It can be written as $\psi(e_i, e_j) = \psi_{i,j} e_{i+j}$. If it is a coboundary then it is also the coboundary $\psi = \delta_1 \phi$ of a linear form ϕ of degree zero. We set $\phi(e_i) = \phi_i e_i$. The systems of $\psi_{i,j}$ and ϕ_i for $i, j \in \mathbb{Z}$ fix ψ and ϕ completely.

We evaluate the cocycle condition for the triple (e_i, e_j, e_k) of elements from $\mathcal{W}$. This yields for the coefficients

$$
\begin{aligned}
0 = (j-i)\psi_{i+j,k} - (k-i)\psi_{i+k,j} + (k-j)\psi_{j+k,i} \\
- (j+k-i)\psi_{j,k} + (i+k-j)\psi_{i,k} - (i+j-k)\psi_{i,j},
\end{aligned}
\tag{5}
$$

and for the coboundary

$$
(\delta\phi)_{i,j} = (j-i)(\phi_{i+j} - \phi_j - \phi_i).
$$

Hence, ψ is a coboundary if and only if there exists a system of $\phi_k \in \mathbb{K}$, $k \in \mathbb{Z}$ such that

$$
\psi_{i,j} = (j-i)(\phi_{i+j} - \phi_j - \phi_i), \quad \forall i,j \in \mathbb{Z}.
$$

A degree zero 1-cochain ϕ will be a 1-cocycle (i.e., $\delta_1\phi = 0$) if and only if

$$
\phi_{i+j} - \phi_j - \phi_i = 0.
$$

This has the solution $\phi_i = i\,\phi_1, \forall i \in \mathbb{Z}$. Hence, given a ϕ we can always find a ϕ' with $(\phi')_1 = 0$ and $\delta_1\phi = \delta_1\phi'$. In the following we will always choose such a ϕ' for our 2-coboundaries.

Lemma 1. *Every 2-cocycle ψ of degree zero is cohomologous to a degree zero cocycle ψ' with*

$$
\psi'_{i,1} = 0, \quad \forall i \in \mathbb{Z}, \quad and \quad \psi'_{-1,2} = \psi'_{2,-1} = 0.
\tag{6}
$$

Proof. Let ψ be the 2-cocycle ψ given by the system of $\psi_{i,j}$. Our goal is to to modify it by adding a coboundary $\delta_1\phi$ (yielding $\psi' = \psi - \delta_1\phi$) with the intention to reach $\psi'_{i,1} = 0$ for all $i \in \mathbb{Z}$.

Hence, ϕ should fulfill

$$
\psi_{i,1} = (1-i)(\phi_{i+1} - \phi_1 - \phi_i) = (1-i)(\phi_{i+1} - \phi_i).
$$

(a) Starting from $\phi_0 := -\psi_{0,1}$ we set in descending order for $i \leq -1$

$$
\phi_i := \phi_{i+1} - \frac{1}{1-i}\,\psi_{i,1}.
\tag{7}
$$

(b) ϕ_2 cannot be fixed in this way, instead we use

$$
\psi_{-1,2} = 3(-\phi_2 - \phi_{-1}), \quad \text{yielding} \quad \phi_2 := -\phi_{-1} - \frac{1}{3}\psi_{-1,2}.
$$

Then we have $\psi'_{-1,2} = 0$.

(c) We use again (7) to calculate recursively (in ascending order) ϕ_i, $i \geq 3$

$$
\phi_{i+1} := \phi_i + \frac{1}{1-i}\,\psi_{i,1}.
$$

For the cohomologous cocycle ψ' we obtain by construction the result (6) □

Lemma 2. *Let ψ be a 2-cocycle of degree zero such that $\psi_{i,1} = 0, \forall i \in \mathbb{Z}$ and $\psi_{-1,2} = 0$, then ψ will be identical zero.*

This says that our original cocycle we started with is cohomologically trivial. Hence, we have proved our main theorem for the Witt algebra.

Proof of the lemma. The coefficient $\psi_{i,m}$ are called of level m (and of level i by antisymmetry). By assumption the cocycle values of level 1 are all zero. We will consider $\psi_{i,m}$ for the values of $|m| \leq 2$ and finally make ascending and descending induction on m.

Specializing the cocycle conditions (5) for the index triple $(i, -1, k)$ gives

$$0 = -(i+1)\psi_{i-1,k} - (k-i)\psi_{i+k,-1} + (k+1)\psi_{k-1,i}$$
$$- (-1+k-i)\psi_{-1,k} + (i+k+1)\psi_{i,k} - (i-1-k)\psi_{i,-1}, \tag{8}$$

and for the triple $(i, 1, k)$

$$0 = (1-i)\psi_{i+1,k} + (k-1)\psi_{k+1,i} + (i+k-1)\psi_{i,k}. \tag{9}$$

Using these two relations we can first show that all values of level $\boxed{m=0}$ and then of level $m = -1$ are zero.

For example for $\boxed{m=-1}$ we set $k = -1$ in (9) and obtain

$$-(i-1)\,\psi_{i+1,-1} + (i-2)\,\psi_{i,-1} = 0.$$

This rewrites as

$$\psi_{i,-1} = \frac{i-1}{i-2}\,\psi_{i+1,-1}, \quad \text{for } i \neq 2,$$

$$\psi_{i+1,-1} = \frac{i-2}{i-1}\,\psi_{i,-1}, \quad \text{for } i \neq 1.$$

Starting from $\psi_{1,-1} = -\psi_{-1,1} = 0$ we get via the first equation $\psi_{i,-1} = 0$, for all $i \leq 1$. From the second equation we get $\psi_{i,-1} = 0$ for $i \geq 3$ and by assumption $\psi_{2,-1} = \psi_{-1,2} = 0$.

For $\boxed{m=-2}$ with similar arguments we get $\psi_{i,-2} = 0$ for $i \neq 2, 3$. The value of $\psi_{2,-2} = -\psi_{3,-2}$ remains undetermined for the moment.

For $\boxed{m=2}$ we get $\psi_{i,2} = 0$ for $i \neq -2, -3$. The value $\psi_{-3,2} = -\psi_{-2,2}$ remains undetermined for the moment.

Next we consider the index triple $(2, -2, 4)$ in the cocycle condition (5) and obtain

$$0 = -2\,\psi_{6,-2} - 8\,\psi_{4,2} + 4\,\psi_{2,-2}.$$

But $\psi_{4,2} = 0$ and $\psi_{6,-2} = 0$ (follows from the $m = 2$ and $m = -2$ discussion). This shows $\psi_{2,-2} = 0$ and consequently all level $m = -2$ and level $m = 2$ values are zero.

The vanishing of all other level m values follow from induction using (8) and (9).
$\qquad\square$

5.3. The Virasoro part

We give only a rough sketch

1. We start from the short exact sequence of Lie algebras (2)

$$0 \longrightarrow \mathbb{K} \longrightarrow \mathcal{V} \longrightarrow \mathcal{W} \longrightarrow 0,$$

defining the central extension.

2. This is also a short exact sequence of Lie modules over $\mathcal{V}$, where $\mathbb{K}$ is the trivial module and $\mathcal{W}$ is a $\mathcal{V}$-module as quotient of $\mathcal{V}$.

3. We obtain the following part of the long exact cohomology sequence

$$\longrightarrow \mathrm{H}^2(\mathcal{V};\mathbb{K}) \longrightarrow \mathrm{H}^2(\mathcal{V};\mathcal{V}) \longrightarrow \mathrm{H}^2(\mathcal{V};\mathcal{W}) \longrightarrow \cdots.$$

4. We showed in [6] that naturally $\mathrm{H}^2(\mathcal{V};\mathcal{W}) \cong \mathrm{H}^2(\mathcal{W};\mathcal{W})$ (i.e., every 2-cocycle of $\mathcal{V}$ with values in $\mathcal{W}$ can be changed by a coboundary such that the restriction to $\mathcal{W} \times \mathcal{W}$ defines a 2-cocycle of $\mathcal{W}$).

5. Also $\mathrm{H}^2(\mathcal{V};\mathbb{K}) = \{0\}$.

6. As we already showed $\mathrm{H}^2(\mathcal{W};\mathcal{W}) = 0$ we obtain $\mathrm{H}^2(\mathcal{V};\mathcal{V}) = 0$ $\qquad\square$

References

[1] Fialowski, A.: *Unpublished Note*, 1989.

[2] Fialowski, A., and Schlichenmaier, M.: *Global deformations of the Witt algebra of Krichever Novikov type*. Comm. Contemp. Math. **5**, 921–945 (2003).

[3] Tsujishita, T.: *On the continuous cohomology of the Lie algebra of vector fields*. Proc. Japan Acad. **53**, Sec. A, 134–138 (1977)

[4] Reshetnikov, V.N.: *On the cohomology of the Lie algebra of vector fields on a circle*. Usp. Mat. Nauk **26**, 231–232 (1971)

[5] Goncharova, I.V.: *Cohomology of Lie algebras of formal vector fields on the line*. Funct. Anal. Appl. **7**, No. 2, 6–14 (1973)

[6] Schlichenmaier, M.: *An elementary proof of the vanishing of the second cohomology of the Witt and Virasoro algebras with values in the adjoint module*, Forum Math. DOI 10.1515/forum-2011-0143 (2012), arXiv:1111.6625.

[7] Fialowski, A.: *Formal rigidity of the Witt and Virasoro algebra*, J. Math. Phys. **53** (2012), arXiv:12023132

[8] Fialowski, A., and Schlichenmaier, M.: *Global Geometric Deformations of Current Algebras as Krichever–Novikov Type Algebras*. Comm. Math. Phys. **260** (2005), 579–612.

[9] Fialowski, A., and Schlichenmaier, M.: *Global Geometric Deformations of the Virasoro algebra, current and affine algebras by Krichever–Novikov type algebras*, International Journal of Theoretical Physics. Vol. **46**, No. 11 (2007) pp. 2708–2724

[10] Gerstenhaber, M.: *On the deformation of rings and algebras I, II, III*. Ann. Math. **79**, 59–103 (1964), **84**, 1–19 (1966), **88**, 1–34 (1968).

[11] Fialowski, A.: *An example of formal deformations of Lie algebras*. In: Proceedings of NATO Conference on Deformation Theory of Algebras and Applications, Il Ciocco, Italy, 1986, pp. 375–401, Kluwer, Dordrecht, 1988.

[12] Fialowski, A., and Fuks, D.: *Construction of miniversal deformations of Lie algebras*. J. Funct. Anal. **161**, 76–110 (1999).

[13] Nijenhuis, A., and Richardson, R.: *Cohomology and deformations of algebraic structures*. Bull. Amer. Math. Soc. **70** (1964), 406–411.

[14] Fuks, D.: *Cohomology of Infinite-dimensional Lie Algebras*, Consultants Bureau, N.Y., London, 1986.

[15] Guieu, L., Roger, C.: *L'algèbre et le groupe de Virasoro*. Les publications CRM, Montreal 2007.

[16] Schlichenmaier, M.: *Degenerations of generalized Krichever–Novikov algebras on tori*, Jour. Math. Phys. **34**, 3809–3824 (1993).

Martin Schlichenmaier
University of Luxembourg
Mathematics Research Unit, FSTC
Campus Kirchberg, 6, rue Coudenhove-Kalergi
L-1359 Luxembourg-Kirchberg, Luxembourg
e-mail: `martin.schlichenmaier@uni.lu`

Geometric Methods in Physics. XXXI Workshop 2012

Trends in Mathematics, 155–162

Invariants for Darboux Transformations of Arbitrary Order for $D_xD_y + aD_x + bD_y + c$

Ekaterina Shemyakova

Abstract. We develop the method of regularized moving frames of Fels and Olver to obtain explicit general formulas for the basis invariants that generate all the joint differential invariants, under gauge transformations, for the operators
$$\mathcal{L} = D_xD_y + a(x,y)D_x + b(x,y)D_y + c(x,y)$$
and an operator of arbitrary order.

The problem appeared in connection with invariant construction of Darboux transformations for $\mathcal{L}$.

Mathematics Subject Classification (2010). Primary 70H06; Secondary 34A26.

Keywords. Joint differential invariants, gauge transformations, bivariate linear partial differential operator of arbitrary order.

1. Introduction

The present paper is devoted to Darboux transformations [1] for the Laplace operator,
$$\mathcal{L} = D_xD_y + aD_x + bD_y + c \;, \tag{1}$$
where $a, b, c \in K$, where K is some differential field (see Sec. 2). Operator $\mathcal{L}$ is transformed into operator $\mathcal{L}_1$ with the same principal symbol (see Sec. 2) by means of operator $\mathcal{M}$ if there is a linear partial differential operator $\mathcal{N}$ such that
$$\mathcal{N}\mathcal{L} = \mathcal{L}_1\mathcal{M} \;.$$
In this case we shall say that there is a *Darboux transformation for pair* $(\mathcal{L}, \mathcal{M})$; we also say that $\mathcal{L}$ admits a *Darboux transformation generated by* $\mathcal{M}$. We define *the order of a Darboux transformation* as the order of the $\mathcal{M}$ corresponding to it.

Given some operator $\mathcal{R} \in K[D]$ and an invertible function $g \in K$, the corresponding *gauge transformation* is defined as
$$\mathcal{R} \to \mathcal{R}^g \;, \quad \mathcal{R}^g = g^{-1}\mathcal{R}g \;.$$

The principal symbol of an operator in $K[D]$ is invariant under the gauge transformations. One can prove [2] that if a Darboux transformation exists for a pair $(\mathcal{L}, \mathcal{M})$, then a Darboux transformation exists for the pair $(\mathcal{L}^g, \mathcal{M}^g)$. Therefore, Darboux transformations can be considered for the equivalence classes of the pairs $(\mathcal{L}, \mathcal{M})$.

A function of the coefficients of an operator in $K[D]$ and of the derivatives of the coefficients of the operator is called a *differential invariant* with respect to the gauge transformations if it is unaltered under the action of the gauge transformations. For several operators, we can consider *joint differential invariants*, which are functions of all their coefficients and of the derivatives of these coefficients. Differential invariants form a differential algebra over K. This algebra may be D-generated over K by some number of basis invariants. We say that these basis invariants form a *generating set of invariants*. For operators of the form (1) known as *Laplace invariants* functions $k = b_y + ab - c$ and $h = a_x + ab - c$ form a generating set of invariants.

In [3] we developed the regularized moving frames of Fels and Olver [4–6] for the individual linear partial differential operators of orders 2 and 3 on the plane under gauge transformations and obtained generating sets of invariants for those operators. In the present paper we extend those ideas and show that there is a finite generating set of invariants for the pairs $(\mathcal{L}, \mathcal{M})$, where $\mathcal{L}$ is of the form (1) and $\mathcal{M} \in K[D]$ is of arbitrary form and of order d. For $\mathcal{M}$ of arbitrary order d but given in its normalized form without mixed derivatives, we find explicit general formulas for the basis invariants for the pairs $(\mathcal{L}, \mathcal{M})$ (Theorem 2).

The existence of such normalized forms for $\mathcal{M}$ is implied by one of the theorems proved in [2], which can be re-formulated as follows.

Theorem 1 ([2]). *Let there be a Darboux transformation for pair $(\mathcal{L}, \mathcal{M})$, where $\mathcal{L}$ be of the form (1) and $\mathcal{M} \in K[D]$ of arbitrary form and order. Then there is a Darboux transformation for pair $(\mathcal{L}, \mathcal{M}')$, where $\mathcal{M}'$ contains no mixed derivatives.*

2. Preliminaries

Let K be a differential field of characteristic zero, equipped with commuting derivations ∂_x, ∂_y. Let $K[D] = K[D_x, D_y]$ be the corresponding ring of linear partial differential operators over K, where D_x, D_y correspond to derivations ∂_x, ∂_y. One can either assume field K to be differentially closed, in other words containing all the solutions of, in general nonlinear, Partial Differential Equations (PDEs) with coefficients in K, or simply assume that K contains the solutions of those PDEs that we encounter on the way.

Let $f \in K$, and $\mathcal{L} \in K[D]$; by $\mathcal{L}f$ we denote the composition of operator $\mathcal{L}$ with the operator of multiplication by a function f, while $\mathcal{L}(f)$ mean the application of operator $\mathcal{L}$ to f. The second lower index attached to a symbol denoting a function means the derivative of that function with respect to the variables listed there. For example, $f_{1,xyy} = \partial_x \partial_x \partial_y f_1$.

In the present paper we use Bell polynomials,

$$B_{n,k}(x_1, x_2, \ldots, x_{n-k+1})$$

$$= \sum \frac{n!}{j_1! j_2! \cdots j_{n-k+1}!} \left(\frac{x_1}{1!}\right)^{j_1} \left(\frac{x_2}{2!}\right)^{j_2} \cdots \left(\frac{x_{n-k+1}}{(n-k+1)!}\right)^{j_{n-k+1}} ,$$

where the sum is over all sequences $j_1, j_2, j_3, \ldots, j_{n-k+1}$ of non-negative integers such that $j_1 + j_2 + \cdots = k$ and $j_1 + 2j_2 + 3j_3 + \cdots = n$. The sum

$$B_n(x_1, \ldots, x_n) = \sum_{k=1}^{n} B_{n,k}(x_1, x_2, \ldots, x_{n-k+1})$$

is called the nth *complete Bell polynomial*, and also it has the following determinant representation:

$$B_n(x_1, \ldots, x_n) = \det \begin{bmatrix} x_1 & \binom{n-1}{1}x_2 & \binom{n-1}{2}x_3 & \binom{n-1}{3}x_4 & \cdots & \cdots & x_n \\ -1 & x_1 & \binom{n-2}{1}x_2 & \binom{n-2}{2}x_3 & \cdots & \cdots & x_{n-1} \\ 0 & -1 & x_1 & \binom{n-3}{1}x_2 & \cdots & \cdots & x_{n-2} \\ \vdots & \vdots & \vdots & \vdots & \ddots & \ddots & \vdots \\ 0 & 0 & 0 & 0 & \cdots & -1 & x_1 \end{bmatrix}.$$

3. Invariants for normalized Darboux transformations of arbitrary order

Theorem 2. *All joint differential invariants[1] for the pairs $(\mathcal{L}, \mathcal{M})$, where $\mathcal{L} = D_x D_y + a D_x + b D_y + c$ and $\mathcal{M} = \sum\limits_{i=1}^{d} m_i D_x^i + m_{-i} D_y^i + m_0$ and where $m_i \in K, i = -d, \ldots, d$ and $a, b, c \in K$, can be generated by the following $2d+3$ basis invariants.*

$$m = a_x - b_y ,$$

$$h = ab - c + a_x ,$$

$$R_j = \sum_{w=j}^{d} m_w \binom{w}{j} B_{w-j}(-b, -\partial_x(b), -\partial_x^2(b), \ldots, -\partial_x^{w-j-1}(b)) ,$$

$$R_{-j} = \sum_{w=j}^{d} m_{-w} \binom{w}{j} B_{w-j}(-a, -\partial_y(a), -\partial_y^2(a), \ldots, -\partial_y^{w-j-1}(a)) ,$$

---

[1]with respect to gauge transformations

$$R_0 = \sum_{w=1}^{d} m_w B_w(-b, -\partial_x(b), -\partial_x^2(b), \ldots, -\partial_x^{w-1}(b))$$
$$+ m_{-w} B_w(-a, -\partial_y(a), -\partial_y^2(a), \ldots, -\partial_y^{w-1}(a)) + m_0 .$$

Proof. We adopt the method of regularized moving frames [4–6]. Possible difficulties with the infinite-dimensional case are addressed in [7]. In this short paper we refer the reader to these works for the rigorous notation and for a justification of the method.

For transformations $\mathcal{L} \mapsto \mathcal{L}^{\exp(\alpha)}$, which implies the following group action on the coefficients of the operator,

$$(a, b, c,) \rightarrow (a + \alpha_y, b + \alpha_x, c + a\alpha_x + b\alpha_y + \alpha_{xy} + \alpha_x\alpha_y) ,$$

consider the prolonged action.

Let us construct a frame

$$\rho : (a_J, b_J, c_J) \mapsto g ,$$

at some regular point (x^0, y^0). Here a_J denotes the jet coefficients of a at (x^0, y^0), and they are to be regarded as the independent group parameters. A moving frame can be constructed through a normalization procedure based on a choice of a cross-section to the group orbits. Here we define a cross-section by normalization equations

$$a_J = 0 ,$$
$$b_X = 0 ,$$

where here J is a string of the form $x \ldots xy \ldots y$, where y has to be present at least once, and there may be no x-s, and X is a string of the form $x \ldots x$, where there can be no y-s. The normalization equations when solved for group parameters produces the moving frame section:

$$a_J = a_{J-y} ,$$
$$b_X = b_{X-x} ,$$

where $J - y$ means that we take one y from the string J, and $X - x$ means that we take one x from the string X. The first two fundamental differential invariants can be then found:

$$(b_1)_y\Big|_\rho = b_y + \alpha_{xy} = b_y - a_x = m = h - k ,$$

$$c_1\Big|_\rho = c - ab - ab - a_x + ab = c - a_x - ab = h .$$

The remaining invariants of the generating set can be obtained using the constructed frame for the group acting on all the coefficients of the second operator in the pair, operator $\mathcal{M}$ since none of them has been used during the normalization process and construction of the frame.

Consider a gauge transformation of each of the terms in the sum $\mathcal{M} = \sum_{i=1}^{d} m_i D_x^i + m_{-i} D_y^i + m_0$:

$$\left(m_i D_x^i\right)^{\exp(\alpha)} = \exp(-\alpha) \cdot m_i \cdot D_x^i \circ \exp(\alpha) \ , \ i \neq 0 \ ,$$

$$\left(m_{-i} D_y^i\right)^{\exp(\alpha)} = \exp(-\alpha) \cdot m_{-i} \cdot D_y^i \circ \exp(\alpha) \ , \ i \neq 0 \ ,$$

$$\left(m_0\right)^{\exp(\alpha)} = m_0 \ .$$

Using the general Leibnitz rule [8], for $i \neq 0$, we have

$$\left(m_i D_x^i\right)^{\exp(\alpha)} = \exp(-\alpha) \cdot m_i \cdot \sum_{k=0}^{i} \binom{i}{k} \frac{\partial^{i-k} \exp(\alpha)}{\partial x^{i-k}} \cdot D_x^k \ ,$$

then applying Faà di Bruno formula [9] we continue

$$\left(m_i D_x^i\right)^{\exp(\alpha)} = \exp(-\alpha) \cdot m_i \cdot \sum_{k=0}^{i} \binom{i}{k} \sum_{t=1}^{i-k} \frac{\partial^t \exp(\alpha)}{\partial \alpha^t}$$

$$\cdot B_{i-k,t}\left(\partial_x(\alpha), \partial_x^2(\alpha), \ldots, \partial_x^{i-k-t+1}(\alpha)\right) \cdot D_x^k$$

$$= m_i \cdot \sum_{k=0}^{i} \binom{i}{k} \sum_{t=1}^{i-k} B_{i-k,t}(\partial_x(\alpha), \partial_x^2(\alpha), \ldots, \partial_x^{i-k-t+1}(\alpha)) \cdot D_x^k \ ,$$

where $B_{i-k,t}$ are Bell polynomials. Since only the terms $B_{i-k,t}$ are summed with respect to t, we can rewrite the expression in terms of complete Bell polynomials:

$$\left(m_i D_x^i\right)^{\exp(\alpha)} = m_i \cdot \sum_{k=0}^{i} \binom{i}{k} B_{i-k}(\partial_x(\alpha), \partial_x^2(\alpha), \ldots, \partial_x^{i-k}(\alpha)) \cdot D_x^k \ . \qquad (2)$$

Now we compute invariants R_j, $j = 1, \ldots, d$ from the statement of the theorem as the coefficients at D_x^j, restricted on the constructed frame ρ. First, equality (2) implies that m_i appears only in R_j with $j \leq i$. Secondly, equality (2) implies that R_j must be a sum of the m_i multiplied by some functional coefficients. The coefficient of m_i in R_j can be found from (2) by the substitution $k = j$. In this way, we can obtain the invariants

$$R_j = \sum_{w=j}^{d} m_w \binom{w}{j} B_{w-j}(\partial_x(\alpha), \partial_x^2(\alpha), \ldots, \partial_x^{w-j}(\alpha))\bigg|_{frame}$$

$$= \sum_{w=j}^{d} m_w \binom{w}{j} B_{w-j}(-b, -\partial_x(b), -\partial_x^2(b), \ldots, -\partial_x^{w-j-1}(b))$$

for $j = 1, \ldots, d$. Analogously, one can compute invariants R_{-j}, $j = 1, \ldots, d$ from the statement of the theorem as the coefficients at D_y^j, restricted on the constructed frame ρ.

We compute invariant R_0 from the statement of the theorem as function $\mathcal{M}(1)$ restricted on the constructed frame ρ. It has to be considered separately as this is the only invariant which contains both a and b.

$$R_0 = \sum_{w=1}^{d} m_w \binom{w}{0} B_w(\partial_x(\alpha), \partial_x^2(\alpha), \ldots, \partial_x^w(\alpha))\Big|_{frame}$$

$$+ \sum_{w=1}^{d} m_w \binom{w}{0} B_w(\partial_y(\alpha), \partial_y^2(\alpha), \ldots, \partial_y^w(\alpha))\Big|_{frame} + m_0$$

$$= \sum_{w=1}^{d} m_w B_w(-b, -\partial_x(b), -\partial_x^2(b), \ldots, -\partial_x^{w-1}(b))$$

$$+ m_{-w} B_w(-a, -\partial_y(a), -\partial_y^2(a), \ldots, -\partial_y^{w-1}(a)) + m_0 . \qquad \square$$

Theorem 3 (Alternative form of Theorem 2). *All joint differential invariants[2] for the pairs $(\mathcal{M}, \mathcal{L})$, where $\mathcal{M} = \sum_{i=1}^{d} m_i D_x^i + m_{-i} D_y^i + m_0$ and $\mathcal{L} = D_x D_y + a D_x + b D_y + c$, where $m_i \in K, i = -d, \ldots, d$ and $a, b, c \in K$, can be generated by the following $2d + 3$ basis invariants.*

$$m = a_x - b_y ,$$
$$h = ab - c + a_x ,$$
$$R_0 = m_0 + \sum_{i=1}^{d} (m_i P_i(b) + m_{-i} P_i(a)) ,$$
$$R_j = \sum_{i=0}^{d-j} \binom{j+i}{j} m_{i|j} P_i(b) , \quad j \geq 1 ,$$
$$R_{-j} = \sum_{i=0}^{d-j} \binom{j+i}{j} m_{-(i+j)} P_i(a) , \quad j \geq 1 ,$$

where

$$P_0(f) = 1 ,$$
$$P_i(f) = -\Omega^i(f) , i \in \mathbb{N}_0 ,$$

and the linear differential operator Ω is defined by

$$\Omega(f) = \begin{cases} (D_x - b)(f) , & if \quad f = b \\ (D_y - a)(f) , & if \quad f = a . \end{cases}$$

---

[2]with respect to gauge transformations

Remark 1. Note the explicit forms taken by the first few operators P_i:

$$P_0(f) = 1 ,$$
$$P_1(f) = -f ,$$
$$P_2(f) = -f_x + f^2 ,$$
$$P_3(f) = -f_{xx} + 3ff_x - f^3 ,$$
$$P_4(f) = -f_{xxx} + 4ff_{xx} - 6f_x f^2 + 3f_x^2 + f^4 ,$$
$$P_5(f) = -f_{xxxx} + 5ff_{xxx} + 10f_{xx}f_x - 15f_x^2 f - 10f_{xx}f^2 + 10f_x f^3 - f^5 .$$

Example. Given the operator $\mathcal{M} = \sum_{i=1}^{5} m_i D_x^i + m_{-i} D_y^i$ and an operator $\mathcal{L}$ in the form (1), with $m_i \in K, i = -5, \ldots, 5$, the following functions form a generating set of invariants.

$$m = a_x - b_y ,$$
$$h = ab - c + a_x ,$$
$$R_1 = -2m_2 b + m_1 + (-3b_x + 3b^2)m_3 + (-4b_{xx} + 12b_x b - 4b^3)m_4$$
$$\quad + (20b_{xx}b + 15b_x^2 - 30b_x b^2 - 5b_{xxx} + 5b^4)m_5 ,$$
$$R_2 = m_2 - 3m_3 b + (-6b_x + 6b^2)m_4 + (-10b_{xx} - 10b^3 + 30b_x b)m_5 ,$$
$$R_3 = m_3 - 4m_4 b + (-10b_x + 10b^2)m_5 ,$$
$$R_4 = m_4 - 5m_5 b ,$$
$$R_5 = m_5 ,$$
$$R_{-5} = m_{-5} ,$$
$$R_{-4} = m_{-4} - 5m_{-5}a ,$$
$$R_{-3} = m_{-3} - 4m_{-4}a + (-10a_y + 10a^2)m_{-5} ,$$
$$R_{-2} = m_{-2} - 3m_{-3}a + (-6a_y + 6a^2)m_{-4} + (-10a_{yy} - 10a^3 + 30a_y a)m_{-5} ,$$
$$R_{-1} = -2m_{-2}a + m_{-1} + (-3a_y + 3a^2)m_{-3} + (-4a_{yy} + 12a_y a - 4a^3)m_{-4}$$
$$\quad + (20a_{yy}a + 15a_y^2 - 30a_y a^2 - 5a_{yyy} + 5a^4)m_{-5} ,$$

and finally

$$R_0 = m_0 - bm_1 - am_{-1} + (-b_x + b^2)m_2 + (a^2 - a_y)m_{-2}$$
$$\quad + (3b_x b - b^3 - b_{xx})m_3 + (3a_y a - a_{yy} - a^3)m_{-3}$$
$$\quad + (b^4 + 3b_x^2 - b_{xxx} + 4b_{xx}b - 6b_x b^2)m_4$$
$$\quad + (-6a_y a^2 + a^4 - a_{yyy} + 3a_y^2 + 4a_{yy}a)m_{-4}$$
$$\quad + (5b_{xxx}b - b_{xxxx} - b^5 - 10b_{xx}b^2 + 10b_x b^3 - 15b_x^2 b + 10b_{xx}b_x)m_5$$
$$\quad + (-15a_y^2 a + 10a_{yy}a_y + 5a_{yyy}a - a_{yyyy} - a^5 - 10a_{yy}a^2 + 10a_y a^3)m_{-5} .$$

References

[1] S. Tsarev. Factorization of linear partial differential operators and Darboux' method for integrating nonlinear partial differential equations. *Theo. Math. Phys.*, 122:121–133, 2000.

[2] E. Shemyakova. Proof of the Completeness of Darboux Wronskian Formulas for Order Two. *Accepted to Canadian Journal of Mathematics, see electronically at http://arxiv.org/abs/1111.1338*, 2012.

[3] E. Shemyakova and E.L. Mansfield. Moving frames for Laplace invariants. In *ISSAC 2008*, pages 295–302. ACM, New York, 2008.

[4] M. Fels and P.J. Olver. Moving coframes. II. Regularization and theoretical foundations. *Acta Appl. Math*, 55:127–208, 1999.

[5] P.J. Olver. Differential invariant algebras. *Contemp. Math.*, 549:95–121, 2011.

[6] E.L. Mansfield. *A Practical Guide to the Invariant Calculus*. Cambridge University Press, 2010.

[7] P.J. Olver and J. Pohjanpelto. Pseudo-groups, moving frames, and differential invariants. In Michael Eastwood and Willard Miller, editors, *Symmetries and Overdetermined Systems of Partial Differential Equations*, volume 144 of *The IMA Volumes in Mathematics and its Applications*, pages 127–149. Springer New York, 2008.

[8] P.J. Olver. *Applications of Lie groups to differential equations*. Graduate texts in mathematics. Springer-Verlag, 1986.

[9] R.P. Stanley and G.C. Rota. *Enumerative Combinatorics:*. Cambridge Studies in Advanced Mathematics. Cambridge University Press, 2000.

Ekaterina Shemyakova
Department of Mathematics
SUNY at New Paltz
1 Hawk Dr.
New Paltz, NY 12561, USA
e-mail: `shemyake@newpaltz.edu`

Geometric Methods in Physics. XXXI Workshop 2012
Trends in Mathematics, 163–168

Decomposition of Weyl Group Orbit Products of $W(A_2)$

Agnieszka Tereszkiewicz

Abstract. Product of two orbits of the Weyl reflection group $W(A_2)$ are decomposed into the union of the orbits.

Mathematics Subject Classification (2010). 20F55, 20H15.

Keywords. Weyl reflection group, product of orbits.

1. Introduction

Weyl group orbits play very important role in the theory of Lie groups. Usually one considers decomposition of the product of two representations and there are known algorithms how to do it, see for example [1]. It is computational problem, because of the weight systems which grow without limits with increasing representations. In this paper decomposition of a tensor product of two orbits of $W(A_2)$ into the union of orbits is presented. The decomposition problem for orbits is a finite one, and it doesn't depend on how large the dominant weights may be, i.e., the problem can be explicitly solved.

To calculate an orbit the dominant point is almost always chosen as the seed. There is only one dominant point in every orbit that has positive coordinates in the ω-basis. The number of points in any orbit cannot exceed the order of the Weyl group $W(A_2)$ and it is known.

The reason why the subject is an interesting is that orbits have been used in description of viruses [2] or symmetries of Clebsch–Gordon coefficients for groups of rank 2, see [3].

2. Preliminaries

The Weyl group $W(A_2)$ is the group of order 6, generated by reflection in two mirrors intersecting at angle $\frac{\pi}{3}$ at the origin of the real Euclidean space $\mathbb{R}^2$, see for example [4, 5].

It is convenient to work in $\mathbb{R}^2$ with a pair of dual bases. The α-basis is simple root basis, defined by the scalar products

$$\langle \alpha_1 \mid \alpha_1 \rangle = 2\,, \ \langle \alpha_2 \mid \alpha_2 \rangle = 2\,, \langle \alpha_1 \mid \alpha_2 \rangle = -1.$$

The ω-basis is defined as a dual to α-basis, $\langle \alpha_j \mid \omega_k \rangle = \frac{\langle \alpha_j \mid \alpha_j \rangle}{2} \delta_{jk}$. Explicitly,

$$\omega_1 = \tfrac{2}{3}\alpha_1 + \tfrac{1}{3}\alpha_2\,, \ \ \omega_2 = \tfrac{1}{3}\alpha_1 + \tfrac{2}{3}\alpha_2\,, \ \ \alpha_1 = 2\omega_1 - \omega_2, \ \ \alpha_2 = -\omega_1 + 2\omega_2.$$

For the group $W(A_2)$ reflections r_1 and r_2 act in $\mathbb{R}^2$ as follows

$$r_k x = x - \frac{2\langle \alpha_k \mid x \rangle}{\langle \alpha_k \mid \alpha_k \rangle} \alpha_k\,, \qquad k = 1,2, \quad x \in \mathbb{R}^2. \tag{1}$$

In particular,

$$r_k \omega_j = \omega_j - \delta_{jk}\alpha_k. \tag{2}$$

An orbit of $W(A_2)$ is the set of distinct points generated from a seed point $x \in \mathbb{R}^2$ by the repeated action of reflections (1). Such an orbit contains at most as many points as is the order of the group, 6 in the case considered in this paper. Each orbit contains precisely one point with non-negative coordinates in the ω-basis. It is called the dominant point. The orbit $O(a,b)$ is specified by its unique dominant point $x = (a,b)$, where $a,b \in \mathbb{Z}^{\geq 0}$.

In the case under consideration there are two kinds of orbits according to the position of the seed point $x \in \mathbb{R}^2$, namely $x \in P^+$ or $x \notin P^+$, where P^+ is the set of all dominant weights in the weight lattice P of $W(A_2)$

$$P^+ = \mathbb{Z}^{\geq 0}\omega_1 + \mathbb{Z}^{\geq 0}\omega_2 \subset P, \qquad P = \mathbb{Z}\omega_1 + \mathbb{Z}\omega_2 \subset \mathbb{R}^2,$$

see [6]. Because of (2), all the points of an orbit $O(x)$ either are in P or not. Let $a,b \in \mathbb{Z}^{>0}$, then one has four types of orbits of $W(A_2)$:

$$O(0,0) = \{(0,0)\},$$
$$O(a,0) = \{(a,0),(-a,a),(0,-a)\},$$
$$O(0,b) = \{(0,b),(b,-b),(-b,0)\},$$
$$O(a,b) = \{(a,b),(-a,a+b),(a+b,-b),$$
$$(-a-b,a),(b,-a-b),(-b,-a)\}\,,$$

The size of an orbit $|O(a,b)|$ is the number of distinct points it contains and it takes three values:

$$|O(a,b)| = |W(A_2)| = 6,$$
$$|O(a,0)| = |O(0,b)| = \tfrac{1}{2}|W(A_2)| = 3, \qquad a,b > 0,$$
$$|O(0,0)| = 1,$$

what could be written in one formula:

$$|O(a,b)| = 2\operatorname{sign}(a) + 2\operatorname{sign}(b) + \operatorname{sign}(ab) + 1, \text{ where } a,b \geq 0.$$

3. Decomposition of Weyl group orbit product of $W(A_2)$

The product of two orbits $W(A_2)$ is the set of points obtained by adding every point of one orbit to every point of the other orbit

$$O(\lambda) \otimes O(\lambda') := \bigcup_{\lambda_i \in O(\lambda), \lambda'_k \in O(\lambda')} (\lambda_i + \lambda'_k),$$

see [6]. The set of these points is invariant with respect to action of the Weyl group, each product of orbits is decomposable into a sum of orbits:

$$O(\lambda) \otimes O(\lambda') = O(\lambda') \otimes O(\lambda) - O(\lambda + \lambda') \cup \cdots \cup mO(\lambda + \overline{\lambda'}), \qquad m \geq 1. \quad (3)$$

Here $\overline{\lambda}$ stands for the lowest weight of the orbit $O(\lambda)$. If the sum $\lambda + \overline{\lambda'}$ is not a dominant weight, it should be reflected into the dominant weight of its orbit. The multiplicity m is strictly positive, $m \geq 1$. Moreover the orbit sizes multiply, $|O(a,b) \otimes O(a',b')| = |O(a,b)| \cdot |O(a',b')|$.

For a group of rank 1 all decompositions are known, but for rank 2 one can find only some special cases, see [6] or [7]. There is no general formula for finding the terms in the decomposition (3). In this paper general formula is found for the orbits of $W(A_2)$, and for the remaining reflection groups of rank 2 such formulas have been found and will be published elsewhere [8].

Proposition 1. *Decomposition of the product* (3) *of two Weyl group orbits of* $W(A_2)$ *with dominant weights* $\lambda = (a_1, a_2)$ *and* $\lambda' = (b_1, b_2)$ *is given by the following formula:*

$$
\begin{aligned}
O(a_1, a_2) &\otimes O(b_1, b_2) \\
&= k_1\, O(a_1 + b_1, a_2 + b_2) \cup k_2\, O\left(|a_1 - b_1|, a_2 + b_2 + \min\{a_1, b_1\}\right) \\
&\quad \cup k_3\, O\left(a_1 + b_1 + \min\{a_2, b_2\}, |a_2 - b_2|\right) \\
&\quad \cup k_4\, O\left(|b_1 + \min\{a_2, b_2 - a_1\}|, |a_2 + \min\{b_1, a_1 - b_2\}|\right) \\
&\quad \cup k_5\, O\left(|a_1 + \min\{b_2, a_2 - b_1\}|, |b_2 + \min\{a_1, b_1 - a_2\}|\right) \\
&\quad \cup k_6\, O\big(\big||a_1 + a_2 - b_1 - b_2| - |\min\{a_1 - b_2, b_1 - a_2, 0\}|\big|, \\
&\qquad\quad \big||a_1 + a_2 - b_1 - b_2| - |\min\{-a_1 + b_2, -b_1 + a_2, 0\}|\big|\big), \\
&\qquad a_1, a_2, b_1, b_2 \in \mathbb{Z}^{\geq 0},
\end{aligned}
$$

where

$$k_1 = \frac{1}{6}\frac{|O(a_1, a_2)||O(b_1, b_2)|}{|O(a_1 + b_1, a_2 + b_2)|}$$

$$k_2 = \frac{1}{6}\frac{|O(a_1, a_2)||O(b_1, b_2)|}{|O\left(|a_1 - b_1|, a_2 + b_2 + \min\{a_1, b_1\}\right)|}$$

$$k_3 = \frac{1}{6}\frac{|O(a_1, a_2)||O(b_1, b_2)|}{|O\left(a_1 + b_1 + \min\{a_2, b_2\}, |a_2 - b_2|\right)|}$$

$$k_4 = \frac{1}{6}\frac{|O(a_1, a_2)||O(b_1, b_2)|}{|O\left(|b_1 + \min\{a_2, b_2 - a_1\}|, |a_2 + \min\{b_1, a_1 - b_2\}|\right)|}$$

$$k_5 = \frac{1}{6}\frac{|O(a_1, a_2)||O(b_1, b_2)|}{|O\left(|a_1 + \min\{b_2, a_2 - b_1\}|, |b_2 + \min\{a_1, b_1 - a_2\}|\right)|}$$

$$k_6 = \frac{1}{6}|O(a_1, a_2)||O(b_1, b_2)|/|O\left(||a_1 + a_2 - b_1 - b_2| - |\min\{a_1 - b_2, b_1 - a_2, 0\}||, \right.$$
$$\left. ||a_1 + a_2 - b_1 - b_2| - |\min\{-a_1 + b_2, -b_1 + a_2, 0\}||\right)| .$$

The formula is proved through direct verification of all cases.

Using general hierarchy of orbits $O(a, b)$ of $W(A_2)$ with integer a and b one finds that they split into three congruence classes according to the value of their congruence number $K(a, b)$

$$K(a, b) = 2a + b \mod 3, \qquad a, b \in \mathbb{Z}.$$

All points of an orbit are in the same congruence class, because difference between two points of the same orbit is an integer linear combination of simple roots, and all simple roots are in the congruence class 0. For the multiplication of orbits, their congruence numbers add up:

$$K(O(a_1, a_2) \otimes O(b_1, b_2)) = K(a_1, a_2) + K(b_1, b_2).$$

All orbits in the decomposition belong to that congruence class. Let us illustrate this fact in an example:

$$O(4, 2) \otimes O(9, 7) = O(3, 11) \cup O(5, 13) \cup O(7, 3) \cup O(11, 1) \cup O(13, 9) \cup O(15, 5)$$

one has

$$K(4, 2) + K(9, 7) = 1 \mod 3 + 1 \mod 3 = 2 \mod 3$$
$$K(3, 11) = 2 \mod 3, \quad K(5, 13) = 2 \mod 3, \quad K(7, 3) = 2 \mod 3,$$
$$K(11, 1) = 2 \mod 3, \quad K(13, 9) = 2 \mod 3, \quad K(15, 5) = 2 \mod 3.$$

In physics, this product can be thought of as a certain interaction between two orbit layers, resulting on the right-hand side in an onion-like structure of several concentric orbit layers, what can be illustrated by example:

$$O(4, 2) \otimes O(1, 1) = O(5, 3) \cup O(2, 3) \cup O(3, 4) \cup O(6, 1) \cup O(3, 1) \cup 2O(5, 0)$$

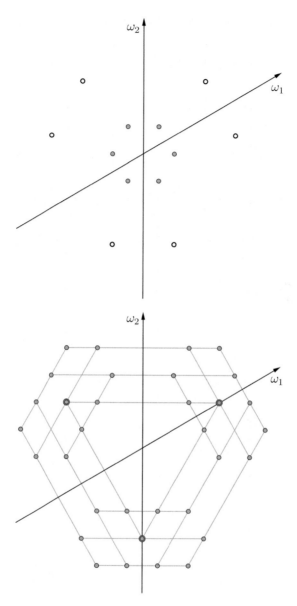

In the first (top) figure, the orbits $O(4,2)$ and $O(1,1)$ are shown. In the second one (bottom) the product of $O(4,2) \otimes O(1,1)$ is presented, points from the same orbit are joint.

By analogy one can calculate the decomposition of the group orbit products for a group of higher rank. Decomposition of Weyl group orbit products could be useful in the calculation of products of orthogonal functions defined on orbits, see [6].

Acknowledgment

The author would like to thank for the hospitality at the Centre de Recherches Mathématiques, Université de Montréal, when the idea of this paper appeared. She is also grateful to J. Patera for helpful comments.

References

[1] M.R. Bremner, R.V. Moody, J. Patera, *Tables of dominant weight multiplicities for representations of simple Lie algebras,* Marcel Dekker, New York 1985, 340 pages, ISBN: 0-8247-7270-9

[2] R. Twarock, *New group structure for carbon onions and nanotubes via affine extension of noncrystallographic Coxeter groups,* Phys. Lett. A **300** (2002) 437–444

[3] R.V. Moody, J. Patera, *General charge conjugation operators in simple Lie groups,* J. Math. Phys., **25** (1984) 2838–2847

[4] J.E. Humphreys, *Reflection groups and Coxeter groups,* Cambridge University Press, 1990.

[5] R. Kane, *Reflection Groups and Invariants,* New York: Springer, 2002

[6] A Klimyk, J. Patera, *Orbit functions,* Symmetry, Integrability and Geometry: Methods and Applications, **2** (2006) 006, 60pp.

[7] L. Háková, M. Larouche, J. Patera, *The rings of n-dimensional polytopes,* J. Phys. A: Math. Theor., **41** (2008) 495202; arXiv:0901.4686.

[8] A. Tereszkiewicz, *Complete Decompositions of Weyl group orbit products of $W(A_2)$, $W(C_2)$, $W(G_2)$ and Coxeter group H_2* – in preparation

Agnieszka Tereszkiewicz
Institute of Mathematics
University in Białystok
Lipowa 41
PL-15-424 Białystok, Poland
e-mail: a.tereszkiewicz@uwb.edu.pl

Part III

General Methods

Geometric Methods in Physics. XXXI Workshop 2012
Trends in Mathematics, 171–178

On Operators Generated by Maps with Separable Dynamics

A. Antonevich and A. Akhmatova

Abstract. Weighted shift operator generated by mapping $\alpha : X \to X$ is an operator of the form $Bu(x) = a_0(x)u(\alpha(x))$, where a_0 is a given function. Spectral properties of such operators are considered.

Mapping α is said to be *compatible* (with one-sided invertibility) if there exists a function a_0 such that operator $B - \lambda I$ is one-sided invertible for a certain spectral value λ. The main results are a dynamical description of the compatible mappings and one-sided invertibility conditions of $B - \lambda I$.

Mathematics Subject Classification (2010). Primary 47B38, 47A10.

Keywords. Weighted shift operator, essential spectrum, one-sided invertibility.

1. Introduction

A bounded linear operator B, acting on a Banach space $F(X)$ of functions on a set X, is called a *weighted shift operator* or *weighted composition operator* if it can be represented in the form

$$Bu(x) = a_0(x)u(\alpha(x)), \quad x \in X,$$

where $\alpha : X \to X$ is a map, $a_0(x)$ is a given function on X. Operator of the form

$$Bu(x) = \sum_k a_k(x)u(\alpha^k(x))$$

is called *functional operator*.

Such operators, operator algebras generated by them and functional equations related to such operators have been studied in application to theory of dynamical systems, integro-functional, differential-functional, functional and difference equations, automorphisms and endomorphisms of Banach algebras, nonlocal boundary value problems, nonclassical boundary value problems for equation of string vibration, the general theory of operator algebras (see the bibliography in [1–3]).

The main problem is to clarify the relationship between the spectral properties of weighted shift operators and the dynamical properties of the map α, i.e., the behavior of the trajectories

$$\{\alpha^n(x), n \in \mathbb{N}\},$$

where $\alpha^n(x) = \alpha(\alpha^{n-1}(x))$.

In this paper we will consider such operators under an assumption that X is a compact topological space and $F(X) = L_2(X, \mu)$, where μ is a Borel measure on X.

Let the following condition hold:

1) $\mu(\omega) \neq 0$ for every open set ω;
2) α is an invertible continuous map which *preserves the class of the measure* μ:

$$\mu(\alpha^{-1}(\omega)) = 0 \iff \mu(\omega) = 0$$

for all measurable ω.

Under these conditions there exists a measurable function ϱ, such that the operator

$$T_\alpha u(x) = \varrho(x)u(\alpha(x)) \tag{1}$$

is unitary. Then the operator B can be written in the form

$$B = aT_\alpha,$$

where

$$a(x) = \frac{1}{\varrho(x)}a_0(x)$$

is the so-called *reduced coefficient*.

2. Example: Symbol of non-local pseudodifferential operator

Let X be a compact smooth manifold and let $A(x, D)$ be a classical pseudodifferential operator. The principal symbol $\sigma(A)$ is a continuous function: $\sigma(A) \in C(S^*X)$, where $C(S^*X)$ is the bundle of unit cotangent vectors of the manifold X.

It is well known

Proposition 1. *Pseudodifferential operator A is Fredholm iff it is elliptic, i.e., symbol $\sigma(A)(x, \xi) \neq 0$ for all $(x, \xi) \in S^*X$.*

Let us consider in the space $L_2(X)$ non-local pseudodifferential operators of the form

$$L = \sum_k A_k(x, D)T_\alpha^k, \tag{2}$$

where $\alpha : X \to X$ is a diffeomorphism, T_α is the unitary operator (1), generated by α, and A_k are classical pseudodifferential operators of the order 0.

In the process of developing a theory of non-local pseudodifferential operators one of the first problems to be solved is that of constructing a symbol for operators (2). The solution looks as follows [1, 3].

For given diffeomorphism $\alpha : X \to X$ we can associate a canonical mapping $\beta : S^*X \to S^*X$ by the rule

$$\beta(\eta) = \frac{1}{\|\partial\alpha(\eta)\|}\partial\alpha(\eta),$$

where $\partial\alpha : T^*X \to T^*X$ is the codifferential of α.

Definition. The symbol $\sigma(L)$ of non-local pseudodifferential operator L (2) is a functional operator in the space $L_2(S^*X)$ given by the expression

$$\sigma(L) = \sum_k \sigma(A_k)T_\beta^k.$$

Proposition 2 ([1, 3]). *The non-local pseudodifferential operator*

$$L = \sum_k A_k(x, D)T_\alpha^k$$

is Fredholm iff the symbol $\sigma(L)$ is invertible as an operator in the space $L_2(S^*X)$.

The operator L is semi-Fredholm iff the symbol $\sigma(L)$ is one-sided invertible operator.

In particular, for binomial non-local pseudodifferential operator

$$L = A_0(x, D) + A_1(x, D)T_\alpha$$

the problem of invertibility of the symbol can be reduced to investigation of the spectrum of weighted shift operator aT_β, where

$$a = \frac{\sigma(A_1)}{\sigma(A_2)}.$$

3. Spectrum of the weighted shift operator

A description of the spectrum of a weighted shift operator in the case of invertible mapping α in classical spaces was obtained in sufficient generality.

Let $EM_\alpha(X)$ be the set of all probability measures on X, invariant and ergodic with respect to α.

For a given function a and given measure ν let us introduce the geometric mean of a with respect to ν:

$$S_\nu(a) = \exp\left[\int_X \ln|a(x)|d\nu\right].$$

Proposition 3 ([1, 2]). *Let $a \in C(X)$ and $B = aT_\alpha$ be the weighted shift operator in the space $L_2(X, \mu)$ with the reduced coefficient a. The following expression is valid for the spectral radius $R(B)$:*

$$R(B) = \max\{S_\nu(a) : \nu \in EM_\alpha(X)\}.$$

By using the expression for spectral radius one can give (under some natural assumptions) the full description of the spectrum $\Sigma(B)$.

The additional conditions are

3) the set of the nonperiodic points with respect to α is a dense subset of X.
4) The space X is α-connected.

Here the space X is called α-*connected* if it cannot be decomposed into two nonempty closed subsets invariant with respect to α.

The description of the spectrum is as follows.

Proposition 4 ([1, 2]). *Let the conditions* 1)–4) *hold,* $a \in C(X)$ *and* $a(x) \neq 0$ *for all* $x \in X$. *Then the spectrum* $\Sigma(B)$ *of the operator* B *is the annulus*

$$\Sigma(B) = \{\lambda \in \mathbb{C} : r(B) \leq |\lambda| \leq R(B)\},$$

where

$$R(B) = \max\{S_\nu(a) : \nu \in EM_\alpha(X)\},$$
$$r(B) = \min\{S_\nu(a) : \nu \in EM_\alpha(X)\}.$$

4. Essential spectra

Let λ belong to the spectrum $\Sigma(B)$. The operator $B - \lambda I$ is not invertible, but it can happen that this operator has some "good" properties. For example, $B - \lambda I$ can be Fredholm, semi-Fredholm or one-sided invertible.

Every "good" property of non-invertible operator $B - \lambda I$ splits the spectrum $\Sigma(B)$ into two parts – the part on which this property holds with the corresponding *essential spectrum* – the part of the spectrum $\Sigma(B)$ on which this property does not hold.

The Fredholm spectrum is used more often:

$$\Sigma_F(B) = \{\lambda \in \mathbb{C} : B - \lambda I \ \text{ is not a Fredholm operator}\}.$$

Other forms of essential spectrum of the operator B are also important for applications:

$$\Sigma^l(B) = \{\lambda \in \mathbb{C} : \nexists \text{ left inverse for } B - \lambda I\} \quad – \text{ left spectrum,}$$
$$\Sigma^r(B) = \{\lambda \in \mathbb{C} : \nexists \text{ right inverse for } B - \lambda I\} – \text{ right spectrum,}$$
$$\Sigma_K(B) = \{\lambda \in \mathbb{C} : Im(B - \lambda I) \ \text{ is not closed}\} – \text{ Kato spectrum.}$$

Our aim is to give a description of $\Sigma^l(B)$ and $\Sigma^r(B)$ for the weighted shift operators B.

In regard to the one-sided invertibility of operators $aT_\alpha - \lambda I$ all the mappings α can be divided into two classes.

Mapping α is said to be *incompatible* (with one-sided invertibility) if for every $a \in C(X)$ operator $aT_\alpha - \lambda I$ is not one-sided invertible for all spectral values λ: for all a holds

$$\Sigma(aT_\alpha) = \Sigma^l(aT_\alpha) = \Sigma^r(aT_\alpha).$$

Mapping α is said to be *compatible* (with one-sided invertibility) if there exists $a \in C(X)$, such that the operator $aT_\alpha - \lambda I$ is one-sided invertible for a certain spectral value λ.

We have two problems. First one is to obtain a characterization of the compatible (and incompatible) mappings in terms of dynamical properties. Second problem is to give one-sided invertibility conditions for $aT_\alpha - \lambda I$.

5. Morse–Smale type mappings

One-sided invertibility condition of the operators $aT_\alpha - \lambda I$ was obtained early for some special classes of mappings with simple dynamics [4–7]. All these mappings are particular case of so-called Morse–Smale type mappings.

Mapping α is said to be a *Morse–Smale type mapping*, if

i) for every point x the trajectory $\alpha^n(x)$ has a limit as $n \to +\infty$ and also has a limit (may be different one) as $n \to -\infty$.

ii) the set $\mathrm{Fix}(\alpha)$ of fixed points is finite:
$$\mathrm{Fix}(\alpha) = \{F_1, F_2, \ldots, F_q\}.$$

For such maps
$$EM_\alpha(X) = \{\delta_{F_1}, \delta_{F_2}, \ldots, \delta_{F_q}\},$$
i.e., the measure is ergodic if it is concentrated in a fixed point F_k. Therefore, the Proposition 4 gives an explicit description of the spectrum for Morse–Smale type mappings α: if $a(x) \neq 0$ for all x, the spectrum of the weighted shift operator $B = aT_\alpha$ is the annulus:
$$\Sigma(B) = \{\lambda \in \mathbb{C} : r(B) \leq |\lambda| \leq R(B)\},$$
where
$$R(B) = \max_{x \in Fix(\alpha)} |a(x)|, \quad r(B) = \min_{x \in Fix(\alpha)} |a(x)|.$$
Circles
$$S_k = \{\lambda : |\lambda| = |a(F_k)|\}, 1 \leq k \leq q,$$
belong to the spectrum $\Sigma(B)$ and they split the spectrum into smaller subrings. The study shows that the properties of the operator $B - \lambda I$ are the same for λ from the same subring, and may be different for λ from different subrings. Therefore, the problem consists of the description of properties of the operator $B - \lambda I$ for different subrings. These properties depend on coefficient a, number λ and dynamical properties of α.

An appropriate dynamical characteristic of a Morse–Smale type map α is an *oriented graph* G_α.

The vertices of the graph G_α are fixed points F_k.

An *oriented edge* $F_k \to F_j$ is included in the graph if and only if there exists a point $x \in X$ such that $\alpha^n(x) \to F_j$ as $n \to +\infty$ and $\alpha^n(x) \to F_k$ as $n \to -\infty$.

The use of the introduced graph permits one to obtain a simple formulation of the important spectral properties of the weighted shift operators generated by α.

Given the coefficient a and number λ one forms two subsets of the set of vertices of the graph

$$G^+(a, \lambda) = \{F_k \in Fix(\alpha) : |a(F_k)| > |\lambda|\},$$
$$G^-(a, \lambda) = \{F_j \in Fix(\alpha) : |a(F_j)| < |\lambda|\}.$$

It is clear that

$$G^+(a, \lambda) \bigcap G^-(a, \lambda) = \emptyset.$$

We say that subsets $G^+(a, \lambda)$ and $G^-(a, \lambda)$ give a *decomposition of the graph*, if the condition

$$G_\alpha = G^+(a, \lambda) \bigcup G^-(a, \lambda)$$

holds, i.e., if

$$|a(F_k)| \neq |\lambda| \quad \text{for all} \quad F_k \in Fix(\alpha).$$

The graph decomposition will be called *oriented to the right* if any edge connecting the point $F_k \in G^-(a, \lambda)$ with the point $F_j \in G^+(a, \lambda)$, is oriented from F_k to F_j.

The decomposition will be called *oriented to the left* if any edge connecting the point $F_k \in G^-(a, \lambda)$ with the point $F_j \in G^+(a, \lambda)$, is oriented from F_j to F_k.

The basic result in this direction is the following.

Proposition 5. *Let α be a Morse–Smale type mapping, $a \in C(X)$, $a(x) \neq 0$ for all x and $B = aT_\alpha$ is the corresponding weighted shift operator. The operator $B - \lambda I$ is invertible from the right (left) if and only if the subsets $G^+(a, \lambda)$ and $G^-(a, \lambda)$ form a decomposition of the graph G_α which is oriented to the right (to the left).*

Corollary. *Morse–Smale type mapping α is compatible iff the graph G_α admits a nontrivial oriented decomposition.*

6. Main results: Mappings with separable dynamics and one-sided invertibility conditions

The first our result is the following global dynamical description of the compatible mappings.

Let $\Omega^+(x)$ be the set of limit points for the positive semi-trajectory $\{x, \alpha(x), \alpha^2(x), \ldots\}$ and similarly $\Omega^-(x)$ is the set of the limit points for the negative semi-trajectory $\{x, \alpha^{-1}(x), \alpha^{-2}(x), \ldots\}$.

The dynamics of an invertible mapping α is said to be *separable*, if there exists a decomposition $X = X^+ \coprod X^-$ into nonempty measurable sets such that

(i) $\alpha(X^+) \subset X^+$, $\quad \alpha^{-1}(X^-) \subset X^-$;
(ii) $\Omega^+(X^+) \bigcap \Omega^-(X^-) = \emptyset$,

where

$$\Omega^+(X^+) = \cup_{x \in X^+} \Omega^+(x),$$
$$\Omega^-(X^-) = \cup_{x \in X^-} \Omega^-(x).$$

Theorem 1. *Mapping α is compatible iff its dynamics is separable.*

For a compatible mapping α it not obvious how to obtain conditions of one-sided invertibility for the operator $aT_\alpha - \lambda I$.

The answer is similar to the case of Morse–Smale type mapping.

A point $x \in X$ is said to be *non wandering*, if for an arbitrary neighborhood U of x there exist $n > 0$ such that

$$\alpha^n(U) \bigcap U \neq \emptyset.$$

Let $\Omega(\alpha)$ be the set of all nonwandering points of α.

Let us assume that there exists a finite number of the α-connected components of the set $\Omega(\alpha)$ and denote these components by $\Omega_k, k = 1, 2, \ldots, p$.

The dynamical characteristic of the map α which was useful in obtaining solutions of the problem in question turned out to be an *oriented graph* G_α, describing the dynamics of the map.

The vertices of the graph G_α are the sets $\Omega_k, k = 1, 2, \ldots, p$.

An *oriented edge* $\Omega_k \to \Omega_j$ is included in the graph if and only if there exists a point $x \in X$ such that its positive semi-trajectory has a limit point from Ω_j and negative semi-trajectory has a limit point from Ω_k.

Note that if ν is an ergodic measure ($\nu \in EM_\alpha(X)$) then the support $supp \, \nu$ belongs to some α-connected component Ω_k.

Denote

$$M_k = \{\nu \in EM_\alpha(X) : \operatorname{supp} \nu \subset \Omega_k\}.$$

For example, in the case of the Morse–Smale type mappings we have $\Omega_k = \{F_k\}$ and $M_k = \{\delta_{F_k}\}$.

Given reduced coefficient a and number λ one forms two subsets of the set of vertices of the graph G_α:

$$G^+(a, \lambda) = \{\Omega_k : S_\nu(a) > |\lambda| \text{ for all } \nu \in M_k\},$$

$$G^-(a, \lambda) = \{\Omega_k : S_\nu(a) < |\lambda| \text{ for all } \nu \in M_k \}.$$

It is clear that

$$G^+(a, \lambda) \bigcap G^-(a, \lambda) = \emptyset.$$

We say that subsets $G^+(a, \lambda)$ and $G^-(a, \lambda)$ give a *decomposition of the graph*, if the condition

$$G_\alpha = G^+(a, \lambda) \bigcup G^-(a, \lambda)$$

holds, i.e., if

$$S_\nu(a) \neq |\lambda| \quad \text{for all} \quad \nu \in EM_\alpha(X).$$

The basic result in this direction is the following.

Theorem 2. *Let the conditions* 1)–4) *hold and* $a(x) \neq 0$ *for all* $x \in X$. *The operator* $B - \lambda I$ *is invertible from the right (left) if and only if the subsets* $G^+(a, \lambda)$ *and* $G^-(a, \lambda)$ *form a decomposition of the graph* G_α *which is oriented to the right (to the left).*

For discrete weighted shift operators similar results were obtained in [8].

Acknowledgment

This work was partially supported by the Grant of the National Science Center (Poland) No DEC-2011/01/B/ST1/03838.

References

[1] A.B. Antonevich. *Linear functional equations. Operator approach.* Universitetskoe, Minsk, 1988 (in Russian); English transl. Birkhäuser, 1996.

[2] A. Antonevich, A. Lebedev. *Functional differential Equations: V.I. C*-theory.* Longman Scientific & Technical. 1994.

[3] A. Antonevich, M. Belousov, A. Lebedev. *Functional differential equations: V.II. C*-applications. Part 1. Equations with continuous coefficients.* Addison Wesley Longman. 1998.

[4] R. Mardiev. *A criterion for semi-Noetherity of a class of singular integral operators with non-Carleman shift.* Dokl. Akad. Nauk UzSSR, **2** (1985), 5–7.

[5] Yu.I. Karlovich,R. Mardiev. *One-sided invertibility of functional operator with non-Carleman shift in Hölder spaces.* Izv. Vyssh. uchebnykh Zaved. Mat., 3 (1987), 77–80 (in Russian). English transl.: Sovet Math. (Iz.VUZ), **31 (3)** (1987), 106–110.

[6] A.Yu. Karlovich, Yu.I. Karlovich. *One-sided invertibility of binomial functional operators with a shift in rearrangement-invariant spaces.* Integral Equations Operator Theory, **42** (2002), 201–228.

[7] Antonevich A., Makowska Yu. *On spectral properties of weighted shift operators generated by mappings with saddle points,* Complex Analysis and Operator theory **2**(2008), 215–240.

[8] Antonevich A.B., Akhmatova A.A. *Spectral properties of discrete weighted shift operators,* Trudy In. Math. NAN Belarusi, Minsk **20(1)** (2012), 14–21.

A. Antonevich
Institute of Mathematics
University of Białystok
Akademicka 2
PL-15-267 Białystok, Poland
e-mail: `antonevich@bsu.by`

A. Akhmatova
Belarus State University
Nezavisimosti 4
220030 Minsk, Belarus
e-mail: `akhmatovaaa@tut.by`

Geometric Methods in Physics. XXXI Workshop 2012
Trends in Mathematics, 179–185

From the Heat Equation to Ordinary Differential Equations

Elena Yu. Bunkova

Abstract. In a recent work we introduced an ansatz for heat equation solutions. The ansatz reduces the problem of solving the heat equation to the problem of solving an ordinary differential equation. In this paper we describe the reduction and construct examples of heat equation solutions and ordinary differential equations corresponding to these solutions. The aim of this work is to present cases important in modern integrable systems theory and mathematical physics.

Mathematics Subject Classification (2010). 35K05; 34M55.

Keywords. Heat equation, polynomial dynamical systems, Chazy equation, Darboux-Halphen system, Painlevé property.

Introduction

In this work a method to construct examples of heat equation solutions starting from ordinary differential equation solutions and vice versa is described. The description presented in this paper is a simplification of the one given in [1].

The classical Darboux-Halphen system is an order 3 dynamical system that has generalizations important in modern integrable systems theory. We show the connections between these generalizations and the heat equation. We obtain analogous systems of order 4 and 5 starting from the heat equation. In this construction special families of ordinary differential equations arise. We pay special attention to ordinary differential equations with the Painlevé property.

This work was completed with the support of RFFI grants 11-01-00197-a, 11-01-12067-ofi-m-2011, RF Government grant 10-220-01-077, ag. 11.G34.31.0005.

1. Special ansatz for solutions of the heat equation

For some $n \in \mathbb{N}$ set

$$\Phi(z; \mathbf{x}) = z^\delta + \sum_{k \geqslant 2} \Phi_k(\mathbf{x}) \frac{z^{2k+\delta}}{(2k+\delta)!},$$

where $\Phi_k(\mathbf{x})$ are degree $-4k$ homogeneous polynomials in $\mathbf{x} = (x_2, \ldots, x_{n+1})$, $\deg x_k = -4k$, and $\delta = 0$ or 1. For $\deg z = 2$ the function $\Phi(z; \mathbf{x})$ is a homogeneous function of degree 2δ. Let $P_n(\mathbf{x})$ be some homogeneous polynomial of degree $-4(n+2)$. Because of the grading, this polynomial does not depend on its last argument.

Our main interest is attracted by the ansatz

$$\psi(z, t) = e^{-\frac{1}{2}h(t)z^2 + r(t)} \Phi(z; \mathbf{x}(t)) \tag{1}$$

for heat equation solutions, where $h(t)$ and $r(t)$ are some functions and $\mathbf{x}(t)$ is a vector-function.

Theorem 1. *Any two of the following conditions imply the third one:*

1) *The function $\psi(z, t)$ solves the heat equation*

$$\frac{\partial \psi}{\partial t} = \frac{1}{2} \frac{\partial^2 \psi}{\partial z^2}. \tag{2}$$

2) *We have the recurrent relations on $\Phi_k(\mathbf{x})$:*

$$\Phi_k = 2 \left(\sum_{l=2}^n x_{l+1} \frac{\partial}{\partial x_l} + P_n(\mathbf{x}) \frac{\partial}{\partial x_{n+1}} \right) \Phi_{k-1}$$
$$+ \frac{(2k+\delta-3)(2k+\delta-2)}{2(1+2\delta)} \Phi_2 \Phi_{k-2} \tag{3}$$

for $k - 4, 5, \ldots$, with initial conditions $\Phi_2(\mathbf{x}) = -2(1+2\delta)x_2$, $\Phi_3(\mathbf{x}) = -4(1+2\delta)x_3$.

3) *We have $r'(t) = -(\delta + \frac{1}{2})h(t)$ and the set of functions $(h(t), \mathbf{x}(t))$ solves the homogeneous polynomial dynamical system*

$$\frac{d}{dt} h = -h^2 + x_2,$$
$$\frac{d}{dt} x_k = x_{k+1} - 2khx_k, \quad for \quad k = 2, \ldots, n,$$
$$\frac{d}{dt} x_{n+1} = P_n(\mathbf{x}) - 2(n+1)hx_{n+1}.$$

Proof. This theorem is a special case of Theorem 3.6 from [1]. Any type of system in the general case can be reduced to the considered one according to results of Section 4 in [1]. $\square$

Remark 1. Condition 3) in Theorem 1 can be replaced by the following condition:

3') We have $r'(t) = -(\delta + \frac{1}{2})h(t)$,

$$x_2 = \frac{d}{dt}h + h^2, \quad x_k = \frac{d}{dt}x_{k-1} + 2(k-1)hx_{k-1} \quad for \quad k = 3, \ldots, n+1, \tag{4}$$

and the function $h(t)$ solves the ordinary differential equation obtained from

$$\frac{d}{dt}x_{n+1} = P_n(\mathbf{x}) - 2(n+1)hx_{n+1} \tag{5}$$

by the substitutions (4).

The substitution of (4) into (5) gives an ordinary differential equation

$$\mathcal{D}_{P_n,n+1}(h) = 0 \tag{6}$$

of order $n+1$ on $h(t)$. It is homogeneous with respect to the grading $\deg h = -4$, $\deg t = 4$. A recurrent formula for this equation can be found in [1]: We have

$$\mathcal{D}_{0,1}(h) = \left(\frac{d}{dt} + h\right)h,$$

$$\mathcal{D}_{0,k}(h) = \left(\frac{d}{dt} + 2kh\right)\mathcal{D}_{0,k-1}(h) \quad \text{for} \quad k > 1, \tag{7}$$

$$\mathcal{D}_{P_k,k+1}(h) = \mathcal{D}_{0,k+1}(h) - P_k(\mathcal{D}_{0,1}(h), \ldots, \mathcal{D}_{0,k-1}(h), 0).$$

Therefore for the heat equation each solution of the form (1) corresponds to an ordinary differential equation of the form (6) with its solution and vice versa. In the next sections we present important examples of this correspondence. An example of such a correspondence for the elliptic theta and sigma functions and the Chazy-3 equation can be found in [2].

2. The Painlevé property

For the definitions given in this chapter we follow [3] and [4].

A singularity of a function is a point in which the function is not analytic (and possibly not defined). A singularity of a function is called a critical singularity if going around this singularity changes the value of the function.

L. Fuchs remarked that differential equation solutions can have movable singularities, that is singularities whose location depends on the initial conditions of the solution. In 1884 L. Fuchs and H. Ponicaré stated the problem of defining new functions by means of non-linear ordinary differential equations. In the same year L. Fuchs proved that among first-order explicit differential equations

$$y' = F(t, y)$$

with F being a rational function of y and a locally-analytical function of t, only the Riccati equation

$$y' = P_0(t) + P_1(t)y + P_2(t)y^2$$

does not have any movable critical singularities. All first-order algebraic differential equations with such property can be transformed into the Riccati equation or the Weierstrass equation

$$(y')^2 = 4y^3 - g_2 y - g_3.$$

Both equations are integrable in terms of previously known special functions.

The next important step in the analytical theory of differential equations was made by S. Kovalevskaya. In 1888 she solved the classical precession of a top under the influence of gravity. Her approach to the problem is based on finding solutions with no movable critical singularities. She proved that there exists only three cases with such solutions: the famous Euler, Lagrange and Kovalevskaya tops. In the third case she found new solutions and thus was first to discover the advantages of solving differential equations whose solutions have no movable critical singularities.

The property of an ordinary differential equation that its general solution has no movable critical singularities is shared by all linear ordinary differential equations but is rare in non-linear equations. It is called the *Painlevé property*. The general solution of equations with this property leads to single-valued function.

Around 1900, P. Painlevé studied second-order explicit non-linear differential equations

$$y'' = F(t, y, y')$$

with F being a rational function of y and y' and a locally-analytical function of t. It turned out that among such equations up to certain transformations only fifty equations have the Painlevé property, and among them six are not integrable in terms of previously known functions. P. Painlevé and B. Gambier have introduced new special functions, now known as Painlevé transcendents, as general solutions to these equations.

In 1910 J. Chazy extended Painlevé's work to higher-order equations, finding some third-order equations with the Painlevé property.

We will show the connection to heat equation solutions of the following explicit autonomous non-linear ordinary differential equations with Painlevé property:

Chazy-3 equation

$$y''' = 2yy'' - 3(y')^2. \tag{8}$$

Chazy-12 equation (see, e.g., [5])

$$y''' = 2yy'' - 3(y')^2 - \frac{4}{k^2 - 36}(6y' - y^2)^2, \tag{9}$$

where $k \in \mathbb{N}$, $k > 1$, $k \neq 6$.

A fourth-order equation (see [6])

$$y'''' + 5yy''' + 10y'y'' + 10y^2y'' + 15y(y')^2 + 10y^3y' + y^5 = 0. \tag{10}$$

The relation of the considered equations to mathematical physics arises from an observation by M.J. Ablowitz and H. Segur that reduction of a partial differential equation of soliton type gives rise to ordinary differential equations that possess the Painlevé property.

3. The case of Chazy-3 and Chazy-12 equations

Let us describe in details the construction of Section 1 in the important case $n = 2$.

We have $\mathbf{x} = (x_2, x_3)$ and $P_2(\mathbf{x})$ is a homogeneous polynomial of degree -16. Thus $P_2(\mathbf{x}) = cx_2^2$ for some constant c. Formulas (7) imply that equation (6) becomes

$$h''' + 12hh'' - 18(h')^2 + (24 - c)(h' + h^2)^2 = 0. \tag{11}$$

After the substitution $y(t) = -6h(t)$ this third-order differential equation for $c = 24$ becomes the Chazy-3 equation (8). For $c = \frac{24k^2}{k^2-36}$ it becomes the Chazy-12 equation (9).

Let us consider the classical Darboux–Halphen system

$$\xi_i' = \xi_j \xi_k - \xi_i(\xi_j + \xi_k), \tag{12}$$

where the indices (i, j, k) run over the three cyclic permutations of $(1, 2, 3)$. Set $h_1 = \xi_1 + \xi_2 + \xi_3$, $h_2 = \xi_1\xi_2 + \xi_1\xi_3 + \xi_2\xi_3$, $h_3 = \xi_1\xi_2\xi_3$. System (12) implies the homogeneous dynamical system

$$h_1' = -h_2,$$
$$h_2' = -6h_3,$$
$$h_3' = h_2^2 - 4h_3 h_1,$$

which reduces to equation (11) with $c = 24$ for $3h(t) = h_1(t)$, that is to Chazy-3 equation for $y(t) = -2h_1(t)$.

Generalizing the dynamical system (12), one can obtain equation (11) for different values of c.

Let us consider a generalization of this system

$$\frac{d\xi_i}{dt} = a\left(\xi_j \xi_k - \xi_i(\xi_j + \xi_k)\right) + b\xi_i^2, \quad c = \frac{6(2a+b)^2}{(a+2b)(a-b)}, \tag{13}$$

where the indices (i, j, k) run over the three cyclic permutations of $(1, 2, 3)$. For $c \neq 0$ it reduces to equation (11) for $h(t) = \frac{a-b}{3}(\xi_1 + \xi_2 + \xi_3)$. This system becomes the classical Darboux–Halphen system in the case $a = 1$, $b = 0$.

Remark that the wide-known generalization of the Darboux–Halphen system (see [7])

$$\zeta_i' = \zeta_j \zeta_k - \zeta_i(\zeta_j + \zeta_k) + \tau^2,$$

where the indices (i, j, k) run over the three cyclic permutations of $(1, 2, 3)$ and $\tau^2 = \alpha^2(\zeta_1 - \zeta_2)(\zeta_3 - \zeta_1) + \beta^2(\zeta_2 - \zeta_3)(\zeta_1 - \zeta_2) + \gamma^2(\zeta_3 - \zeta_1)(\zeta_2 - \zeta_3)$ in the case $\alpha^2 = \beta^2 = \gamma^2$ is the case $b = a - 1$ of the same system (13) in coordinates

$$\zeta_i = a\xi_i - \frac{1}{2}(a-1)(\xi_j + \xi_k), \quad i \neq j \neq k$$

with $\alpha^2 = (a - 1)^2/(3a - 1)^2$.

Given a solution $h(t)$ of (11), we get

$$x_2 = h' + h^2, \quad x_3 = h'' + 6hh' + 4h^3.$$

Equation (3) takes the form

$$\Phi_k = 2\left(x_3\frac{\partial}{\partial x_2} + cx_2^2\frac{\partial}{\partial x_3}\right)\Phi_{k-1} + \frac{(2k+\delta-3)(2k+\delta-2)}{2(1+2\delta)}\Phi_2\Phi_{k-2}$$

for $k = 4, 5, \ldots$, with initial conditions $\Phi_2(\mathbf{x}) = -2(1+2\delta)x_2$, $\Phi_3(\mathbf{x}) = -4(1+2\delta)x_3$.

This gives us all functions $\Phi_k(t)$ as functions of t recurrently. The equation $r'(t) = -(\delta+\frac{1}{2})h(t)$ gives the function $r(t)$ up to a constant of integration.

Thus, up to a constant factor, we get two functions $\psi(z,t)$ determined by (1) (one even for $\delta = 0$ and one odd for $\delta = 1$), each being a solution of the heat equation (2).

4. The case of fourth-order equations

For $n = 3$ we have $\mathbf{x} = (x_2, x_3, x_4)$ and $P_3(\mathbf{x})$ is a homogeneous polynomial of degree -20. Thus $P_3(\mathbf{x}) = cx_2x_3$ for some constant c. Formulas (7) imply that equation (6) becomes

$$h'''' + 20hh''' - 24h'h'' + 96h^2h'' - 144h(h')^2$$
$$+ (48-c)(h'+h^2)(h''+6hh'+4h^3) = 0. \tag{14}$$

After the substitution $y(t) = 4h(t)$ this equation for $c = -16$ becomes equation (10).

A generalization of system (13) is

$$\frac{d\xi_i}{dt} = a\Big(\xi_j\xi_k + \xi_j\xi_l + \xi_k\xi_l\Big) - 2a\xi_i(\xi_j+\xi_k+\xi_l)\Big) + b\xi_i^2, \quad c = \frac{16(3a+b)^2}{(a+b)(3a-b)},$$

where the indices (i,j,k,l) run over the four cyclic permutations of $(1,2,3,4)$. For $c \neq 0$ it reduces to equation (14) for $h(t) = \frac{3a-b}{4}(\xi_1+\xi_2+\xi_3+\xi_4)$.

Given a solution $h(t)$ of (14), one can construct solutions to the heat equation the same way as illustrated in section 3.

5. The case of fifth-order equations

For $n = 4$ we have $\mathbf{x} = (x_2, x_3, x_4, x_5)$ and $P_3(\mathbf{x})$ is a homogeneous polynomial of degree -24. Thus $P_4(\mathbf{x}) = c_1x_3^2 + c_2x_2^3 + c_3x_4x_2$ for some constants c_1, c_2, c_3. Formulas (7) imply that equation (6) becomes

$$h''''' + 30hh'''' - 300h'h''' + 300(h'')^2$$
$$- (276+c_1)(h''+6hh'+4h^3)^2 - (1920+c_2)(h'+h^2)^3$$
$$+ (344-c_3)(h'+h^2)(h'''+12hh''+6(h')^2+48h^2h'+24h^4) = 0. \tag{15}$$

A generalization of system (13) is

$$\frac{d\xi_i}{dt} = a\Big(\xi_j\xi_k + \xi_j\xi_l + \xi_k\xi_l + \xi_j\xi_m + \xi_k\xi_m + \xi_l\xi_m$$

$$- 3a\xi_i\big(\xi_j + \xi_k + \xi_l + \xi_m\big)\Big) + b\xi_i^2,$$

where the indices (i, j, k, l, m) run over the five cyclic permutations of $(1, 2, 3, 4, 5)$. For $c_2 = -\frac{45}{676}c_1^2$, $c_3 = \frac{31}{26}c_1$ and $c_1 = \frac{52(4a+b)^2}{(3a+2b)(6a-b)} \neq 0$ it reduces to equation (15) for $h(t) = \frac{6a-b}{5}(\xi_1 + \xi_2 + \xi_3 + \xi_4 + \xi_5)$.

References

[1] V.M. Buchstaber, E.Yu. Bunkova, *Polynomial dynamical systems and ordinary differential equations associated with the heat equation*, Functional Analysis and Its Applications, 46, 3, (2012).

[2] E.Yu. Bunkova, V.M. Buchstaber, *Heat Equations and Families of Two-Dimensional Sigma Functions*, Geometry, Topology, and Mathematical Physics. II, Collected papers. Dedicated to Academician Sergei Petrovich Novikov on the occasion of his 70th birthday, Tr. Mat. Inst. Steklova, 266, MAIK Nauka/Interperiodica, Moscow, (2009), 5–32.

[3] N.A. Kudryashov, *The Painlevé property in the theory of differential equations*, Soros Educational Journal, 1999, 9, 118–122 (in Russian).

[4] P.A. Clarkson, P.J. Olver, *Symmetry and the Chazy equation*, J. Diff. Eq. 124, (1996), 225–246.

[5] C.M. Cosgrove, *Chazy Classes IX–XI Of Third-Order Differential Equations*, Stud. Appl. Math. 104, 3, (2000), 171–78.

[6] N.A. Kudryashov, *Some fourth-order ordinary differential equations which pass the Painlevé test*, Journal of Nonlinear Mathematical Physics, (2001), v. 8, Supplement, 172–177, Proceedings: Needs'99.

[7] M.J. Ablowitz, S. Chakravarty, R. Halburd, *The generalized Chazy equation from the self-duality equations*, Stud. Appl. Math. 103, (1999), 75-7.

Elena Yu. Bunkova
Steklov Mathematical Institute
Moscow, Russia
e-mail: bunkova@mi.ras.ru

Geometric Methods in Physics. XXXI Workshop 2012
Trends in Mathematics, 187–193

4-planar Mappings of Quaternionic Kähler Manifolds

Irena Hinterleitner

Abstract. In this paper we study fundamental equations of 4-planar mappings of almost quaternionic manifolds with respect to the smoothness class of metrics. We show that 4-planar mappings preserve the smoothness class of metrics.

Mathematics Subject Classification (2010). Primary 53B20; Secondary 53B21; 53B30; 53B35; 53C26.

Keywords. 4-planar mappings, almost quaternionic manifolds.

1. Introduction

4-planar and 4-quasiplanar mappings of almost quaternionic spaces have been studied in [1, 2] and [3]. These mappings generalize the geodesic, quasigeodesic and holomorphically projective mappings of Riemannian and Kählerian spaces, see [4–23]. Almost quaternionic structures were studied by many authors for example [24–26]. Generalisations of the above-introduced mappings were studied in [27–32].

First we study the general dependence of 4-planar mappings of almost quaternionic manifolds in dependence on the smoothness class of the metric. We present well-known facts, which were proved by Kurbatova, see [1], without stress on details about the smoothness class of the metric. They were formulated "for sufficiently smooth" geometric objects. In the present article we want to make this issue more precise.

2. Almost quaternionic and quaternionic Kähler manifolds

Under an *almost quaternionic* space we understand a differentiable manifold M_n with almost complex structures $\overset{1}{F}$ and $\overset{2}{F}$ defined on it, satisfying

$$\overset{1}{F}{}^h_\alpha \overset{1}{F}{}^\alpha_i = -\delta^h_i; \quad \overset{2}{F}{}^h_\alpha \overset{2}{F}{}^\alpha_i = -\delta^h_i; \quad \overset{1}{F}{}^h_\alpha \overset{2}{F}{}^\alpha_i + \overset{2}{F}{}^h_\alpha \overset{1}{F}{}^\alpha_i = 0,$$

where δ^h_i is the Kronecker symbol, see, e.g., [4, 24].

The tensor $\overset{3}{F}{}^h_i \equiv \overset{1}{F}{}^\alpha_i \overset{2}{F}{}^h_\alpha$ is further an almost complex structure. The relations among the tensors $\overset{1}{F}$, $\overset{2}{F}$ and $\overset{3}{F}$ are the following

$$\overset{1}{F}{}^h_i = \overset{2}{F}{}^\alpha_i \overset{3}{F}{}^h_\alpha = -\overset{3}{F}{}^\alpha_i \overset{2}{F}{}^h_\alpha; \quad \overset{2}{F}{}^h_i = \overset{3}{F}{}^\alpha_i \overset{1}{F}{}^h_\alpha = -\overset{1}{F}{}^\alpha_i \overset{3}{F}{}^h_\alpha; \quad \overset{3}{F}{}^h_i = \overset{1}{F}{}^\alpha_i \overset{2}{F}{}^h_\alpha = -\overset{2}{F}{}^\alpha_i \overset{1}{F}{}^h_\alpha.$$

Each pair chosen from the three structures $\overset{1}{F}$, $\overset{2}{F}$ and $\overset{3}{F}$ determines an *almost quaternionic structure*. The tensors $*\overset{1}{F}$, $*\overset{2}{F}$, $*\overset{3}{F}$ and $\overset{1}{F}$, $\overset{2}{F}$, $\overset{3}{F}$ define the same almost quaternionic structure if $\quad *\overset{\sigma}{F} = \sum_{\rho=1}^{3} \alpha_\rho \overset{\rho}{F}$ where α_ρ are some functions.

An almost quaternionic manifold A_n is called a *quaternionic Kähler manifold*, if there exists a metric g such that $(g, \overset{s}{F})$, $s = 1, 2, 3$ are Kähler spaces, so that

$$g(X, \overset{s}{F}X) = 0, \quad \text{and} \quad \nabla \overset{s}{F} = 0,$$

for any $X \in TA_n$ and $s = 1, 2, 3$. Here and in the following ∇ is an affine connection with components Γ on A_n.

Let $A_n(\Gamma, \overset{1}{F}, \overset{2}{F}, \overset{3}{F})$ be a space with affine connection Γ without torsion with almost quaternionic structures $(\overset{1}{F}, \overset{2}{F}, \overset{3}{F})$.

Definition 1. A curve ℓ in A_n which is given by the equation $\ell = \ell(t)$, $\lambda = d\ell/dt$, $(\neq 0)$, $t \in I$, where t is a parameter, is called 4-*planar*, if under the parallel translation along the curve, the tangent vector λ belongs to the four-dimensional distribution $D = \mathrm{Span}\{\lambda, \overset{1}{F}\lambda, \overset{2}{F}\lambda, \overset{3}{F}\lambda\}$, that is, it satisfies

$$\nabla_t \lambda = a(t)\lambda + b(t)\overset{1}{F}\lambda + c(t)\overset{2}{F}\lambda + d(t)\overset{3}{F}\lambda,$$

where $a(t)$, $b(t)$, $c(t)$ and $d(t)$ are some functions of the parameter t.

Particularly, in the case $b(t) = c(t) = d(t) = 0$, a 4-planar curve is a geodesic. Evidently, a 4-planar curve with respect to the structure $(\overset{1}{F}, \overset{2}{F}, \overset{3}{F})$ is 4-planar with respect to the structure $(*\overset{1}{F}, *\overset{2}{F}, *\overset{3}{F})$, too.

3. 4-planar mappings

Consider two almost quaternionic manifolds with affine connections without torsion A_n and $\bar{A}_n$ with connection components Γ and $\bar{\Gamma}$, respectively. Let an almost quaternionic structure $(\overset{1}{F}, \overset{2}{F}, \overset{3}{F})$ be defined on A_n.

Definition 2. A diffeomorphism $f: A_n \to \bar{A}_n$ is called a 4-*planar mapping*, if it maps any 4-planar curve in A_n onto a 4-planar curve in $\bar{A}_n$.

Assume a 4-planar mapping $f: A_n \to \bar{A}_n$. Since f is a diffeomorphism, we can introduce local coordinate charts on M or $\bar{M}$, respectively, such that locally $f: A_n \to \bar{A}_n$ maps points onto points with the same coordinates, and $\bar{M} = M$.

A manifold A_n admits a 4-planar mapping onto $\bar{A}_n$ if and only if the following equations [3]:

$$\bar{\nabla}_X Y = \nabla_X Y + \sum_{s=0}^{3} \left\{ \underset{s}{\psi}(X) \underset{s}{\overset{s}{F}} Y + \underset{s}{\psi}(Y) \underset{s}{\overset{s}{F}} X \right\} \tag{1}$$

hold for any tangent fields X, Y and where $\underset{s}{\psi}$ are differential forms; $\overset{0}{F} = \mathrm{Id}$. If $\underset{s}{\psi} \equiv 0$, $(s = 0, 1, 2, 3)$ then f is *affine*.

Beside these facts it was proved [3] that the quaternionic structure of A_n and $\bar{A}_n$ is preserved; for this reason we can assume that $\overset{s}{\bar{F}} = \overset{s}{F}$. This was a priori assumed in the definition and results by Kurbatova [1].

Equation (1) in the common coordinate system x with respect to the mapping, has the following form

$$\bar{\Gamma}_{ij}^{h}(x) = \Gamma_{ij}^{h}(x) + \sum_{s=0}^{3} \underset{s}{\psi}_{(i} \overset{s}{\underset{}{F}}{}_{j)}^{h}$$

where Γ_{ij}^{h} and $\bar{\Gamma}_{ij}^{h}$ are components of ∇ and $\bar{\nabla}$, $\underset{s}{\psi}_i(x)$ are components of $\underset{s}{\psi}$, $(i\,j)$ denotes a symmetrization without division by 2.

Finally we will assume that the space $A_n(\Gamma, \overset{1}{F}, \overset{2}{F}, \overset{3}{F})$ is mapped onto the (pseudo-) Riemannian space $\bar{V}_n(\bar{g})$. A *mapping* $f: A_n \to \bar{V}_n$ *is 4-planar if and only if the metric tensor* $\bar{g}_{ij}(x)$ *satisfies the following equations:*

$$\bar{g}_{ij,k} = \sum_{s=0}^{3} \left(\underset{s}{\psi}_k \, \bar{g}_{\alpha(i} \, \overset{s}{\underset{}{F}}{}_{j)}^{\alpha} + \underset{s}{\psi}_{(i} \, \bar{g}_{j)\alpha} \, \overset{s}{\underset{}{F}}{}_{k}^{\alpha} \right) \tag{2}$$

where the comma is the covariant derivative in A_n *(see* [3]*),*

4. 4-planar mapping theory for $K_n \to \bar{K}_n$ of class C^1

Let us consider the quaternionic Kähler manifolds $K_n = (M, g, F)$ and $\bar{K}_n = (\bar{M}, \bar{g}, \bar{F})$ with metrics g and $\bar{g}$, structures $F = (\overset{1}{F}, \overset{2}{F}, \overset{3}{F})$ and $\bar{F} = (\overset{1}{\bar{F}}, \overset{2}{\bar{F}}, \overset{3}{\bar{F}})$, Levi–Civita connections ∇ and $\bar{\nabla}$, respectively. Here $K_n, \bar{K}_n \in C^1$, i.e., $g, \bar{g} \in C^1$ which means that their components $g_{ij}, \bar{g}_{ij} \in C^1$.

We further assume that K_n admits a 4-planar mapping onto $\bar{K}_n$. Then we can consider $\bar{M} = M$ and $\overset{s}{\bar{F}} = \overset{s}{F}$ for $s = 1, 2, 3$.

In the present case we can simplify formula (2) as follows:

$$\bar{g}_{ij,k} = 2\psi_k \, \bar{g}_{ij} + \sum_{s=1}^{3} \left(\underset{s}{\psi}_i \, \bar{g}_{j\alpha} \, \overset{s}{\underset{}{F}}{}_{k}^{\alpha} + \underset{s}{\psi}_j \, \bar{g}_{i\alpha} \, \overset{s}{\underset{}{F}}{}_{k}^{\alpha} \right). \tag{3}$$

Here and in the following $\psi_k \equiv \underset{0}{\psi}_k$. When $n > 4$, it was proved in [1] that

$$\underset{s}{\psi}_i = -\frac{n}{n-4}\psi_\alpha \, \overset{s}{F}_i^\alpha, \quad s = 1, 2, 3.$$ Moreover, ψ is gradient-like, that is

$$\psi_i = \partial\Psi/\partial x^i \quad \text{and} \quad \Psi = \frac{n^2 - 4}{2(n-4)} \ln \left| \frac{\det \bar{g}}{\det g} \right|.$$

Kurbatova [9] proved that equations (3) are equivalent to

$$a_{ij,k} = \lambda_\alpha \check{Q}_{(ij)}^{\alpha\beta} \, g_{\beta k}, \tag{4}$$

where

$$\check{Q}_{ij}^{\alpha\beta} = \delta_i^\alpha \delta_j^\beta + \frac{n}{n-4} \sum_{s=1}^{3} \overset{s}{F}_i^\alpha \, \overset{s}{F}_j^\beta,$$

and

$$\text{(a)} \quad a_{ij} = e^{2\Psi} \bar{g}^{\alpha\beta} g_{\alpha i} g_{\beta j}; \quad \text{(b)} \quad \lambda_i = -e^{2\Psi} \bar{g}^{\alpha\beta} g_{\beta i} \psi_\alpha.$$

In addition, the formula

$$a_{\alpha\beta} \, \overset{s}{F}_i^\alpha \, \overset{s}{F}_j^\beta = a_{ij}$$

holds. From (4) follows $\lambda_i = \partial_i \lambda = \partial_i(\text{const} \cdot a_{\alpha\beta} g^{\alpha\beta})$. On the other hand

$$\bar{g}_{ij} = e^{2\Psi} \tilde{g}_{ij}, \quad \Psi = \frac{1}{2} \ln \left| \frac{\det \tilde{g}}{\det g} \right|, \quad \|\tilde{g}_{ij}\| = \|g^{i\alpha} g^{j\beta} a_{\alpha\beta}\|^{-1}. \tag{5}$$

The above formulas are the criterion for 4-planar mappings $K_n \to \bar{K}_n$, globally as well as locally.

5. 4-planar mapping theory for $K_n \to \bar{K}_n$ of class C^2

Let K_n and $\bar{K}_n \in C^2$ be quaternionic Kähler manifolds, then the integrability conditions of equations (4) have the following form

$$a_{ij,kl} - a_{ij,lk} \equiv a_{i\alpha} R_{jkl}^\alpha + a_{j\alpha} R_{ikl}^\alpha = \lambda_{\alpha l} \check{Q}_{(ij)}^{\alpha\beta} \, g_{\beta k} - \lambda_{\alpha k} \check{Q}_{(ij)}^{\alpha\beta} \, g_{\beta l}.$$

Here R_{ijk}^h are components of the Riemann tensor.

After contraction with g^{jk} we get [1]:

$$n\lambda_{i,k} = \mu g_{ik} + a_{\alpha\beta} B_{ik}^{\alpha\beta}, \tag{6}$$

where

$$\mu = \lambda_{\alpha\beta} g^{\alpha\beta}, \quad B_{il}^{\alpha\beta} = \hat{Q}_{(i\delta)}^{\beta\gamma} R_{.\,\gamma\,.\,l}^{\alpha\,\,\,\delta}, \quad R_{.\,\gamma\,.\,l}^{\alpha\,\,\,\delta} = g^{\beta\delta} R_{i\delta l}^\alpha,$$

$$\hat{Q}_{i\delta}^{\beta\gamma} = \frac{n(n-4)}{16(n-1)} \left(\frac{4-3n}{n} \delta_i^\beta \delta_\delta^\gamma + \sum_{s=1}^{3} \overset{s}{F}_i^\beta \, \overset{s}{F}_\delta^\gamma \right).$$

6. 4-planar mappings between $K_n \in C^r$ $(r > 2)$ and $\bar{K}_n \in C^2$

We demonstrate the following theorem

Theorem 1. *If $K_n \in C^r$ $(r > 2)$ admits 4-planar mappings onto $\bar{K}_n \in C^2$, then $\bar{K}_n \in C^r$.*

The proof of this theorem follows from the following lemmas.

Lemma 1 (see [7]). *Let $\lambda^h \in C^1$ be a vector field and ρ a function. If*

$$\partial_i \lambda^h - \rho\,\delta_i^h = f_i^h \in C^1 \tag{7}$$

then $\lambda^h \in C^2$ and $\rho \in C^1$.

Lemma 2. *If $K_n \in C^3$ admits a 4-planar mapping onto $\bar{K}_n \in C^2$, then $\bar{K}_n \in C^3$.*

Proof. In this case equations (4) and (6) hold. According to the foregoing assumptions, $g_{ij} \in C^3$ and $\bar{g}_{ij} \in C^2$. By a simple check-up we find $\Psi \in C^2$, $\psi_i \in C^1$, $a_{ij} \in C^2$, $\lambda_i \in C^1$ and $R_{ijk}^h \in C^1$.

From the above-mentioned conditions we easily convince ourselves that we can write equation (6) in the form (7), where

$$\lambda^h = g^{h\alpha}\lambda_\alpha \in C^1, \quad \rho = \mu/n \text{ and } f_i^h = \tfrac{1}{n}\,g^{hl}a_{\alpha\beta}B_{il}^{\alpha\beta} \in C^1.$$

From Lemma 1 it follows that $\lambda^h \in C^2$, $\rho \in C^1$, and evidently $\lambda_i \in C^2$. Differentiating (4) twice we show that $a_{ij} \in C^3$. From this and formula (5) follows that also $\Psi \in C^3$ and $\bar{g}_{ij} \in C^3$. $\qquad\square$

Further we notice that for 4-planar mappings between quaternionic Kähler manifolds K_n and $\bar{K}_n$ of class C^3 holds the following third set of equations (after simple modifications of [1]):

$$(n-1)\mu_{,k} = \lambda_\alpha C_{\gamma[\delta k]}^\alpha g^{\gamma\delta} + a_{\alpha\beta}B_{\gamma[\delta,k]}^{\alpha\beta}g^{\gamma\delta}, \tag{8}$$

where $C_{ilk}^\alpha = \check{Q}_{\gamma\beta}^{\alpha\delta}B_{i[l}^{\gamma\beta}g_{k]\delta}$.

If $K_n \in C^r$ and $\bar{K}_n \in C^2$, then by Lemma 2, $\bar{K}_n \in C^3$ and (8) holds. Because the system (4), (6) and (8) is closed, we can differentiate equations (4) $(r-1)$ times. So we convince ourselves that $a_{ij} \in C^r$, and also $\bar{g}_{ij} \in C^r$ $(\equiv \bar{K}_n \in C^r)$.

Remark 1. Moreover, in this case from equation (8) follows that the function $\mu \in C^{r-1}$.

Acknowledgment

The paper was supported by the project FAST-S-12-25/1660 of the Brno University of Technology.

References

[1] I.N. Kurbatova, *4-quasi-planar mappings of almost quaternion manifolds.* Sov. Math. **30** (1986), 100–104; transl. from Izv. Vyssh. Uchebn. Zaved., Mat. (1986), 75–78.

[2] J. Mikeš, O. Pokorná, J. Němčíková, *On the theory of the 4-quasiplanar mappings of almost quaternionic spaces.* Suppl. Rend. Circ. Mat. Palermo, II. Ser. **54** (1998), 75–81.

[3] J. Mikeš, O. Pokorná, J. Bělohlávková, *On special 4-planar mappings of almost Hermitian quaternionic spaces.* Proc. of the 2nd meeting on quaternionic structures in mathematics and physics, Roma, Italy, Sept. 6–10, 1999. Rome: Univ. di Roma La Sapienza", 265-271, electronic only (2001).

[4] D.V. Beklemishev, *Differential geometry of spaces with almost complex structure.* Akad. Nauk SSSR Inst. Nauchn. Informacii, Moscow, 1965 Geometry, (1963), 165–212.

[5] V.V. Domashev, J. Mikeš, *Theory of holomorphically projective mappings of Kählerian spaces.* Math. Notes **23** (1978), 160–163; transl. from Mat. Zametki **23** (1978), 297–303.

[6] I. Hinterleitner, J. Mikeš, *On F-planar mappings of spaces with affine connections.* Note Mat. **27** (2007), 111–118.

[7] I. Hinterleitner, J. Mikeš, *Geodesic Mappings and Einstein Spaces.* Geometric Methods in Physics, XXX Workshop, Białowieża, Poland, June 26 to July 2, 2011, Series: Trends in Mathematics, Kielanowski, P.; Ali, S.T.; Odzijewicz, A.; Schlichenmaier, M.; Voronov, T. (eds.), Birkhäuser, Springer Basel (2013), 331–335.

[8] I. Hinterleitner, J. Mikeš, J. Stránská, *Infinitesimal F-planar transformations.* Russ. Math. **52** (2008), 13–18; transl. from Izv. Vyssh. Uchebn. Zaved., Mat. (2008), 16–22.

[9] I.N. Kurbatova, *HP-mappings of H-spaces.* (Russian) Ukr. Geom. Sb. **27** (1984), 75–83.

[10] J. Mikeš, *On holomorphically projective mappings of Kählerian spaces.* (Russian) Ukr. Geom. Sb. **23** (1980), 90–98.

[11] J. Mikeš, *Special F-planar mappings of affinely connected spaces onto Riemannian spaces.* Mosc. Univ. Math. Bull. **49** (1994), 15–21; transl. from Vestn. Mosk. Univ., Ser. 1 (1994), 18–24.

[12] J. Mikeš, *Geodesic mappings on affine-connected and Riemannian spaces.* J. Math. Sci. New York **78** (1996), 311–333.

[13] J. Mikeš, *Holomorphically projective mappings and their generalizations.* J. Math. Sci., New York **89** (1998) 1334–1353.

[14] J. Mikeš, M. Shiha, A. Vanžurová, *Invariant objects by holomorphically projective mappings of Kähler spaces.* 8th Int. Conf. on Applied Mathematics (APLIMAT 2009), Bratislava, Feb. 03-06, 2009, APLIMAT 2009: 8th Int. Conf. Proc. (2009), 439–444.

[15] J. Mikeš, N.S. Sinyukov, *On quasiplanar mappings of space of affine connection.* Sov. Math. **27** (1983), 63–70; transl. from Izv. Vyssh. Uchebn. Zaved., Mat. (1983), 55–61.

[16] J. Mikeš, A. Vanžurová, I. Hinterleitner, *Geodesic mappings and some generalizations.* Palacky University Press, Olomouc, 2009.

[17] A.Z. Petrov, *Simulation of physical fields*. Grav. i teor. Otnos., Kazan State Univ. Press, Kazan **4–5** (1968), 7–21.

[18] M. Prvanović, *Holomorphically projective transformations in a locally product space*. Math. Balk. **1** (1971), 195–213.

[19] N.S. Sinyukov, *Geodesic mappings of Riemannian spaces*. Moscow, Nauka, 1979.

[20] N.S. Sinyukov, *On almost geodesic mappings of affine-connected and Riemannian spaces*. Itogi nauki i techn., Geometrija. Moskva, VINITI **13** (1982), 3–26.

[21] N.S. Sinyukov, I.N. Kurbatova, J. Mikeš, *Holomorphically projective mappings of Kähler spaces*. Odessa, Odessk. Univ., 1985.

[22] S.-I. Tachibana, S. Ishihara, *On infinitesimal holomorphically projective transformations in Kählerian manifolds*. Tohoku Math. J. II **12** (1960), 77–101.

[23] K. Yano, Differential geometry of complex and almost complex spaces. Pergamon Press, 1965.

[24] D.V. Alekseevsky, S. Marchiafava, *Transformation of a quaternionic Kählerian manifold*. C.R. Acad. Sci., Paris, Ser. I **320** (1995), 703–708.

[25] M. Jukl, L. Juklová, J. Mikeš, *The decomposition of tensor spaces with quaternionic structure*. APLIMAT 2007, 6th Int. Conf., **I** (2007), 217–222.

[26] M. Jukl, L. Juklová, J. Mikeš, *Some results on traceless decomposition of tensors*. J. of Math. Sci. **174** (2011) 627–640.

[27] J. Hrdina, *Almost complex projective structures and their morphisms*. Arch. Math., Brno **45** (2009), 255–264.

[28] J. Hrdina, *Remarks on F-planar curves and their generalizations*. Banach Center Publications **93** (2011), 241–249.

[29] J. Hrdina, J. Slovák, *Morphisms of almost product projective geometries*. Proc. 10th Int. Conf. on Diff. Geom. and its Appl., DGA 2007, Olomouc, Czech Republic, Aug. 27–31, 2007. Hackensack, NJ: World Sci. (ISBN 978-981-279-060-6/hbk) (2008), 253–261.

[30] M.S. Stanković, M.L. Zlatanović, L.S. Velimirović, *Equitorsion holomorphically projective mappings of generalized Kählerian space of the second kind*. Int. Electron. J. Geom. **3** (2010), 26–39.

[31] M.S. Stanković, M.L. Zlatanović, L.S. Velimirović, *Equitorsion holomorphically projective mappings of generalized Kählerian space of the first kind*. Czech. Math. J. **60** (2010), 635–653.

[32] M.S. Stanković, S.M. Minčić, L.S. Velimirović, *On equitorsion holomorphically projective mappings of generalized Kählerian spaces*. Czech. Math. J. **54** (2004), 701–715.

Irena Hinterleitner
Department of Mathematics
Faculty of Civil Engineering
Brno University of Technology
Žižkova 17
CZ-602 00 Brno, Czech Rep.
e-mail: `hinterleitner.irena@seznam.cz`

Geometric Methods in Physics. XXXI Workshop 2012
Trends in Mathematics, 195–202
© 2013 Springer Basel

Unduloid-like Equilibrium Shapes of Carbon Nanotubes Subjected to Hydrostatic Pressure

Ivaïlo M. Mladenov, Mariana Ts. Hadzhilazova,
Vassil M. Vassilev and Peter A. Djondjorov

Abstract. The aim of this work is to obtain numerically unduloid-like equilibrium shapes of carbon nanotubes subjected to external pressure.

Mathematics Subject Classification (2010). Primary 82D80; Secondary 74G15, 74G65.

Keywords. Carbon nanotubes, equilibrium shapes, unduloid-like shapes.

1. Introduction

Carbon nanotubes are carbon molecules in the shape of hollow cylindrical fibers of nanometer-size diameter and length-to-diameter ratio of up to 107 : 1. Carbon nanotubes exhibit extraordinary strength, unique electrical properties, and are efficient conductors of heat. For this reason, carbon nanotubes have many practical applications in electronics, optics and other fields of material science. If the tube wall is composed by one layer of carbon atoms, then the tube is referred to as a single-walled one (SWNT). Otherwise, the tube is called multi-walled (MWNT).

The predominating opinion among the scientists working in this field is that they are discovered in 1991 by Sumio Iijima [1]. However, carbon nanotubes have been produced and observed prior to 1991. In 1952 appeared a paper in the Soviet Journal of Physical Chemistry (in Russian) by Radushkevich and Lukyanovich [2] where images of 50 nanometer diameter tubes made of carbon are presented. Oberlin, Endo and Koyama [3] reported observations of hollow carbon fibers (SWNT) with nanometer-scale diameters in 1976. In 1987, Howard G. Tennent of Hyperion Catalysis was issued a US patent for the production of "... *cylindrical discrete carbon fibrils with a constant diameter between about* 3.5 *and about* 70 *nanometers ..., length* 102 *times the diameter ...*"

Shortly after the experimental discovery of multi-wall [1] and single-wall [1, 4] carbon nanotubes and the reported progress in their large-scale synthesis [5], a remarkable mechanical properties of this carbon alotrops was observed. The findings provided by high-resolution transmission electron microscopy [6] demonstrated that these nano-structures can sustain large deformations of their initial circular-cylindrical shape without occurrence of irreversible atomic lattice defects. As noticed in [6] "*Thus, within a wide range of bending, the tube retains an all-hexagonal structure and reversibly returns to its initial straight geometry upon removal of the bending force.*"

One of the most widely used approaches for determining the mechanical response of CNS's is the molecular dynamic simulation. Within this approach, a CNS is considered as a multibody system in which the interaction of a given atom with the neighbouring ones is regarded. The energy of this interaction is modelled through certain empirical interatomic potentials. In 1988, Tersoff [7] suggested a general approach for derivation of such potentials and applied it to silicon. In 1990, Brenner [8] adapted and modified Tersoff's results and suggested an interatomic potential for carbon atomic bonds. Another potential of such kind was introduced in 1992 by Lenosky *et al.* [9].

On the other hand, the observed elastic behaviour of the carbon nanotubes, their essentially two-dimensional atomic lattice structure and the intrinsic hexagonal symmetry of the latter gave firm arguments to Yakobson *et al.* [10] to develop a continuum mechanics approach, based on the classical theory of isotropic thin elastic shells [11], for explanation of the mechanical properties and exploration of the deformed configurations of these carbon molecules. The advantage of such an approach, in comparison with the ones based directly on the interatomic interactions, is that the continuum mechanics models are amenable to analytical calculations and allow efficient numeric simulations. Therefore, it is not surprising that since the pioneering work by Yakobson *et al* [10], the application of continuum mechanics to the study of mechanical behaviour of carbon nano-structures has become common practice although, as noted there "... *its relevance for a covalent-bonded system of only a few atoms in diameter is far from obvious* ... " The easiest way of introducing a continuum model of the regarded type of atomistic systems is to emulate the basic idea of the above-mentioned work, that is to adopt a standard continuum theory (some of the well-known shell theories [11–13], for instance) and to adjust the material parameters (such as the Young's modulus, Poisson's ratio, bending rigidity, shell thickness, etc.) to the data available by atomistic simulations. This approach has been used by various authors and turned out to be quite successful.

2. Modeling

In the case of a curved two-dimensional continuum, such as that required within the continuum modeling of the carbon nanotubes, the geometry of the deformed lattice is expressed in terms of the invariants of the strain (first fundamental form) and curvature (second fundamental form) of the deformed surface.

In what follows, we adopt the continuum theory developed by Tu and Ou-Yang in [14], which is based on the continuum limit of the Lenosky potential [9], an extra term being added to the corresponding deformation energy to take into account the screw dislocation core-like deformation as it was suggested by Xie *et al.* in [15].

According to Lenosky *et al.* [9], the deformation energy of a single layer of curved graphite carbon has the form

$$\mathcal{F} = \epsilon_0 \sum_{(ij)} \frac{1}{2} (r_{ij} - r_0)^2 + \epsilon_1 \sum_i \left(\sum_{(j)} \mathbf{u}_{ij} \right)^2$$
$$+ \epsilon_2 \sum_{(ij)} (1 - \mathbf{n}_i \cdot \mathbf{n}_j)^2 + \epsilon_3 \sum_{(ij)} (\mathbf{n}_i \cdot \mathbf{u}_{ij})(\mathbf{n}_j \cdot \mathbf{u}_{ji})$$

where r_{ij} is the bond length between atoms i and j after the deformation, r_0 is the initial bond length of planar graphite; $\mathbf{u}_{ij}$ is a unit vector pointing from carbon atom i to its neighbor j, $\mathbf{n}_i$ is a unit vector normal to the plane determined by the three neighbors of atom i, ϵ_0, ϵ_1, ϵ_2 and ϵ_3 are the so-called bond-bending parameters. The summation $\sum_{(j)}$ is taken over the three nearest-neighbor atoms j to i atom and $\sum_{(ij)}$ is taken over all nearest-neighbor atoms.

The continuum limit of the Lenosky potential $\mathcal{F}$ yields the following expression for the deformation energy [14]

$$\mathcal{F}_{cl} = \int_{\mathcal{S}} \left[\frac{k_c}{2} (2H)^2 + k_G K + \frac{k_d}{2} (2J)^2 + \tilde{k}Q \right] \mathrm{d}A$$

where $\mathcal{S}$ is the deformed surface; H and K are its mean and Gaussian curvatures; $\mathrm{d}A$ is the area element on the surface $\mathcal{S}$, J and Q are the first and second invariants of the in-plane deformation tensor, which are often referred to as the "mean" and "Gaussian" strains, respectively, and the constants k_c, k_G, k_d and $\tilde{k}$ are

$$k_c = 1.62 \, \mathrm{eV}, \qquad k_G = -0.72 \, \mathrm{eV}$$
$$k_d = 22.97 \, \mathrm{eV/\mathring{A}^2}, \qquad \tilde{k} = 19.19 \, \mathrm{eV/\mathring{A}^2}.$$

These "material" parameters and the functional $\mathcal{F}_{cl}$ describe the deformation of a single-wall carbon nanotube as that of a two-dimensional isotropic elastic continuous media.

Actually, the last two terms in the functional $\mathcal{F}_{cl}$ accounting for the in-plane deformation can be neglected since the contribution of the bond stretching to the deformation energy is less than 1%, see [9]. Instead of this, the graphene sheet can be assumed to be inextensible under the bending related to the remaining two terms in the functional $\mathcal{F}_{cl}$. If, in addition, it is assumed that a uniform hydrostatic pressure p is applied and the term accounting for the screw dislocation core-like deformation suggested in [15] is included, one arrives at the conclusion

that the equilibrium shapes of a single-wall carbon nanotube are determined by
the extremals of the functional

$$\mathcal{F}_{cb} = \frac{k_c}{2} \int_S (2H + \text{lh})^2 \mathrm{d}A + k_G \int_S K \mathrm{d}A + \lambda \int_S \mathrm{d}A + p \int \mathrm{d}V$$

where λ is a Lagrange multiplier corresponding to the constraint of fixed total
area, which can be interpreted as the chemical potential associated with the atoms
located at the cross section of carbon nanotube, lh is a constant that accounts for
the non-planarity of the stress-free configuration of the tube and the last term
is the work done by the hydrostatic pressure p. The Euler–Lagrange equation
associated with the foregoing functional is [16]

$$2k_c \Delta H + k_c (2H^2 - \text{lh} H - 2K) (2H + \text{lh}) - 2\lambda H + p = 0. \tag{1}$$

It is usually called the shape equation, and is a nonlinear fourth-order partial
differential equation with respect to the components of the position vector.

The shape equation admits exact solutions determining cylindrical equilibrium shapes as is reported in [17–19]. Here, axisymmetric equilibrium shapes,
predicted by equation (1) are sought. The quantum-mechanical aspects of such
systems are treated in [20].

3. Axisymmetric equilibrium shapes of CNS's

For an axially symmetric equilibrium shape of a carbon nano-structure, as it is
assumed in the present paragraph, the curvature energy functional $\mathcal{F}_{cb}$ takes the
form

$$\mathcal{F}_{ca} = 2\pi k_c \int_0^L \frac{1}{2} \left(\frac{\mathrm{d}\psi}{\mathrm{d}s} + \frac{\sin\psi}{x} + \text{lh} \right)^2 x \, \mathrm{d}s + 2\pi k_G \int_0^L \frac{\mathrm{d}\psi}{\mathrm{d}s} \sin\psi \, \mathrm{d}s$$

$$+ 2\pi\lambda \int_0^L x \, \mathrm{d}s + 2\pi \frac{p}{3} \int_0^L \left(x \frac{\mathrm{d}z}{\mathrm{d}s} - z \frac{\mathrm{d}x}{\mathrm{d}s} \right) x \, \mathrm{d}s$$

since the mean H and Gaussian K curvatures of a surface in revolution are given
by the expressions

$$H = \frac{1}{2} \left(\frac{\mathrm{d}\psi}{\mathrm{d}s} + \frac{\sin\psi}{x} \right), \qquad K = \frac{\mathrm{d}\psi}{\mathrm{d}s} \frac{\sin\psi}{x}.$$

Here, s is the arc length of the profile curve of the surface, which is assumed
to lie in the XOZ-plane (see Figure 1) and to be determined by the parametric
equations $X = x(s)$, $Z = z(s)$ while $\psi(s)$ is the slope angle defined by the relations

$$\frac{\mathrm{d}x}{\mathrm{d}s} = \cos\psi, \qquad \frac{\mathrm{d}z}{\mathrm{d}s} = \sin\psi.$$

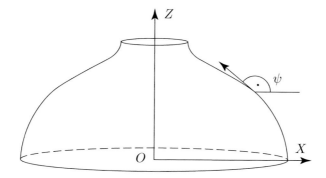

FIGURE 1. A surface of revolution and its geometrical description.

Then, setting to zero the first variation of the functional $\mathcal{F}_{ca}$ one obtains the following system of Euler-Lagrange equations

$$\frac{\mathrm{d}^2\psi}{\mathrm{d}s^2} = -\frac{\mathrm{d}\psi}{\mathrm{d}s}\frac{\cos\psi}{x} + \frac{\sin 2\psi}{2x^2} + \mu\frac{\sin\psi}{x} - \eta\frac{\cos\psi}{x}$$

$$\frac{\mathrm{d}\mu}{\mathrm{d}s} = \frac{1}{2}\left(\frac{\mathrm{d}\psi}{\mathrm{d}s} + \mathrm{lh}\right)^2 - \frac{1}{2}\left(\frac{\sin\psi}{x}\right)^2 - \lambda + px\sin\psi \qquad (2)$$

$$\frac{\mathrm{d}\eta}{\mathrm{d}s} = -px\cos\psi, \qquad \frac{\mathrm{d}x}{\mathrm{d}s} = \cos\psi, \qquad \frac{\mathrm{d}z}{\mathrm{d}s} = \sin\psi, \qquad \frac{\mathrm{d}\lambda}{\mathrm{d}s} = 0$$

and natural boundary conditions

$$\left\{\hat{M}\delta\psi + \hat{\mu}\delta x + \hat{\eta}\delta z + \mathcal{H}\delta s\right\}_0^L + Q_0\delta s(0) + Q_L\delta s(L)$$

$$+ \sigma_0\delta x(0) + \left(\frac{1}{2}\mathrm{lh}^2 x(L) + \sigma_L\right)\delta x(L) = 0 \qquad (3)$$

for determination of the profile curves each of which yields an extremum of the functional $\mathcal{F}_{ca}$, where L is the length of the profile curve, and

$$\hat{M} = \left[\frac{\mathrm{d}\psi}{\mathrm{d}s} + \left(1 + \frac{k_G}{k_c}\right)\frac{\sin\psi}{x} + \mathrm{lh}\right]x$$

$$\mathcal{H} = \frac{1}{2}\left[\left(\frac{\mathrm{d}\psi}{\mathrm{d}s}\right)^2 - \left(\frac{\sin\psi}{x} + \mathrm{lh}\right)^2\right]x + \lambda x + \mu\cos\psi + \eta\sin\psi$$

$$\hat{\mu} = \mu - \frac{1}{3}pxz, \qquad \hat{\eta} = \eta + \frac{1}{3}px^2 \qquad (4)$$

$$Q_0 = \sigma_0\cos\psi(0), \qquad Q_L = \left(\frac{1}{2}\mathrm{lh}^2 x(L) + \sigma_L\right)\cos\psi(L).$$

Here, σ_0 and σ_L denote line tensions in the circles $s = 0$ and $s = L$ if these circles are free ends of the nanostructure, otherwise $\sigma_0 = \sigma_L = 0$.

Observing expressions (3), one can immediately interpret

$$M = 2\pi k_c \hat{M}, \qquad \mathbf{F} = 2\pi k_c (\hat{\mu}\mathbf{i} + \hat{\eta}\mathbf{k})$$

where $\mathbf{i}$ and $\mathbf{k}$ denote the unit vectors along the coordinate axes X and Z, as the bending moment (couple resultant) and force (stress resultant) at any point of the membrane, except for the points with jump discontinuities.

Alternatively, for axysimmetric equilibrium shapes with $\mathbb{h} = 0$ the shape equation (1) reads

$$\frac{d^3\psi}{ds^3} = -\frac{2\cos\psi}{x}\frac{d^2\psi}{ds^2} - \frac{1}{2}\left(\frac{d\psi}{ds}\right)^3 + \frac{3\sin\psi}{2x}\left(\frac{d\psi}{ds}\right)^2$$

$$+ \frac{2\sigma x^2 + 2 - 3\sin^2\psi}{2x^2}\frac{d\psi}{ds} + \frac{2\sigma x^2 - 2 + \sin^2\psi}{2x^3}\sin\psi - q \qquad (5)$$

where $\sigma = \lambda/k_c$ and $q = p/k_c$. Besides Jülicher and Seifert [21] have already shown, that $\mathcal{H} = 0$ is the necessary and sufficient condition for the shape equations (1) and (5) to be equivalent. Since $\mathcal{H}$ is a conserved quantity on the smooth solutions of the Euler–Lagrange equations (2) due to the invariance of the functional $\mathcal{F}_{ca}$ under the translations of the independent variable s, henceforward we assume $\mathcal{H} = 0$.

4. Unduloid-like equilibrium shapes

Consider an infinitely long carbon nanotube subjected to external pressure p. In this Section, numerical solutions to equation (5) that describe unduloid-like equilibrium shapes of this tube are determined. Consider a representative part of the profile curve whose length is L and let the arc length corresponds to $0 \leq s \leq L$. Then, for symmetry reasons we adopt the following values of the slope angle

$$\psi(0) = \psi(L) = \frac{\pi}{2}.$$

On the other hand, the relation $\mathcal{H} = 0$ implies boundary values of $\eta(s)$ of form

$$\eta\big|_{s=0,\,L} = \frac{x}{2}\left[\left(\frac{1}{x} + \mathbb{h}\right)^2 - \left(\frac{d\psi}{ds}\right)^2 - 2\lambda\right]\bigg|_{s=0,\,L}.$$

An example of a unduloid-like equilibrium shape of a carbon nanotube subjected to external pressure $p = 130$ is presented in Figure 2. At that, the bending moment $M(s)$ and stress resultant $\hat{\eta}(s)$ are periodic functions of period L. The stress resultant $\hat{\mu}(s)$ is a sum of the periodic function $\mu(s)$ (its derivative is periodic, as the equations (2) imply) and the term $-(1/3)pxz$ that increases with z, see equations (4).

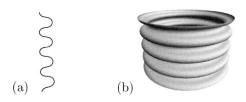

FIGURE 2. Profile curve (a) and the corresponding surface of revolution (b).

Acknowledgment

This work has been done within the bilateral scientific project *Nonlinear Geometry of Membranes, Filaments and Drops* between the Bulgarian and Polish Academies of Sciences.

References

[1] S. Iijima, *Helical microtubules of graphitic carbon.* Nature **354** (1991), 56–58.

[2] L. Radushkevich, V. Lukyanovich, *About carbon structures formed during thermal decomposition of carbon oxides at iron surface* (in Russian). Soviet Journal of Physical Chemistry **XXVI** (1952), 88–95.

[3] A. Oberlin, M. Endo, T. Koyama, *High-resolution electron-microscope observations of graphitized carbon-fibers.* Carbon **14** (1976), 133–135.

[4] D. Bethune, C. Kiang, M. de Vries, G. Gorman, R. Savoy, J. Vazquez, R. Beyers, *Cobalt-catalysed growth of carbon nanotubes with single-atomic-layer walls.* Nature **363** (1993), 605–607.

[5] T. Ebbesen, P. Ajayan, *Large-scale synthesis of carbon nanotubes.* Nature **358** (1992), 220–222.

[6] S. Iijima, Ch. Brabec, A. Maiti, J. Bernholc, *Structural flexibility of carbon nanotubes.* J. Chem. Phys. **104** (1996), 2089–2090.

[7] J. Tersoff, *New empirical approach for the structure and energy of covalent systems.* Phys. Rev. B **37** (1988), 6991–7000.

[8] D. Brenner, *Empirical potential for hydrocarbons for use in simulating the chemical vapor deposition of diamond films.* Phys. Rev. B **42** (1990), 9458–9471.

[9] T. Lenosky, H. Gonze, M. Teter, V. Elser, *Energetics of negatively curved graphitic carbon.* Nature **355** (1992), 333–335.

[10] B. Yakobson, C. Brabec, J. Bernholc, *Nanomechanics of carbon tubes: instabilities beyond linear response.* Phys. Rev. Lett. **76** (1996), 2511–2514.

[11] P. Naghdi, *Foundations of elastic shell theory*, Progress in Solid Mechanics, vol. 4 (I.N. Sneddon and R. Hill, eds.), North-Holland, Amsterdam, 1963, pp. 1–90.

[12] P. Naghdi, *The theory of shells and plates*, Handbuch der Physik, vol. VI a-2, Springer, New York, 1972, pp. 425–640.

[13] F. Niordson, *Shell Theory*, North-Holland, New York, 1985.

[14] Z.-C. Tu, Z.-C. Ou-Yang, *Elastic theory of low-dimensional continua and its applications in bio- and nano-structures.* J. Comput. Theoret. Nanoscience **5** (2008), 422–448.

[15] S.S. Xie, W.Z. Li, L.X. Qian, B.H. Chang, C.S. Fu, R.A. Zhao, W.Y. Zhou, G. Wang, *Equilibrium shape equation and possible shapes of carbon nanotubes.* Phys. Rev. B **54** (1996), 16436–16439.

[16] Z.-C. Ou-Yang, W. Helfrich, *Bending energy of vesicle membranes: general expressions for the first, second, and third variation of the shape energy and applications to spheres and cylinders.* Phys. Rev. A **39** (1989), 5280–5288.

[17] V. Vassilev, P. Djondjorov, I. Mladenov, *Cylindrical equilibrium shapes of fluid membranes.* J. Phys. A: Math. & Theor. **41** (2008), 435201 (16pp).

[18] V. Vassilev, P. Djondjorov, I. Mladenov, *Application of continuum mechanics approach to the shape analysis of carbon nanotubes and their junctions.* In: Nanoscience & Nanotechnology IX (Eds. E. Balabanova and J. Dragieva), BPS Ltd., Sofia, 2009, pp. 11–16.

[19] P. Djondjorov, V. Vassilev, I. Mladenov, *Analytic description and explicit parametrisation of the equilibrium shapes of elastic rings and tubes under uniform hydrostatic pressure.* Int. J. Mech. Sci. **53** (2011), 355–364.

[20] J. Smotlacha, *Numerical solutions of some models in quantum mechanics,* PhD Thesis, Czech Technical University 2012.

[21] F. Jülicher, U. Seifert, *Shape equations for axisymmetric vesicles: A clarification.* Phys. Rev. E **49** (1994), 4728–4731.

Ivaïlo M. Mladenov and Mariana Ts. Hadzhilazova
Institute of Biophysics
Bulgarian Academy of Sciences
Acad. G. Bonchev Street, Block 21
1113 Sofia, Bulgaria
e-mail: mladenov@obzor.bio21.bas.bg
        murryh@obzor.bio21.bas.bg

Vassil M. Vassilev and Peter A. Djondjorov
Institute of Mechanics
Bulgarian Academy of Sciences
Acad. G. Bonchev Street, Block 4
1113 Sofia, Bulgaria
e-mail: vasilvas@imbm.bas.bg
        padjon@imbm.bas.bg

Geometric Methods in Physics. XXXI Workshop 2012
Trends in Mathematics, 203–207
© 2013 Springer Basel

Resonance Phenomenon for Potentials of Wigner–von Neumann Type

Barbara Pietruczuk

Abstract. The paper discusses the behavior of solutions of the second-order differential equations possessing a resonance effect known for the Wigner–von Neumann potential. A class of potentials generalizing that of Wigner–von Neumann is presented.

Mathematics Subject Classification (2010). Primary 34E10; Secondary 34F15.

Keywords. Resonance, Wigner–von Neumann potential, asymptotic behavior.

1. Introduction

Consider a second-order differential equation

$$y'' + (\lambda - q(x))y = 0, \quad 0 < x < \infty. \tag{1}$$

If $q(x)$ decays sufficiently rapidly as $x \to \infty$, for example $q(x) \to 0$ and $q(x) \in L^1(0, \infty)$, then (see [1]) the spectrum of the corresponding operator

$$-\frac{d^2}{dx^2} + q(x)$$

is absolutely continuous in $\lambda \in \mathbb{R}_+$. The example of von Neumann and Wigner

$$y'' + (1 - 2ax^{-1} \sin kx)y = 0, \tag{2}$$

where a and k are real parameters, illustrates the phenomenon of geometrically large and small oscillating solutions, which is called resonance. The asymptotic form of the solutions of this equation depends on whether k has values ± 2 or not.

Fundamental system of solutions of equation (2) has the following asymptotic behavior as $x \to \infty$

$$
\begin{array}{lll}
y_1(x) = \cos x + o(1), & y_2(x) = \sin x + o(1), & (k \neq \pm 2), \\
y_1(x) = x^{-\frac{1}{2}a}(\cos x + o(1)), & y_2(x) = x^{\frac{1}{2}a}(\sin x + o(1)), & (k = \pm 2).
\end{array}
$$

In non-resonant case $k \neq \pm 2$, equation (2) has no non-trivial $L^2(0, +\infty)$ solution.

The case $k = \pm 2$ can be treated as a resonance phenomenon and here $y_1(x) \in L^2(0, +\infty)$, if $a > 1$. Thus the Wigner–von Neumann potential

$$q(x) = 2ax^{-1} \sin 2x$$

gives an example of the Schrödinger operator $-\dfrac{d^2}{dx^2} + q(x)$ having an eigenvalue embedded into continuous spectrum.

2. A construction yielding square-integrable solutions

The Wigner–von Neumann potential was first constructed in Ref. [2] by the procedure in which an equation with the L^2 solution is generated from the initial equation with an explicitly known solution that is not square integrable.

The general idea of this method is the following. Let us consider an equation

$$y'' + (\mu - q_0(x))y = 0 \tag{3}$$

and let $\phi(x)$ be a solution of this equation for some given μ and $q_0(x)$. Now, let us define

$$\psi(x) = \frac{\phi(x)}{G(x)},$$

where $G(x)$ is chosen in such a way, that $\psi(x)$ is square integrable. Then $\psi(x)$ is a $L^2(0, \infty)$ solution of the equation

$$y'' + (\mu - q(x))y = 0$$

with

$$q(x) = q_0(x) - \frac{G''(x)}{G(x)} + \frac{2G'^2(x)}{G^2(x)} - \frac{2G'(x)\phi'(x)}{G(x)\phi(x)}.$$

A. Wintner [3] used the above construction for $q_0(x) = 0$ and $\phi(x) = \cos x\sqrt{\mu}$. The choice

$$G(x) = \exp\left\{ \int_0^x g(t)\phi(t)dt \right\}, \quad g(t) = at^{-1} \cos t\sqrt{\mu}$$

gives $\psi(x) \in L^2(0, \infty)$, if $a > 1$. The corresponding potential is of the form

$$q(x) = 2a\sqrt{\mu}x^{-1} \sin(2x\sqrt{\mu}) + x^{-2}(a\cos^2 x\sqrt{\mu} + a^2 \cos^4 x\sqrt{\mu}),$$

which is in a sense a perturbation of the Wigner–von Neumann potential.

This approach gives an explicit expression for the square integrable solution of the generated equation. When asymptotic behavior is of importance further methods should be developed.

3. Asymptotics

The asymptotic analysis for equations of which (2) is an example, and the identification of resonant and non-resonant situations is due to F.V. Atkinson, A. Lutz
and W.A. Harris [4].

Theorem 1. *Suppose that the infinite integrals*

$$q_1(x) := \int_x^\infty q(t)dt, \qquad q_2(x) := \int_x^\infty q(t)\cos(2t\sqrt{\lambda})dt$$

converge for $\lambda > 0$.

(i) (*Non-resonance case*). *Assume that the integral*

$$q_3(x) := \int_x^\infty q(t)\sin(2t\sqrt{\lambda})dt$$

*exists for $x > x_0 > 0$, and let $q(x)q_i(x) \in L(x_0, \infty)$, $i = 2,\ 3$. Then equa-
tion (1) has solutions $y_1(x)$ and $y_2(x)$ such that, as $x \to \infty$,*

$$y_1(x) = \sin x\sqrt{\lambda} + o(1), \qquad y_2(x) = \cos x\sqrt{\lambda} + o(1).$$

(ii) (*Resonance case*). *Let $q(x)q_i(x) \in L(x_0, \infty)$, $i = 1,\ 2$. If the infinite integral
$q_3(x)$ diverges and for a certain C, inequality*

$$\kappa(x_2) - \kappa(x_1) > C$$

holds for all x_1 and x_2 such that $x_0 < x_1 < x_2 < \infty$, where

$$\kappa(x) = \frac{1}{2}\lambda^{-\frac{1}{2}}\int_{x_0}^x q(t)\sin(2t\sqrt{\lambda})dt.$$

Then equation (1) has solutions $y_1(x)$ and $y_2(x)$ such that, as $x \to \infty$,

$$y_1(x) = \exp\{\kappa(x)\}\left[\sin x\sqrt{\lambda} + o(1)\right],$$
$$y_2(x) = \exp\{-\kappa(x)\}\left[\cos x\sqrt{\lambda} + o(1)\right].$$

The Wigner–von Neumann potential $2a\sqrt{\mu}x^{-1}\sin(2x\sqrt{\mu})$ satisfies the conditions of Part (i) of Theorem 1 if $\lambda \neq \mu$ and the conditions of Part (ii) if $\lambda = \mu$.
Thus it provides the existence of an $L^2(0, \infty)$ solution $y_2(x)$ when $\lambda = \mu$ and $a > 1$.

4. Potentials of Wigner–von Neumann type

Methods elaborated in [1] can be applied to a wider class of potentials possessing
resonance phenomenon.

Theorem 2. *Assume that $|\phi(x)| < M$ and let the integrals*

$$\int_0^s \phi(t)\cos 2tdt, \qquad \int_0^s \phi(t)dt$$

be bounded. Moreover suppose that $\phi(s)\sin 2s = A + f(s)$, where f is bounded and
$$\int_0^\infty f(t)(1+t)^{-1}dt < \infty.$$
　　Then differential equation
$$y'' + \left(1 - \frac{\phi(x)}{1+x}\right)y = 0$$
has fundamental system of solutions with asymptotic behavior as $x \to \infty$
$$y_1(x) = (1+x)^{\frac{A}{2}}(\sin x + o(1)),$$
$$y_2(x) = (1+x)^{-\frac{A}{2}}(\cos x + o(1)).$$

Sketch of the proof. Transformation
$$\begin{pmatrix} y \\ y' \end{pmatrix} = \begin{pmatrix} \cos x & \sin x \\ -\sin x & \cos x \end{pmatrix} \begin{pmatrix} u_1 \\ u_2 \end{pmatrix},$$
reduces the equation to the system
$$\begin{cases} u_1'(x) + \dfrac{\phi(x)}{2(1+x)}\sin 2x\, u_1(x) = \dfrac{\phi(x)}{2(1+x)}(\cos 2x - 1)\,u_2(x), \\[2mm] u_2'(x) - \dfrac{\phi(x)}{2(1+x)}\sin 2x\, u_2(x) = \dfrac{\phi(x)}{2(1+x)}(\cos 2x + 1)\,u_1(x). \end{cases}$$
Further we pass to integral equations
$$\begin{cases} u_1(x) = u_1(0)e^{-Q(x)} + e^{-Q(x)}\displaystyle\int_0^x \dfrac{\phi(s)}{2(1+s)}(\cos 2s - 1)e^{Q(s)}u_2(s)ds, \\[3mm] u_2(x) = u_2(0)e^{Q(x)} + e^{Q(x)}\displaystyle\int_0^x \dfrac{\phi(s)}{2(1+s)}(\cos 2s + 1)e^{-Q(s)}u_1(s)ds, \end{cases}$$
where
$$Q(x) = \int_0^x \frac{\phi(s)}{2(1+s)}\sin 2s\, ds$$
is an integration factor.
Eliminate $u_1(x)$ and solve by iterations the equation
$$u_2(x) = H(x) + \int_0^x K_x(r)u_2(r)dr,$$
where
$$H(x) = u_2(0) + u_1(0)\int_0^x \frac{\phi(s)}{2(1+s)}(\cos 2s + 1)e^{-2Q(s)}ds$$
and
$$K_x(r) = \frac{\phi(r)}{2(1+r)}(\cos 2r - 1)e^{2Q(r)}\int_r^x \frac{\varphi(s)}{2}(\cos 2s + 1)e^{-2Q(s)}ds.$$
This gives the desired information about the asymptotic behavior of solutions of the equation in question. $\qquad\square$

Example. Potential

$$q(x) = \frac{\psi(x)}{1 + x},$$

where ψ is bounded and π-periodic function with zero mean $\int_0^\pi \psi(t)dt = 0$, shows that the class described by Theorem 2 supplements the one from Theorem 1.

Acknowledgment

The author is indebted to prof. S. Stepin for his suggestions, help and illuminating discussions, especially about resonance phenomenon for solutions of second-order differential equations.

References

[1] D.B. Hinton, M. Klaus, J.K. Shaw, *Embedded half-bound states for potentials of Wigner–von Neumann type*, Proc. London Math. Soc. (3) 62 (1991), no. 3, 607–646.

[2] M.S.P. Eastham, H. Kalf, *Schrödinger-type operators with continuous spectra* (1982).

[3] A. Wintner, *Asymptotic integration of the adiabatic oscillator* Amer. Journ. Math., vol. 69 (1947), pp. 251–272.

[4] W.A. Harris, D.A. Lutz, *Asymptotic integration of adiabatic oscillator* Amer. Journ. Math. Anal. Appl. 51 (1975), 76–93.

Barbara Pietruczuk
Institute of Mathematics
University of Białystok
Akademicka 2
PL-15-267 Białystok, Poland
e-mail: bpietruczuk@math.uwb.edu.pl

Geometric Methods in Physics. XXXI Workshop 2012
Trends in Mathematics, 209–220
© 2013 Springer Basel

Form Factor Approach to the Calculation of Correlation Functions of Integrable Models

N.A. Slavnov

Abstract. We describe form factor approach to the study of correlation functions of quantum integrable models in the critical regime. We illustrate the main features of this method using the example of impenetrable bosons. We introduce dressed form factors and show that they are well defined in the thermodynamic limit.

Mathematics Subject Classification (2010). 81Q80; 82C22.

Keywords. Form factors, correlation functions.

1. Introduction

Form factor approach can be applied for calculation of correlation functions in various quantum models. Let us briefly describe the main idea of this method. Suppose that we have a quantum model with the Hamiltonian H and a system of its eigenfunctions $|\psi\rangle$

$$H|\psi\rangle = E|\psi\rangle. \tag{1}$$

The form factor of an operator $\mathcal{O}$ between the states $\langle\psi|$ and $|\psi'\rangle$ is the matrix element $\mathcal{F}^{\mathcal{O}}_{\psi,\psi'}$

$$\mathcal{F}^{\mathcal{O}}_{\psi,\psi'} = \frac{\langle\psi|\mathcal{O}|\psi'\rangle}{\|\psi\| \, \|\psi'\|}.$$

Suppose that for some operators $\mathcal{O}_1$ and $\mathcal{O}_2$ all form factors are known. Then we can calculate the two-point correlation function

$$G^{\mathcal{O}_1,\mathcal{O}_2} = \frac{\langle\psi|\mathcal{O}_1\mathcal{O}_2|\psi\rangle}{\langle\psi|\psi\rangle},$$

This work was supported by Program of RAS Basic Problems of the Nonlinear Dynamics and Grants SS-4612.2012.1, CDO-2012.33.

by inserting the complete set of the Hamiltonian eigenfunctions between two operators

$$G^{\mathcal{O}_1,\mathcal{O}_2} = \sum_{|\psi'\rangle} \frac{\langle\psi|\mathcal{O}_1|\psi'\rangle\langle\psi'|\mathcal{O}_2|\psi\rangle}{\langle\psi|\psi\rangle\langle\psi'|\psi'\rangle} = \sum_{|\psi'\rangle} \mathcal{F}^{\mathcal{O}_1}_{\psi,\psi'}\mathcal{F}^{\mathcal{O}_2}_{\psi',\psi}. \tag{2}$$

The sum in (2) is called the form factor series. Thus, the calculation of two-point correlation functions can be reduced to the computation of form factor series. Clearly this method can be applied to the calculation of multi-point correlation functions as well. We, however, restrict our selves with the two-point functions only.

The form factor approach appears to be especially powerful for quantum integrable models. Many of them can be solved by Bethe Ansatz [1–4]. This method provides us with a convenient description of the Hamiltonian spectrum. The Quantum Inverse Scattering Method [5–7] allows to study the problem of form factors and correlation functions. A variety of works were devoted to the calculation of form factors of local operators (see, e.g., [8–14]). Thus, the problem of calculation of correlation functions reduces to the summation of the form factor series. For some quantum integrable models there exist effective tools to work with the series (2). However for critical (gapless) models there are still several unsolved questions [12, 15–17]. In the present paper we consider the problems appearing in the form factor approach to the critical models and describe a way to solve them. As an example we consider one of the simplest integrable model, namely the model of impenetrable bosons.

2. Quantum Nonlinear Schrödinger equation

We start with the Quantum Nonlinear Schrödinger equation. This is $(1 + 1)$-dimensional model. The Hamiltonian is given by

$$H = \int_0^L \left(\partial_x\phi^\dagger\partial_x\phi + c\phi^\dagger\phi^\dagger\phi\phi - h\phi^\dagger\phi \right) dx, \tag{3}$$

where $\phi(x,t)$ and $\phi^\dagger(x,t)$ are Bose-fields with canonical equal-time commutation relations

$$[\phi(x,t), \phi^\dagger(y,t)] = \delta(x-y).$$

The operators $\phi(x,t)$ and $\phi^\dagger(x,t)$ are defined on a finite interval $[0, L]$, and we assume that they enjoy periodically boundary conditions. Later we will take the limit $L \to \infty$. The field $\phi(x,t)$ acts in the Fock space as annihilation operator $\phi(x,t)|0\rangle = 0$, the conjugated field $\phi^\dagger(x,t)$ acts in the dual space $\langle 0|\phi^\dagger(x,t) = 0$. The evolution of these operators is defined in a standard way

$$\phi(x,t) = e^{iHt}\phi(x,0)e^{-iHt}, \qquad \phi^\dagger(x,t) = e^{iHt}\phi^\dagger(x,0)e^{-iHt}.$$

The Hamiltonian (3) depends also on a coupling constant c and a chemical potential h. We consider the case $c > 0$, what corresponds to the repulsive interaction.

We also set $h > 0$, as in this case the model has a non-trivial ground state (see below).

The Quantum Nonlinear Schrödinger equation is completely integrable model. There exists infinitely many integrals of motion commuting with the Hamiltonian H. For example, the charge operator $\hat{Q}$ and the momentum operator $\hat{P}$ have the form

$$\hat{Q} = \int_0^L \phi^\dagger \phi \, dx, \qquad \hat{P} = \frac{i}{2} \int_0^L \left(\partial_x \phi^\dagger \cdot \phi - \phi^\dagger \cdot \partial_x \phi \right) dx.$$

2.1. Eigenfunctions of the Hamiltonian

The Hamiltonian eigenfunctions $|\psi\rangle$ and their eigenvalues E can be found by Bethe Ansatz. All the integrals of motion possess a common system of eigenstates. It is clear that the eigenstates of the charge operator $\hat{Q}$ have the form

$$|\psi\rangle = \int_0^L \chi_N(x_1, \dots, x_N) \prod_{k=1}^N \phi^\dagger(x_k) \, dx_k |0\rangle, \qquad N = 0, 1, \dots, \tag{4}$$

where $\chi_N(x_1, \dots, x_N)$ are some coefficients, satisfying periodic boundary conditions with respect to every x_k. Then the eigenvalues of $\hat{Q}$ corresponding to such $|\psi\rangle$ are equal to N. Obviously, one can look for the Hamiltonian eigenfunctions in the form (4). Substituting (4) into the equation for the eigenvalues (1) we find that $\chi_N(x_1, \dots, x_N)$ should satisfy the following differential equation

$$\left(-\sum_{k=1}^N \frac{\partial^2}{\partial x_k^2} + 2c \sum_{j>k}^N \delta(x_j - x_k) - Nh \right) \chi_N = E \chi_N. \tag{5}$$

The form of this equation coincides with Schrödinger equation describing a model of N one-dimensional particles interacting by delta-function potential. Therefore the model of Quantum Nonlinear Schrödinger equation is often called one-dimensional Bose-gas [18–21].

There is no a big problem to solve the equation (5). It turns out that different solutions of (5) can be parameterized by different sets of real variables $\lambda_1, \dots, \lambda_N$. Namely,

$$\chi_N(x_1, \dots, x_N) = \sum_P (-1)^{[P]} \prod_{j>k}^N \left(\lambda_{P_j} - \lambda_{P_k} - ic \, \mathrm{sgn}(x_j - x_k) \right) \prod_{m=1}^N e^{ix_m \lambda_{P_m}}, \tag{6}$$

where the sum is taken over permutations P of the set $\lambda_1, \dots, \lambda_N$. Hereby the parameters λ_j satisfy the system of Bethe equations

$$e^{iL\lambda_j} = -\prod_{k=1}^N \frac{\lambda_j - \lambda_k + ic}{\lambda_j - \lambda_k - ic}, \qquad j = 1, \dots, N. \tag{7}$$

System (7) provides the periodicity of the function $\chi_N(x_1, \dots, x_N)$. Every solution of Bethe equations with pair-wise distinct λ_j determines a function $\chi_N(x_1, \dots, x_N)$, which in its turn gives an eigenfunction of the Hamiltonian via (4). One can show that the set of states constructed in this way is complete.

Knowing the solutions of Bethe equations on can find the eigenvalues E of the Hamiltonian and the eigenvalues P of the momentum operator

$$E = \sum_{j=1}^{N} (\lambda_j^2 - h), \qquad P = \sum_{j=1}^{N} \lambda_j.$$

2.2. Form factors

Consider now form factors of local operators in this model. For definiteness we will focus on the form factor of the filed ϕ (and the conjugated form factor of the field $\phi^\dagger$):

$$\mathcal{F}_{\psi,\psi'}^{\phi} = \frac{\langle\psi|\phi(0,0)|\psi'\rangle}{\|\psi'\| \, \|\psi\|}, \qquad \mathcal{F}_{\psi',\psi}^{\phi^\dagger} = \frac{\langle\psi'|\phi^\dagger(0,0)|\psi\rangle}{\|\psi'\| \, \|\psi\|} = \left(\mathcal{F}_{\psi,\psi'}^{\phi}\right)^*. \qquad (8)$$

Let the state $|\psi\rangle$ be parameterized by the set of N Bethe roots $\{\lambda\} = \lambda_1, \ldots, \lambda_N$. Then the state $|\psi'\rangle$ should be parameterized by the set of $N+1$ Bethe roots $\{\mu\} = \mu_1, \ldots, \mu_{N+1}$. Otherwise the matrix elements (8) vanish.

It is easy to check that

$$\mathcal{F}_{\psi,\psi'}^{\phi}(x,t) = \frac{\langle\psi|\phi(x,t)|\psi'\rangle}{\|\psi'\| \, \|\psi\|} = \mathcal{F}_{\psi,\psi'}^{\phi} e^{ix\mathcal{P}-it\mathcal{E}},$$

where $\mathcal{P}$ and $\mathcal{E}$ are the excitation momentum and energy

$$\mathcal{P} = P_\psi - P_{\psi'} = \sum_{j=1}^{N} \lambda_j - \sum_{j=1}^{N+1} \mu_j,$$

$$\mathcal{E} = E_\psi - E_{\psi'} = \sum_{j=1}^{N} (\lambda_j^2 - h) - \sum_{j=1}^{N+1} (\mu_j^2 - h).$$

Thus, we can restrict our selves with the computation of $\mathcal{F}_{\psi,\psi'}^{\phi}$ only. Using (4) one can express this matrix element as a multiple integral

$$\langle\psi|\phi(0,0)|\psi'\rangle = N! \int_0^L \chi_N^*(\{\lambda\}|z_1, \ldots, z_N) \chi_{N+1}(\{\mu\}|0, z_1, \ldots, z_N) \prod_{m=1}^{N} dz_m. \qquad (9)$$

The norms of the states $|\psi\rangle$ and $|\psi'\rangle$ also can be written in terms of integrals of the functions χ_N. In particular,

$$\|\psi\|^2 = N! \int_0^L |\chi_N(\{\lambda\}|z_1, \ldots, z_N)|^2 \prod_{m=1}^{N} dz_m, \qquad (10)$$

and similarly for $\|\psi'\|^2$.

Equations (9), (10) formally give us possibility to calculate form factors of the field ϕ. In practice, however, the explicit evaluation of the above integrals was done only for small N. For large N this evaluation turns into a very complicated combinatorial problem. Therefore the calculation of form factors needs more advanced methods. We refer the readers to the works [12–14, 22], where this problem was solved in the frameworks of the Algebraic Bethe Ansatz.

2.3. Form factor series

Let us describe now the problems, which appear in the summation of the form factor series. First of all we should specify the correlation function we deal with.

We will consider ground state two-point correlation function of the fields ϕ and $\phi^\dagger$ in the thermodynamic limit. This means that the state $|\psi\rangle$ corresponds to the minimal energy. Thus, we should fix the set of Bethe roots $\{\lambda\}$ in such a way that the corresponding eigenvalue $E = \sum_{j=1}^N (\lambda_j^2 - h)$ approaches its minimum. We will do it in the next section.

The thermodynamic limit means that the volume of the gas L goes to infinity, while the average density of the gas $D = N/L$ remains fixed and finite.

Suppose that we have computed the form factor of the field ϕ. Then it is a function of the sets $\{\lambda\}$ and $\{\mu\}$, which parameterize the eigenstates $|\psi\rangle$ and $|\psi'\rangle$ respectively

$$\mathcal{F}^\phi_{\psi,\psi'} = \mathcal{F}^\phi_{\psi,\psi'}(\{\lambda\};\{\mu\}).$$

Knowing the explicit expression for this function we can substitute it into the form factor series

$$G^\phi(x,t) \equiv \frac{\langle\psi|\phi(x,t)\phi^\dagger(0,0)|\psi\rangle}{\langle\psi|\psi\rangle} = \sum_{\{\mu\}} |\mathcal{F}^\phi_{\psi,\psi'}|^2 e^{ix\mathcal{P}-it\mathcal{E}}. \tag{11}$$

Thus, the sum over the excited states actually turns into the sum over all possible Bethe roots $\mu_1,\ldots,\mu_{N+1}$.

There exists a way to sum up form factor series for L finite, and then to take the limit $L \to \infty$ in the result obtained. This way is rather complicated from the technical viewpoint. Eventually it leads to various multiple integral representations for correlation functions [23, 24].

Another way is to take the thermodynamic limit directly in the series (11). This method seems to be more simple. The matter is that in the $L \to \infty$ limit the excited states (and the corresponding form factors) can be described in terms of so-called 'particles' and 'holes' and their rapidities (see the next section). Then the summation over the solutions of Bethe equations $\{\mu\}$ can be replaced by the integration over particle and hole rapidities $\hat{\mu}_p$ and $\hat{\mu}_h$. Symbolically one can write

$$\sum_{\{\mu\}} |\mathcal{F}^\phi_{\psi,\psi'}|^2 e^{ix\mathcal{P}-it\mathcal{E}} \to \int |\mathcal{F}^\phi_{\psi,\psi'}|^2 e^{ix\mathcal{P}-it\mathcal{E}} \, d\hat{\mu}_p \, d\hat{\mu}_h, \qquad L \to \infty.$$

Thus, following this way we can get rid of the summation over solutions of the system of transcendental equations (7).

This method perfectly works in the massive models. However, its application to critical models deals with several difficulties. First of all, it turns out that in

the thermodynamic limit all form factors scale to zero as some fractional power of
L [12, 25]

$$|\mathcal{F}^{\phi}_{\psi,\psi'}|^2 \to L^{-\theta} C(\{\hat{\mu}_p\}; \{\hat{\mu}_h\}), \qquad L \to \infty,$$

where $C(\{\hat{\mu}_p\}; \{\hat{\mu}_h\})$ is a finite part of the form factor. The second problem is that
integrals over particle/hole rapidities of the finite part $C(\{\hat{\mu}_p\}; \{\hat{\mu}_h\})$ are divergent

$$\int C(\{\hat{\mu}_p\}; \{\hat{\mu}_h\}) e^{ix\mathcal{P}-it\mathcal{E}}\, d\hat{\mu}_p\, d\hat{\mu}_h \to \infty.$$

Such a senseless result arises, because form factors in critical models have no uniform thermodynamic limit for all values of the particle/hole rapidities. Therefore
generically one can not take the limit $L \to \infty$ in separate terms of the series (11).

In order to solve the problem we consider the series (11) for L large but finite.
Then we split the excited states into special classes **P** and re-order the form factor
series as follows

$$\sum_{|\psi'\rangle} |\mathcal{F}^{\phi}_{\psi,\psi'}|^2 e^{ix\mathcal{P}-it\mathcal{E}} = \sum_{\mathbf{P}} \sum_{|\psi'\rangle \in \mathbf{P}} |\mathcal{F}^{\phi}_{\psi,\psi'}|^2 e^{ix\mathcal{P}-it\mathcal{E}}.$$

The sum of form factors within one class of the excited states can be computed
explicitly. It gives dressed form factor

$$|\mathcal{F}^{\phi}_{\mathbf{P}}|^2 e^{ix\mathcal{P}_{\mathbf{P}}-it\mathcal{E}_{\mathbf{P}}} = \sum_{|\psi'\rangle \in \mathbf{P}} |\mathcal{F}^{\phi}_{\psi,\psi'}|^2 e^{ix\mathcal{P}-it\mathcal{E}},$$

where $\mathcal{P}_{\mathbf{P}}$ and $\mathcal{E}_{\mathbf{P}}$ are the excitation momentum and energy of the class **P**. It turns
out that the dressed form factor has a finite value in the thermodynamic limit.
This value can be substituted into the integrals over the particle/hole rapidities.

All the properties of the form factors announced above will be shown in the
next section in the special limit of the model of one-dimensional Bose-gas.

3. Impenetrable bosons

Consider one-dimensional bosons in the limit when the coupling constant c goes
to infinity. Such the limit is called impenetrable one-dimensional Bose-gas. Then
the expression (6) simplifies

$$\chi_N(x_1, \ldots, x_N) = \prod_{j>k}^{N} \mathrm{sgn}(x_j - x_k) \cdot \det_N e^{ix_j \lambda_k},$$

and the system of Bethe equations becomes trivial

$$e^{iL\lambda_j} = (-1)^{N-1}.$$

It is clear that the roots of these system are $\lambda_j = \frac{2\pi}{L}\ell_j$, where ℓ_j are integers or
half-integers and $\ell_j \neq \ell_k$ for $j, k = 1, \ldots, N$. Below without loss of generality we
assume that N is even.

Let us find the ground state of this model. Let $q = \sqrt{h}$. It is easy to see
that the minimum of the functional $E = \sum_{j=1}^{N}(\lambda_j^2 - h)$ approaches if all vacancies

$\frac{2\pi}{L}\ell_j$ within the interval $[-q, q]$ are occupied. Indeed, if we add to this set some λ with $|\lambda| > q$, then the energy E obtains a positive contribution. Similarly, if we create a hole in the uniform distribution of λ_j, then we subtract a negative quantity from E. Thus, the set of roots $\lambda_j = \frac{2\pi}{L}\left(j - \frac{N+1}{2}\right)$ does correspond to the minimal eigenvalue E, provided $N/L = [q/\pi]$. The interval $[-q, q]$ is called Fermi zone.

Consider now excited states. Recall that all the excited states giving rise to the field form factor are parameterized by $N + 1$ Bethe roots $\mu_{\ell_j} = \frac{2\pi}{L}\ell_j$. On the other hand, as we have already mentioned, the excited states can be described in terms of particles and holes.

We begin with the following set of integers $\ell_j = j - (N+2)/2$. We say that this state does not contain particles and holes. The state with one particle and one hole is constructed as follows. Let $\ell_j = j - \frac{N+2}{2}$ for $j \neq h$, where h is some integer from the set $\{1, \ldots, N+1\}$. For $j = h$ we set $\ell_h = p - \frac{N+2}{2}$, where $p \notin \{1, \ldots, N+1\}$. The integers p and h are called the quantum numbers of the particle and the hole. We say that this state has the particle with the quantum number p and the hole with the quantum number h. The quantities $\hat{\mu}_p = \frac{2\pi}{L}\left(p - \frac{N+2}{2}\right)$ and $\hat{\mu}_h = \frac{2\pi}{L}\left(h - \frac{N+2}{2}\right)$ are called the rapidities of the particle and the hole. Enumerating all possible quantum numbers p and h we run through all excited states with one particle and one hole.

Similarly we can consider states with two particles and two holes and so on. Thus, instead of original parametrization of the excited states in terms of integers ℓ_j we can parameterize them in terms of the quantum numbers p_j, h_j and the corresponding rapidities $\hat{\mu}_{p_j}$ and $\hat{\mu}_{h_j}$. Observe that the distance between two nearest values of the particle/hole rapidities is $2\pi/L$. In the thermodynamic limit it goes to zero, therefore sums over particle/hole rapidities have the sense of integral sums. That is why one can hope to replace the form factor series by the integrals over $\hat{\mu}_{p_j}$ and $\hat{\mu}_{h_j}$ in the thermodynamic limit.

For the model of impenetrable bosons the field form factors can be calculated via the formulas (9)–(10) [26]. We have

$$\left|\mathcal{F}_{\psi,\psi'}^{\phi}\right|^2 = \frac{1}{L}\left(\frac{2}{L}\right)^{2N} \frac{\displaystyle\prod_{j>k}^{N}(\lambda_j - \lambda_k)^2 \prod_{j>k}^{N+1}(\mu_j - \mu_k)^2}{\displaystyle\prod_{j=1}^{N}\prod_{k=1}^{N+1}(\lambda_j - \mu_k)^2}. \tag{12}$$

This result is valid for arbitrary sets of Bethe roots $\{\lambda\}$ and $\{\mu\}$. Let the set $\{\lambda\}$ correspond to the ground state, while the set $\{\mu\}$ describes the excited state without particles and holes. Substituting the explicit values of λ_j and μ_j into (12) we obtain after simple algebra

$$\left|\mathcal{F}_{\psi,\psi'}^{\phi}\right|^2 = \frac{\pi^2 G^4(1/2)}{L}\frac{G^2(N+1)G^2(N+2)}{G^4(N+3/2)},$$

where $G(z)$ is Barnes function: $G(z+1) = \Gamma(z)G(z)$. Using the asymptotic behavior of the Barnes function we immediately find that in the thermodynamic limit

$$\left|\mathcal{F}^\phi_{\psi,\psi'}\right|^2 \to C \cdot L^{-1/2}, \qquad L, N \to \infty, \qquad N/L = D,$$

where C is a finite constant. Thus, as we have mentioned in the previous section, we see that the form factor goes to zero in the thermodynamic limit.

Form factors corresponding to other excited states behave similarly. Consider, for example, an excited state with one particle and one hole with quantum numbers p $(p > N + 1)$ and h $(1 \le h \le N + 1)$ respectively. Simple calculation shows that for L large enough this form factor behaves as

$$\left|\mathcal{F}^\phi_{\psi,\psi'}\right|^2 = \frac{C \cdot L^{-1/2}}{(p-h)^2} \frac{\Gamma^2(p)\Gamma^2(p-N-\frac{1}{2})}{\Gamma^2(p-\frac{1}{2})\Gamma^2(p-N-1)} \cdot \frac{\Gamma^2(h-\frac{1}{2})\Gamma^2(N-h+\frac{3}{2})}{\Gamma^2(h)\Gamma^2(N-h+2)}. \quad (13)$$

One can use the equation (13) in order to show non-uniform behavior of the form factor with respect to the particles and hole rapidities. Suppose that $\hat{\mu}_p$ and $\hat{\mu}_h$ are separated from the Fermi boundaries $\pm q$ by a finite gap in the thermodynamic limit. This means that $p \gg N + 1$ and $1 \ll h \ll N + 1$. Then all arguments of the Γ-functions in (13) go to infinity, and we can apply the Stirling formula. Then we obtain

$$|\mathcal{F}^\phi_{\psi,\psi'}|^2 \to \frac{C \cdot L^{-1/2}}{(2\pi)^2} \frac{1}{L^2(\hat{\mu}_p - \hat{\mu}_h)^2} \frac{\hat{\mu}_p^2 - q^2}{q^2 - \hat{\mu}_h^2}, \qquad L \to \infty. \quad (14)$$

In order to find the contribution of such form factors into the correlation function we should take the sum over $\hat{\mu}_p$ and $\hat{\mu}_h$. For $L \to \infty$ this sum can be replaced by the integral. Hereby the factor L^{-2} absorbs into the integration measure. Then we obtain

$$\sum_{|\psi'\rangle} e^{ix\mathcal{P}-it\mathcal{E}} |\mathcal{F}^\phi_{\psi,\psi'}|^2 \to \frac{C \cdot L^{-1/2}}{(2\pi)^2} \int\limits_{-q+\epsilon}^{q-\epsilon} d\hat{\mu}_h \int\limits_{q+\epsilon}^{\infty} d\hat{\mu}_p \frac{e^{ix\mathcal{P}-it\mathcal{E}}}{(\hat{\mu}_p - \hat{\mu}_h)^2} \frac{\hat{\mu}_p^2 - q^2}{q^2 - \hat{\mu}_h^2},$$

where ϵ is an arbitrary small positive number. We see that for $\epsilon > 0$ this contribution vanishes due to the prefactor $L^{-1/2}$. On the other hand one can not set $\epsilon = 0$, since in this case the integral becomes divergent. Moreover, we can not consider the limit $\epsilon = 0$, because in this case $\hat{\mu}_p$ and $\hat{\mu}_h$ can approach the Fermi boundaries, while the formula (14) was obtained under the condition that $\hat{\mu}_p$ and $\hat{\mu}_h$ were separated from $\pm q$.

Consider now the case when particle or hole rapidity goes to the Fermi boundary in the thermodynamic limit. Let for definiteness $\hat{\mu}_h \to -q$. Then the quantum number h is of order 1. Hence, we can not apply the Stirling formula to $\Gamma(h-\frac{1}{2})$ and $\Gamma(h)$ in (13), therefore we do not reproduce (14). We also see that the thermodynamic limit of the form factor can not be described in terms of the rapidities $\hat{\mu}_p$ and $\hat{\mu}_h$ only. It depends on the quantum number h as well.

It is easy to see that similar effects occur, when $\hat{\mu}_h \to q$ or $\hat{\mu}_p \to \pm q$. Thus, we come to conclusion that particle-hole form factors have no uniform thermodynamic limit. One should consider separately:

- particle/hole rapidities are separated from the Fermi boundaries in the limit $L \to \infty$;
- particle/hole rapidities are on the Fermi boundaries $\hat{\mu}_p = \pm q$, $\hat{\mu}_h = \pm q$ at $L \to \infty$.

This difference is the basis for the division of the excited states into classes. We say that two excited states belong to the same class **P** if:

- they have the same excitation momentum $\mathcal{P}$ and energy $\mathcal{E}$ in the thermodynamic limit;
- they have the same number of particles and holes with the same rapidities $\hat{\mu}_{p_a}$ and $\hat{\mu}_{h_a}$, which are separated from the Fermi boundaries in the thermodynamic limit.

Pay attention that the total number of particles and holes in the states of the same class is not fixed. One can add arbitrary number of excitations at the Fermi boundaries. Indeed, adding particles or holes, whose rapidities coincide with $\pm q$, we do not change the excitation energy, as $q^2 - h = 0$. In order to preserve the excitation momentum we should add equal number of particles and holes to the right Fermi boundary q and equal number of particles and holes to the left Fermi boundary $-q$. Then the excitation momentum does not change and, hence, the obtained states belong to the same class.

As an example we consider an excited state depending on n rapidities $\hat{\mu}_{p_a}$ and n rapidities $\hat{\mu}_{h_a}$. Suppose that all of them are separated from $\pm q$. This state is a representative of a class **P**. Other representatives of this class have additional excitations at the Fermi boundaries. Calculating form factors for all these states and taking their sum we obtain a dressed form factor corresponding to the class **P**

$$\sum_{|\psi'\rangle \in \mathbf{P}} e^{ix\mathcal{P}-it\mathcal{E}} |\mathcal{F}^\phi_{\psi,\psi'}|^2 = e^{ix\mathcal{P}_\mathbf{P}-it\mathcal{E}_\mathbf{P}} |\mathcal{F}^\phi_\mathbf{P}|^2.$$

Using the equation (12) we can find the explicit form of this dressing for L large enough

$$|\mathcal{F}^\phi_\mathbf{P}|^2 = L^{-1/2} C_\mathbf{P}(\{\hat{\mu}_p\}; \{\hat{\mu}_h\}) R_+(\nu(+q)) R_-(\nu(-q)).$$

Here $C_\mathbf{P}(\{\hat{\mu}_p\}; \{\hat{\mu}_h\})$ is a finite amplitude, which is common for all form factors of the **P** class, the factors $R_\pm(\nu(\pm q))$ describe the dressing at the right and left Fermi boundaries respectively. They have the following form [27, 28]

$$R_\pm(\nu) = \sum_{n=0}^{\infty} \frac{1}{(n!)^2} \sum_{\substack{p_1,\ldots,p_n=1 \\ h_1,\ldots,h_n=1}}^{\infty} \left(\det_n \frac{1}{p_j + h_k - 1} \right)^2 \prod_{k=1}^{n} z^{p_k+h_k-1}$$

$$\left(\frac{\sin \pi \nu}{\pi} \right)^{2n} \prod_{k=1}^{n} \frac{\Gamma^2(p_k \pm \nu)\Gamma^2(h_k \mp \nu)}{\Gamma^2(p_k)\Gamma^2(h_k)}. \qquad (15)$$

where $z = e^{-2\pi i(2qt \mp x)/L}$. In the model of impenetrable bosons $\nu(\pm q) = \pm 1/2$. The series (15) is absolutely convergent if $|z| < 1$, therefore one should understand it as the limit $t \to t - i0$.

It is remarkable that the multiple series (15) can be summed up explicitly [27, 29–31]. The result reads

$$R_\pm(\nu) = (1-z)^{-\nu^2}.$$

Substituting here $z = e^{-2\pi i(2qt \mp x)/L}$ we obtain for the dressed form form factor

$$|\mathcal{F}^{\phi}_{\mathbf{P}}|^2 = \frac{C_{\mathbf{P}}(\{\hat{\mu}_p\}; \{\hat{\mu}_h\})}{L^{1/2}\left(1 - e^{-2\pi i(2qt-x)/L}\right)^{1/4}\left(1 - e^{-2\pi i(2qt+x)/L}\right)^{1/4}}.$$

We see that now we can easily proceed to the thermodynamic limit $L \to \infty$:

$$|\mathcal{F}^{\phi}_{\mathbf{P}}|^2 = \frac{C_{\mathbf{P}}(\{\hat{\mu}_p\}; \{\hat{\mu}_h\})}{\sqrt{2\pi}(x - 2qt)^{1/4}(x + 2qt)^{1/4}}, \qquad L \to \infty.$$

Thus, as we have claimed above, the dressed form factor has a finite thermodynamic limit.

Conclusion

All the properties of the form factors, as described above, persist in more complex critical models, in particular in the model of one-dimensional Bose-gas at finite coupling constant. In this case one can not solve the Bethe equations explicitly. However, the description of excited states and form factors in terms of particles and holes is still possible. Division of the excited states into classes is exactly the same as for the impenetrable bosons. Summation of the form factors within the same class gives dressed form factor, which has a well-defined thermodynamic limit. Dressing comes with the formula (15). The only difference from the model of impenetrable bosons is that there are new constants in (15). These constants can be found from a system of linear integral equations. For example, $\nu(\pm q)$ makes sense of the shift function on the Fermi boundary. The argument z also changes to $z = e^{-2\pi i(vt \mp x)/L}$, where v is the speed of sound in the gas at the Fermi boundary. Therefore dressed form factors at finite coupling constant have the following form:

$$|\mathcal{F}^{\phi}_{\mathbf{P}}|^2 = \frac{C_{\mathbf{P}}(\{\hat{\mu}_p\}; \{\hat{\mu}_h\})}{(x - vt)^{\nu^2(q)}(x + vt)^{\nu^2(-q)}},$$

where $C_{\mathbf{P}}$ is a finite amplitude, which is common for all form factors of $\mathbf{P}$ class.

In order to calculate correlation functions one should integrate dressed form factors over rapidities of particles and holes separated from the Fermi boundary. Generically analytical evaluation of these integrals is hardly possible because of very complicated dependence of $C_{\mathbf{P}}$ on $\{\hat{\mu}_p\}$ and $\{\hat{\mu}_h\}$. However, in many important cases the integrals over $\{\hat{\mu}_p\}$ and $\{\hat{\mu}_h\}$ can be simplified. In particular, in the asymptotic regime ($x \to \infty$, $t \to \infty$) the integrals over particle/hole rapidities are localized in the vicinities of the Fermi boundaries $\pm q$ and in the vicinity of the saddle point (if the last one exists). In this case the amplitude $C_{\mathbf{P}}$ actually can be treated as a constant. Then, using the form factor approach we reproduce the conformal field theory prediction for the asymptotics of the correlation function [32–34] and some additional contributions coming from the saddle point [28]. The last ones appear to be dominant for certain correlation functions.

Finally, there exists always a possibility to compute the integrals over particle/hole rapidities numerically. This was done in various works [35–38].

Acknowledgment

The author thanks the conference organizers for the opportunity to present this work, N. Kitanine, K. Kozlowskii, J.M. Maillet, V. Terras for a long and enjoyable collaboration.

References

[1] H. Bethe, Zeitschrift für Physik, **71** (1931) 205.

[2] R. Orbach, Phys. Rev., **112** (1958) 309.

[3] L.R. Walker, Phys. Rev., **116** (1959) 1089.

[4] M. Gaudin, *La Fonction d'Onde de Bethe*, Paris: Masson, 1983.

[5] L.D. Faddeev, E.K. Sklyanin and L.A. Takhtajan, Theor. Math. Phys. **40** (1980) 688.

[6] L.A. Takhtajan and L.D. Faddeev, Russ. Math. Surveys. **34** (1979) 11.

[7] L.D. Faddeev, Les Houches 1982, *Recent advances in field theory and statistical mechanics*, edited by J.B. Zuber and R. Stora, Elsevier Science Publ., 1984, 561.

[8] F.A. Smirnov, Form Factors in Completely Integrable Models of Quantum Field Theory (World Scientific, Singapore, 1992).

[9] J.L. Cardy, G. Mussardo, Nucl. Phys. B 340 (1990) 387.

[10] A. Koubek and G. Mussardo, Phys. Lett. B 311 (1993) 193.

[11] M. Jimbo, K. Miki, T. Miwa, A. Nakayashiki, Phys. Lett. A **168** (1992) 256.

[12] N.A. Slavnov, Theor. Math. Phys. **82** (1990) 273.

[13] T. Kojima, V.E. Korepin and N.A. Slavnov, Comm. Math. Phys. **188** (1997) 657.

[14] N. Kitanine, J.M. Maillet and V. Terras, Nucl. Phys. B **554** (1999) 647.

[15] F. Lesage, H. Saleur and S. Skorik, Nucl. Phys. B **474** (1996) 602.

[16] F. Lesage and H. Saleur, J. Phys. A: Math. Gen. **30** (1997) L457.

[17] A. Koutouza, F. Lesage and H. Saleur, Phys. Rev. B **68** (2003) 115422.

[18] E. Lieb, T. Shultz and D. Mattis, Ann. Phys. **16** (1961) 407.

[19] E.H. Lieb, W. Liniger, Phys. Rev. **130** (1963) 1605.

[20] E.H. Lieb, Phys. Rev. **130** (1963) 1616.

[21] E. Lieb and D. Mattis (eds.), *Mathematical Physics in One Dimension*, New York: Academic Press, 1966.

[22] N. Kitanine, K.K. Kozlowski, J.M. Maillet, N.A. Slavnov and V. Terras, J. Stat. Mech. (2011) P05028.

[23] N. Kitanine, J.M. Maillet, N.A. Slavnov and V. Terras, Nucl. Phys. B **641** (2002) 487.

[24] N. Kitanine, K.K. Kozlowski, J.M. Maillet, N.A. Slavnov and V. Terras, J. Stat. Mech. (2007) P01022.

[25] N. Kitanine, K.K. Kozlowski, J.M. Maillet, N.A. Slavnov and V. Terras, J. Math. Phys. **50** (2009) 095209.

[26] V.E. Korepin, N.A. Slavnov, Commun. Math. Phys. **129** (1990) 103.

[27] N. Kitanine, K.K. Kozlowski, J.M. Maillet, N.A. Slavnov and V. Terras, J. Stat. Mech. (2011) P12010.

[28] N. Kitanine, K.K. Kozlowski, J.M. Maillet, N.A. Slavnov and V. Terras, J. Stat. Mech. (2012) P09001.

[29] S. Kerov, G. Olshanski and A. Vershik, Comptes Rend. Acad. Sci. Paris, Ser. I **316** (1993) 773.

[30] A. Borodin, and G. Olshanski, Electron. J. Combin. **7** (2000) R28.

[31] A. Borodin and G. Olshanski, Comm. Math. Phys. **211** (2000) 335.

[32] A.A. Belavin, A.M. Polyakov and A.B. Zamolodchikov, Nucl. Phys. B **241** (1984) 333. Phys. Rev. Lett. **56** (1986) 742.

[33] J.L. Cardy, J. Phys. A: Math. Gen. **17** (1984) L385.

[34] J.L. Cardy, J. Phys. A: Math. Gen. **17** (1984) L385.

[35] J.S. Caux and J.M. Maillet, Phys. Rev. Lett. **95** (2005) 077201.

[36] J.S. Caux, R. Hagemans and J.M. Maillet, J. Stat. Mech. (2005) P09003.

[37] R.G. Pereira, J. Sirker, J.S. Caux, R. Hagemans, J.M. Maillet, S.R. White and I. Affleck, Phys. Rev. Lett. **96** (2006) 257202.

[38] J.S. Caux, P. Calabrese and N.A. Slavnov, J. Stat. Mech. (2007) P01008.

N.A. Slavnov
Steklov Mathematical Institute
8 Gubkina str.
Moscow 119991, Russia
e-mail: nslavnov@mi.ras.ru

Geometric Methods in Physics. XXXI Workshop 2012
Trends in Mathematics, 221–228
© 2013 Springer Basel

On Space-like Hypersurfaces in a Space-time

Sergey Stepanov and Josef Mikeš

Abstract. In the present paper we study the global geometry of convex, totally umbilical and maximal space-like hypersurfaces in space-times and, in particular, in de Sitter space-times.

Mathematics Subject Classification (2010). Primary 53C50; Secondary 83C75.

Keywords. Space-times, de Sitter space, space-like hypersurface.

1. Introduction

Space-like hypersurfaces with special second fundamental forms have been an important tool in the study of space-times (see the proof of the positive mass theorem in [1] and the analysis of Cauchy problem for the Einstein equation in [2]).

In the present paper we study the global geometry of convex, totally umbilical and maximal space-like hypersurfaces in space-times and, in particular, in de Sitter space-times.

2. Definitions and notations

2.1 Let (M, g) be an $(n + 1)$-dimensional $(n \geq 3)$ Lorentzian manifold with the Levi–Civita connection ∇. If we denote by $\Lambda^2 M$ the vector bundle of bivectors over M with its pseudo-Riemannian metric $\langle \, , \, \rangle$, then we can define on (M, g) a symmetric curvature operator $\mathcal{R} \colon \Lambda^2 M \to \Lambda^2 M$ of (M, g) by the equality $\langle \mathcal{R}(X_x \wedge Y_x), Z_x \wedge W_x \rangle := g(R(X_x, Y_x), W_x, Z_x)$ for the curvature tensor R of (M, g) and any $X_x, Y_x, Z_x, W_x \in T_x M$.

Let (M, g) be an $(n + 1)$-dimensional Lorentzian manifold of a non-zero constant curvature c. If $c > 0$, then (M, g) is called a *de Sitter space* $S_1^{n+1}(c)$ and here $\mathcal{R}$ is positive and if $c < 0$, then (M, g) is called an *anti-de Sitter space* $H_1^{n+1}(c)$ and here $\mathcal{R}$ is negative.

2.2 Let M' be an n-dimensional $(n \geq 3)$ connected C^∞-manifold in M with imbedding map $f\colon M' \to M$. We call the image $f(M')$ a *hypersurface* in M and identify it with M'.

Consider a neighborhood U with a local coordinate system $\{y^1, \ldots, y^{n+1}\}$ on M and a neighborhood U' with a local coordinate system $\{x^1, \ldots, x^n\}$ on M' such that $f(U') \subset U$.

In the local coordinates, the imbedding map f is given by $y^\alpha = y^\alpha(x^i)$ where $i, j, k, \ldots = 1, 2, \ldots, n$ and $\alpha, \beta, \gamma, \ldots = 1, 2, \ldots, n+1$.

The differential df of the imbedding map $f\colon M' \to M$ will be denoted by f_*, so that a vector field X' in TM' corresponds to a vector field f_*X' in TM. Thus if X' has local expression $X' = X'^i\, \partial/\partial x^i$, then f_*X' has the local expression $f_*X' = f_i^\alpha X'^i\, \partial/\partial y^\alpha$, where $f_i^\alpha = \partial y^\alpha/\partial x^i$.

We denote by $g' = g(f_*, f_*)$ the metric tensor $g' = f^*g$ induced in M' from g by f, where f^* is the mapping conjugate to f_*. The hypersurface M' is called *space-like* if the metric tensor g' is positive definite. We denote by ∇' the covariant differential operator corresponding to the Riemannian metric g'. Then the *second fundamental form Q of the submanifold* (M', g') is defined by the formula (see [3])

$$(\nabla'_{X'} f_*)\, Y' = \nabla_{f_*X'} f_*Y' - f_*(\nabla'_{X'}Y') = Q(X', Y')$$

for any $X', Y' \in C^\infty TM'$. The Gauss curvature equation for (M', g') has the form

$$\begin{aligned}
R'(X', Y', V', W') = \ & R'(f_*X', f_*Y', f_*V', f_*W') \\
& + g(Q(X', W'), Q(Y', V')) \\
& - g(Q(Y', W'), Q(X', V'))
\end{aligned} \tag{1}$$

for all $X', Y', Z', W' \in C^\infty TM'$. Here R' is the curvature tensor of (M', g').

The mean curvature vector H is given by $n \cdot H = \mathrm{trace}_{g'} Q$. The hypersurface (M', g') is *totally geodesic* in (M, g) if $Q = 0$, and *maximal* if $H = 0$ (in contrast to the Riemannian case). If $Q = g' \cdot H$, then (M', g') is *totally umbilical*.

2.3 Let a continuous everywhere non-vanishing unit time-like vector field ξ be given on (M, g). It means that ξ is defining the *future direction* at each point of (M, g). In this case, we call (M, g) the $(n+1)$-dimensional *space-time* (see [3]).

A space-time (M, g) is said to satisfy the *strong energy condition* (the time-like convergence condition in Hawking–Ellis [4]) if $\mathrm{Ric}(\xi, \xi) \geq 0$ for all time-like future-directed vectors fields ξ.

Let (M', g') be a space-like hypersurface in an $(n+1)$-space-time (M, g). We will denote by $\mathcal{N}$ the (globally defined) unitary time-like vector field $\mathcal{N} \in (TM')^\perp$ in the same time-orientation as ξ, that is $g(\xi, \mathcal{N}) = -1$ at any point $x \in M'$, and we will say that (M', g') is *time oriented by $\mathcal{N}$* (which is also called as a *time oriented space-like hypersurface*).

3. Convex space-like hypersurfaces in a space-time

3.1 For a space-like hypersurface (M', g') in a space-time (M, g) we have

$$Q(X', Y) = h(X', Y') \cdot \mathcal{N}.$$

Then we can rewrite (1) in the new form

$$R'(X', Y', V', W') = (f^* R)(X', Y', V', W') \\ - (h(X', W')h(Y', V') - h(Y', W')h(X', V')), \qquad (2)$$

where we denote by $f^* R$ the tensor induced in M' from R by f. The new tensor $f^* R$ has the all symmetry properties of the curvature tensor R'.

Let the curvature operator $\mathcal{R}$ of (M, g) satisfy the following inequality

$$\langle \mathcal{R}(f^* \theta), f^* \theta \rangle \geq s^* \cdot \langle f^* \theta, f^* \theta \rangle \qquad (3)$$

for any $\theta \in \Lambda^2 M'$, some $\delta > 0$ and $s^* = \frac{1}{n(n-1)} \sum_{i<j} (f^* R)(X_i, X_j, X_j, X_i)$. Then from (2) we obtain

$$\langle \mathcal{R}'(\theta), \theta \rangle' = \langle \mathcal{R}(f^* \theta), f^* \theta \rangle - 2 \sum_{i<j} \lambda_i \lambda_j (\theta^{ij})^2$$

$$\geq \delta s^* \cdot \langle f^* \theta, f^* \theta \rangle - 2 \sum_{i<j} \lambda_i \lambda_j (\theta^{ij})^2$$

$$= \delta s' \cdot \langle \theta, \theta \rangle' + 2 \delta \sum_{i<j} \lambda_i \lambda_j \langle \theta, \theta \rangle' - 2 \sum_{i<j} \lambda_i \lambda_j (\theta^{ij})^2, \qquad (4)$$

where $\theta = \Sigma_{i<j} \theta^{ij} X_i' \wedge X_j'$ and $s^* = s' + 2\Sigma_{i<j} \lambda_i \lambda_j$ for a local orthonormal basis $\{X_1', \ldots, X_n'\}$ of vector fields on (M', g') such that $h(X_i', X_j') = \lambda_i \delta_{ij}$ where δ_{ij} is the Kronecker delta.

If (M', g') is a time oriented hypersurface by $\mathcal{N} \in (TM')^\perp$, then its second fundamental form $Q(X', Y') = h(X', Y') \cdot \mathcal{N}$ is said to be *semi-definite* at an arbitrary point $x \in M'$ if $h(X_x', X_x') \geq 0$ or $h(X_x', X_x') \leq 0$ for all non-zero vectors $X_x \in T_x M'$. It is well known that if (M', g') is *convex* at $x \in M'$, then the h is semi-definite at x. It is obvious that $\lambda_i \lambda_j \geq 0$ if (M', g') is convex at each point $x \in M'$. In this case we obtain from (4)

$$\langle \mathcal{R}'(\theta), \theta \rangle' \geq \delta s' \cdot \langle \theta, \theta \rangle' + 4\delta \sum_{i<j} \lambda_i \lambda_j \cdot \sum_{k<l} (\theta^{kl})^2 - 2 \sum_{i<j} \lambda_i \lambda_j (\theta^{ij})^2 \geq s' \cdot \langle \theta, \theta \rangle'. \quad (5)$$

On the other hand, in [5] was showed that a complete Riemannian manifold (M', g') with $n \geq 3$ must be compact if its curvature operator $\mathcal{R}'$ satisfies the following inequalities

$$\langle \mathcal{R}'(\theta), \theta \rangle' \geq \delta \cdot \frac{s'}{n(n-1)} \cdot \langle \theta, \theta \rangle > 0 \qquad (6)$$

for some $\delta > 0$. Hence, if the equalities (3) and $s' > 0$ hold, then from the above result (for $\delta^{-1} = 2n(n-1)$) we conclude that complete and convex hypersurface (M', g') must be compact. In this case, from (5) we conclude that $\mathcal{R}'$ is positive definite and (M', g') is diffeomorphic to a spherical space form (see [6]).

Here we recall that (M', g') is a *spherical space form* if M' is a compact manifold which admits another metric of positive constant sectional curvature; these manifolds have been classified by Wolf (see [7]). Summarizing, we have

Theorem 1. *Let (M', g') be a space-like convex and complete hypersurface with positive scalar curvature in an $(n+1)$-dimensional $(n \geq 3)$ space-time (M, g). If the curvature operator $\mathcal{R}$ of (M, g) satisfies the following inequalities $\langle \mathcal{R}(f^*\theta), f^*\theta \rangle \geq \delta s^* \cdot \langle f^*\theta, f^*\theta \rangle > 0$ for an arbitrary $\theta \in \Lambda^2 M'$ and some $\delta > 0$, then (M', g') is diffeomorphic to a spherical space form.*

Topologically, it is known, if the manifold (M', g') is simply-connected and its dimension n is even then (M', g') is diffeomorphic to the sphere. Then we have

Corollary 1. *Let (M', g') be a space-like convex and complete hypersurface with positive scalar curvature in a space-time (M, g) of odd dimension. If $\mathcal{R}$ of (M, g) satisfies the following inequalities $\langle \mathcal{R}(f^*\theta), f^*\theta \rangle \geq \delta s^* \cdot \langle f^*\theta, f^*\theta \rangle$ for an arbitrary $\theta \in \Lambda^2 M'$ and some $\delta > 0$, then (M', g') is diffeomorphic to a sphere.*

From (2) we have $\sec'(e_i \wedge e_j) = \sec'(f_* e_i \wedge f_* e_j) - \lambda_i \lambda_j$ for any $i < j$ and an orthonormal basis $\{e_1, \ldots, e_n\}$ in $T_x M'$ at an arbitrary $x \in M'$ such that $h(e_i, e_j) = \lambda_i \delta_{ij}$ where δ_{ij} is the Kronecker delta. Let (M', g') be a convex and compact hypersurface in (M, g) then $\lambda_i \lambda_j \geq 0$ and hence there exists $\lambda^2 = \max\limits_{x \in M'} \{\lambda_i(x) \lambda_j(x)\}$ for all $i, j = 1, \ldots, n$.

It is well known, that in three-dimensional compact simply-connected manifold with positive sectional curvatures is diffeomorphic to the sphere (*sphere theorem*). Therefore if $\dim M' = 3$ and $\sec(\pi) > \lambda^2$ for any 2-space π in $T_x M'$ at each point $x \in M$ then (M', g') must be diffeomorphic to the sphere. Summarizing, we formulate the following corollary.

Corollary 2. *Let (M', g') be a space-like convex and compact hypersurface in a four-dimensional space-time (M, g). If the sectional curvature of (M, g) satisfies the inequality $\sec(\pi) > \lambda^2$ for any 2-space π in $T_x M'$ at each point $x \in M$ and $\lambda^2 = \max\limits_{x \in M'} \{\lambda_i(x) \lambda_j(x)\}$ for all eigenvalues λ_i of the second fundamental form Q of the hypersurface (M', g') then (M', g') is diffeomorphic to a sphere.*

3.2 Let (M', g') be a space-like convex hypersurface in a de Sitter space-time $S_1^{n+1}(c)$. If we assume that $c > \lambda_i \lambda_j$ for all eigenvalues λ_i of the second fundamental form Q of (M', g') and $i, j = 1, \ldots, n$ then we can rewrite (6) in the following form

$$\mathcal{R}'(\theta), \theta\rangle' = \langle \mathcal{R}(f^*\theta), f^*\theta \rangle - 2\sum_{i<j} \lambda_i \lambda_j (\theta^{ij})^2$$

$$= c \cdot \langle f^*\theta, f^*\theta \rangle - 2\sum_{i<j} \lambda_i \lambda_j (\theta^{ij})^2 = 2c\sum_{k<l}(\theta^{kl})^2 - 2\sum_{i<j} \lambda_i \lambda_j (\theta^{ij})^2$$

It is clear that the following corollary holds.

Corollary 3. *Let (M', g') be a space-like convex and compact hypersurface in a de Sitter space-time $S_1^{n+1}(c)$. If $c > \lambda^2 = \max_{x \in M'}\{\lambda_i(x), \lambda_j(x)\}$ for all eigenvalues λ_i of the second fundamental form Q of the hypersurface (M', g') then (M', g') is diffeomorphic to a spherical space form.*

From (2) we obtain $s' = s + 2\,\mathrm{Ric}(\mathcal{N}, \mathcal{N}) - 2\sum_{i<j}\lambda_i\lambda_j$ where Ric is the Ricci curvature of (M, g); s and s' are the scalar curvatures of (M, g) and (M', g'), respectively. If (M', g') is a space-like convex hypersurface in a de Sitter space-time $S_1^{n+1}(c)$ then we can rewrite the equality in the following form $s' - n(n-1) \cdot c = -2\sum_{i<j}\lambda_i\lambda_j \leq 0$ because in this case we have $\mathrm{Ric}(\mathcal{N}, \mathcal{N}) = n \cdot c \cdot g(\mathcal{N}, \mathcal{N}) = -n \cdot c$ and $s = n(n+1) \cdot c$. Then the following proposition is true.

Proposition 1. *Let (M', g') be a space-like convex hypersurface in a de Sitter space-time $S_1^{n+1}(c)$ then its scalar curvature satisfies the following inequality $s' \leq n(n-1) \cdot c$. The equality attained on a totally geodesic hypersurface.*

Remark 1. It is obvious, that a space-like convex hypersurface (M', g') in an anti-de Sitter space-time $H_1^{n+1}(c)$ has a negative definite scalar curvature, i.e., $s' < 0$.

4. Totally umbilical hypersurfaces in a space-time

4.1 Let (M', g') be a n-dimensional space-like totally umbilical hypersurface, then we rewrite the Gauss curvature equation (1) in the following form

$$R'(X', Y', V', W') = (f^*R)(X', Y', V', W')$$
$$- |H^2|\,(g'(X', W')g'(Y', V') - g'(Y', W')g'(X', V')) \quad (7)$$

for all $X', Y', V', W' \in C^\infty TM'$. From (7) we have $\sec'(\pi) = \sec(\pi) - |H^2|$ for any 2-space π in $T_x M'$ at each point $x \in M$. By the "sphere theorem" we have

Proposition 2. *Let (M', g') be a space-like compact totally umbilical hypersurface in a four or three-dimensional space-time (M, g). If the sectional curvature of (M, g) satisfies the inequality $\sec(\pi) > |H^2|$ for any 2-space π in $T_x M'$ at each point $x \in M$, then (M', g') is diffeomorphic to a sphere.*

4.2 Let (M', g') be a space-like totally umbilical hypersurface in $S_1^{n+1}(c)$, then we can rewrite the Gauss curvature equation (7) in the following form

$$R('X', Y', V', W') = (c - |H^2|) \cdot (g'(X', W')g'(Y', V') - g'(Y', W')g'(X', V')). \quad (8)$$

By (8) we conclude that (M', g') is a Riemannian manifold of constant curvature $c' = c - |H^2|$. On the other hand, from the above equality we obtain $\langle \mathcal{R}'(\theta), \theta \rangle' = \frac{s'}{n(n-1)} \cdot \langle \theta, \theta \rangle'$ for any $\theta \in \Lambda^2 M'$, the Riemannian metric $\langle\ ,\ \rangle'$ on $\Lambda^2 M'$ and $s' = n(n-1)(c - |H^2|)$. In this case, if (M', g') is complete and $|H^2| < c$ then (M', g') is necessarily compact (see [5]). In addition, if we assume that M' is simple connected then (M', g') must be isometric to the sphere S^n, equipped with its standard metric (see [8]; [9]). We have

Proposition 3. *Let (M', g') be a space-like totally umbilical hypersurface in $S_1^{n+1}(c)$ then it is a Riemannian manifold of constant curvature $c' = c - |H^2|$. If (M', g') is complete, simple connected and $|H^2| < c$ then it is isometric to a sphere.*

Ramanathan proved (see [10]) that if the mean curvature of a space-like complete surface in $S_1^3(c)$ is constant and satisfies the inequality $|H^2| < c$, then the surface is totally umbilical. From Ramanathan's result and our Proposition 3 we obtain

Corollary 4. *Let (M', g') be a space-like complete simple connected surface with constant mean curvature in $S_1^3(c)$. If $|H^2| < c$ then it is isometric to a sphere.*

In [11] it was shown that a compact space-like hypersurface with constant mean curvature is totally umbilical in the de Sitter space $S_1^{n+1}(c)$. From this result and our Proposition 3 we have

Corollary 5. *Let (M', g') be a space-like compact simple connected hypersurface with constant mean curvature in $S_1^{n+1}(c)$. If $|H^2| < c$ then it is isometric to a sphere.*

5. Maximal space-like hypersurfaces of a space-time

5.1 Let (M', g') be a space-like maximal hypersurface is a space-time (M, g). Then from (2) we obtain

$$s' = s + 2\operatorname{Ric}(\mathcal{N}, \mathcal{N}) + \sum_{i=1,\cdots,n} (\lambda_i)^2. \qquad (9)$$

Therefore the following theorem holds.

Theorem 2. *If a space-time (M, g) satisfies the strong energy condition and there exists a space-like, time oriented and maximal hypersurface (M', g') then $s' \geq s$ for the scalar curvatures s' and s of (M', g') and (M, g), respectively. The equality is attained on a totally geodesic hypersurface.*

If (M', g') is a space-like maximal hypersurface in $S_1^{n+1}(c)$, then we can rewritten (9) in the following form $\quad s' - n(n-1) \cdot c = \sum_{i=1,\cdots,n} (\lambda_i)^2$. Then the following proposition holds.

Proposition 4. *Let (M', g') be a space-like maximal hypersurface in $S_1^{n+1}(c)$ then its scalar curvature satisfies the following inequality $s' \geq n(n-1) \cdot c$. The equality is attained on a totally geodesic hypersurface.*

The tensor Ricci Ric' of a space-like hypersurface (M', g') in a de Sitter space-time $S_1^{n+1}(c)$ has the form $\operatorname{Ric}'(X', Y') = (n-1) \cdot c \cdot g'(X', Y') - (\operatorname{trace}_{g'} h \cdot h(X', Y') - h^2(X', Y'))$. Then from this equality for the maximal hypersurface (M', g') in $S_1^{n+1}(c)$ we have

$$\operatorname{Ric}'(X', Y') = (n-1) \cdot c \cdot g'(X', Y') + \lambda_i^2 (X^i)^2 \geq (n-1) \cdot c \cdot g'(X', X') > 0, \quad (10)$$

where X^i are local components of an arbitrary non-zero vector field $X' \in TM'$. We recall that (10) is a necessary condition for a complete Riemannian manifold to be compact (see [12]). This proves the following

Proposition 5. *If (M', g') is a space-like, complete maximal hypersurface in a de Sitter space-time $S_1^{n+1}(c)$ then (M', g') is compact.*

Hamilton proved in [13] that any compact and simply connected three-dimensional Riemannian manifold (M', g') with positive Ricci curvature Ric$'$ is diffeomorphic to a Euclidean sphere S^3. As a corollary of our proposition and Hamilton's result we obtain

Theorem 3. *If (M', g') is a space-like, complete, simply connected and maximal hypersurface in a de Sitter space-time $S_1^4(c)$ then (M', g') is diffeomorphic to a Euclidean sphere S^3.*

Moreover, in [11] was proved the following: Let (M', g') be a space-like compact hypersurface in the de Sitter space $S_1^{n+1}(c)$, $n \geq 2$, then (M', g') is diffeomorphic to a Euclidean sphere S^n (see also [14] and [15]). Therefore, we can formulate a theorem which is a generalization of Theorem 3.

Theorem 4. *If (M', g') is a space like, complete, simply connected and maximal hypersurface in a de Sitter space-time then (M', g') is diffeomorphic to a Euclidean sphere S^n.*

Acknowledgment

The paper was supported by grant P201/11/0356 of The Czech Science Foundation.

References

[1] R. Schoen, S.T. Yau, *Proof of the positive mass theorem I.* Comm. Math. Phys. **65** (1979), 45–76.

[2] Y. Choquet-Bruhat, J. York, *The Cauchy problem.* General Relativity and Gravitation **1**, Plenum Press, 1980.

[3] J. Beem, P. Ehrlich, *Global Lorentzian Geometry.* Marcel Dekker, NY, 1981.

[4] S.W. Hawking, G.F. Ellis, *The large scale structure of space-time.* Cambridge University Press, United Kingdom, 1973.

[5] L. Ni, B. Wu, *Complete manifolds with nonnegative curvature operator.* Proc. Amer. Math. Soc. **135** (2007), 3021–3028.

[6] C. Bohm, B. Wilkig, *Manifolds with positive curvature operators are space forms.* Annals of Mathematics **167** (2008), 1079–1097.

[7] J.A. Wolf, *Space of constant curvature.* McGraw-Hill, NY, 1967.

[8] H. Hopf, *Zum Clifford-Kleinschen Raumproblem.* Math. Ann. **95** (1926), 313–339.

[9] H. Hopf, *Differentialgeometrie und topologische Gestalt.* Jahresber. Deutsch. Math.-Verein. **41** (1932), 209–229.

[10] J. Ramanathan, *Complete space-like hypersurface of constant mean curvature in the de Sitter space.* Indiana Univ. Math. J. **36** (1987), 349–359.

[11] S. Montiel, *An integral inequality for compact space-like hypersurface in the de Sitter space and applications to the case of constant mean curvature.* Indiana Univ. Math. J. 37 (1988), 909–917.

[12] D. Gromoll, W. Meyer, *On complete open manifolds of positive curvature.* Ann. of Math. **90** (1969), 75–90.

[13] R.S. Hamilton, *Three-manifold with positive Ricci curvature.* J. Diff. Geom. **17** (1982), 255–306.

[14] J.A. Aledo, L.J. Alías, *On the volume and the Gauss map image of spacelike hypersurfaces in de Sitter space.* Proc. Amer. Math. Soc. **130** (2002), 1145–1151.

[15] J. Lv, *Compact space-like hypersurfaces in de Sitter space.* Int. J. Math. and Math. Sci. **13** (2005), 2053–2069.

Sergey Stepanov
Department of Mathematics
Finance University under the Government of Russian Federation
Moscow, Russia
e-mail: `s.e.stepanov@mail.ru`

Josef Mikeš
Department Algebra and Geometry
Palacky University
17. listopadu 27
CZ-77146 Olomouc, Czech Republic
e-mail: `josef.mikes@upol.cz`

Geometric Methods in Physics. XXXI Workshop 2012
Trends in Mathematics, 229–237
© 2013 Springer Basel

On Bogomolny Decompositions
for the Baby Skyrme Models

Ł.T. Stępień

Abstract. We derive the Bogomolny decompositions (Bogomolny equations) for the baby Skyrme models: restricted (also called, as extreme or pure baby Skyrme model), and full one, in $(2+0)$ dimensions, by using so-called, concept of strong necessary conditions.

Mathematics Subject Classification (2010). 35Q51, 37K10, 37K40.

Keywords. Bogomolny equations, Bogomol'nyi equations, baby Skyrme model.

1. Introduction

The baby Skyrme model appeared for the first time as an analogical model (on plane) to the Skyrme model in three-dimensional space (introduced in [1], provides good description of low-energy physics of strong interactions [2]). The target space of baby Skyrme model is S^2. In both these models static field configurations can be classified topologically by their winding numbers. The baby Skyrme model includes the terms, being the analogons to the terms of Skyrme model. Although the presence of potential is necessary for existence of static solutions with finite energy in baby Skyrme model, the form of this potential is not restricted and different forms of the potential were investigated in [3, 4], some recent results are, among others in [5–7]. This model can be applied for the description of the quantum Hall effect [8]. In [9] noncommutative baby Skyrmions were studied. In [10] spinning baby Skyrmions in a restricted baby Skyrme model were investigated. The energy functional of the full baby Skyrme model in $(2 + 0)$ dimensions, has the form [5]:

$$H = \frac{1}{2} \int d^2x \left(\partial_i \vec{S} \cdot \partial_i \vec{S} + \frac{1}{4} (\epsilon_{ij} \partial_i \vec{S} \times \partial_j \vec{S})^2 + \gamma^2 V(\vec{S}) \right), \tag{1}$$

where $\vec{S}$ is a three-component vector field, such that $\mid \vec{S} \mid^2 = 1$ and $\gamma > 0$. However, on the other hand, baby Skyrme model is still complicated, non-integrable, topologically non-trivial and nonlinear field theory. We can simplify the problem of solving of the Euler–Lagrange equations of this model, for example, by deriving Bogomolny equations (sometimes called as Bogomol'nyi equations) for the model, mentioned above. All the solutions of Bogomolny equations satisfy the Euler–Lagrange equations, which order is bigger than the order of the Bogomolny equations.

In this paper we derive Bogomolny equations (we call them the *Bogomolny decomposition*) for the baby Skyrme models: restricted and full one, in $(2+0)$ dimensions. In the first one, the $O(3)$ term in (1) is absent. The Bogomolny equations for ungauged restricted baby Skyrme model in $(2+0)$ dimensions, but for some special class of the potentials, were derived in [11] and [5] (some more general results were obtained in [6]), by using the technique, for the first time applied by Bogomolny in [12], among others, for the non abelian gauge theory (although historically earlier, it was applied in [13]). We will call this technique the traditional one. In [14], so-called improved Bogomolny bound for the full baby Skyrme model was found. In contrary to the papers, where Bogomolny equations were derived for Skyrme-like models by using traditional technique, in this paper we derive Bogomolny equations by applying the so-called concept of strong necessary conditions (we call it here shortly, as CSNC), presented for the first time in [15] and developed in [16]. This paper is organized, as follows. In Section 2 we briefly describe the baby Skyrme models, mentioned above, and the concept of strong necessary conditions. In Section 3, we derive the Bogomolny decompositions for the baby Skyrme models, mentioned above, by using the CSNC. Section 4 contains some conclusions.

2. Skyrme models in $(2+0)$ dimensions

2.1. Restricted baby Skyrme model

In this paper we consider the energy functional for restricted baby Skyrme model in $(2+0)$ dimensions of the following form [5] (we introduce here the real constant β):

$$H = \frac{1}{2}\int d^2x \mathcal{H} = \frac{1}{2}\int d^2x\left(\frac{\beta}{4}(\epsilon_{ij}\partial_i\vec{S}\times\partial_j\vec{S})^2 + \gamma^2 V(\vec{S})\right),\quad x_1 = x,\ x_2 = y, \quad (2)$$

where $\gamma > 0$ and the potential V depends only on $\vec{S}$ (we assume **nothing** more about the form of the potential V (of course, $V \in \mathcal{C}^1$)). We make the stereographic projection (where ω is twice differentiable field variable):

$$\vec{S} = \left[\frac{\omega + \omega^*}{1 + \omega\omega^*}, \frac{-i(\omega - \omega^*)}{1 + \omega\omega^*}, \frac{1 - \omega\omega^*}{1 + \omega\omega^*}\right],\ \omega = \omega(x,y)\in\mathbb{C}\ ,\ x,y\in\mathbb{R}. \quad (3)$$

Then, the density of energy functional (2) has the form [17]:

$$\mathcal{H} = -4\beta\frac{(\omega_{,x}\omega^*_{,y} - \omega_{,y}\omega^*_{,x})^2}{(1 + \omega\omega^*)^4} + V(\omega,\omega^*),$$

where γ is included in $V(\omega, \omega^*)$ and $\omega_{,x} \equiv \frac{\partial \omega}{\partial x}$, etc. The Euler–Lagrange equations for this model are as follows [17]:

$$16\beta \frac{(\omega_{,x}\omega^*_{,y} - \omega_{,y}\omega^*_{,x})^2 \omega^*}{(1 + \omega\omega^*)^5} + 8\beta \frac{2\omega_{,xy}\omega^*_{,x}\omega^*_{,y} + \omega_{,x}\omega^*_{,x}\omega^*_{,yy} + \omega_{,y}\omega^*_{,y}\omega^*_{,xx}}{(1 + \omega\omega^*)^4}$$

$$- 8\beta \frac{\omega_{,xx}(\omega^*_{,y})^2 + \omega_{,yy}(\omega^*_{,x})^2 + (\omega_{,x}\omega^*_{,y} + \omega_{,y}\omega^*_{,x})\omega^*_{,xy}}{(1 + \omega\omega^*)^4} - V_{,\omega} = 0,$$

and the corresponding equation, obtained by varying the functional with respect to ω^*.

2.2. Full baby Skyrme model

The Hamiltonian, in $(2 + 0)$ dimensions, is given in (1). If we make stereographic projection (3), then the density of the functional of energy is in $(2 + 0)$ dimensions as follows (we include the constant γ in V):

$$\mathcal{H} = 4\alpha \frac{\omega_{,x}\omega^*_{,x} + \omega_{,y}\omega^*_{,y}}{(1 + \omega\omega^*)^2} - 4\beta \frac{(\omega_{,x}\omega^*_{,y} - \omega_{,y}\omega^*_{,x})^2}{(1 + \omega\omega^*)^4} + V(\omega, \omega^*),$$

where we have introduced the real coupling constants α and β. It is convenient to write this energy density in terms of the real field variables u, v $(u, v \in \mathcal{C}^2)$: $\omega = u + iv, \omega^* = u - iv$, and introduce some new constants λ_1, λ_2 [17]:

$$\mathcal{H} = \frac{\lambda_1}{2} \frac{u^2_{,x} + u^2_{,y} + v^2_{,x} + v^2_{,y}}{(1 + u^2 + v^2)^2} + \lambda_2 \frac{(u_{,x}v_{,y} - u_{,y}v_{,x})^2}{(1 + u^2 + v^2)^4} + V(u, v), \qquad (4)$$

where $\lambda_1 = 8\alpha$, $\lambda_2 = 16\beta$. The Euler–Lagrange equations of this model have the following form [17]:

$$\lambda_1 \frac{u_{,xx} + u_{,yy}}{(1 + u^2 + v^2)^2} - 2\lambda_1 \frac{u(u^2_{,x} + u^2_{,y} - v^2_{,x} - v^2_{,y}) + 2v(u_{,x}v_{,x} + u_{,y}v_{,y})}{(1 + u^2 + v^2)^3}$$

$$+ 2\lambda_2 \frac{u_{,xx}v^2_{,y} + u_{,yy}v^2_{,x} + (u_{,x}v_{,y} + u_{,y}v_{,x})v_{,xy}}{(1 + u^2 + v^2)^4}$$

$$- 2\lambda_2 \frac{2u_{,xy}v_{,x}v_{,y} + u_{,x}v_{,x}v_{,yy} + u_{,y}v_{,y}v_{,xx}}{(1 + u^2 + v^2)^4} - 8\lambda_2 \frac{(u_{,x}v_{,y} - u_{,y}v_{,x})^2 u}{(1 + u^2 + v^2)^5} - V_{,u} = 0,$$

and the corresponding equation, obtained by varying the functional with respect to v.

2.3. The concept of strong necessary conditions

The idea of the strong necessary conditions (CSNC) is such that instead of considering of the Euler–Lagrange equations:

$$F_{,u} - \frac{d}{dx}F_{,u_{,x}} - \frac{d}{dt}F_{,u_{,t}} = 0, \qquad (5)$$

following from the extremum principle, applied to the functional:

$$\Phi[u] = \int_{E^2} F(u, u_{,x}, u_{,t}) \, dx dt, \qquad (6)$$

we consider the strong necessary conditions (they generate so-called *dual equations* – the equations of lower order than the order of (5)) [15, 16]:

$$\tilde{F}_{,u} = 0, \ \tilde{F}_{,u_{,t}} = 0, \ \tilde{F}_{,u_{,x}} = 0, \quad \text{where} \quad \tilde{F}_{,u} \equiv \frac{\partial \tilde{F}}{\partial u}, \quad \text{etc.}, \tag{7}$$

applied to the gauged functional $\Phi \to \Phi + \text{Inv} = \int_{E^2} \tilde{F}(u, u_{,x}, u_{,t}) \, dx dt$ (where Inv is such functional that its local variation, with respect to $u(x,t)$, vanishes: $\delta \text{Inv} \equiv 0$). Owing to the fact that all solutions of the system of equations (7) satisfy the Euler–Lagrange equation (5) and owing to the gauge transformation of (6), we have a possibility of finding nontrivial solutions of (5). The procedure of deriving of the Bogomolny decomposition from the extended concept of strong necessary conditions, was presented in [18, 19] and developed in [20].

3. Bogomolny decompositions of baby Skyrme models in (2 + 0) dimensions

3.1. The case of the restricted baby Skyrme model

Now, we will find Bogomolny decomposition for the restricted baby Skyrme model. We make the following gauge transformation [17]:

$$\mathcal{H} \longrightarrow \tilde{\mathcal{H}} = -4\beta \frac{(\omega_{,x}\omega^*_{,y} - \omega_{,y}\omega^*_{,x})^2}{(1 + \omega\omega^*)^4} + V(\omega, \omega^*) + \sum_{k=1}^{3} I_k, \tag{8}$$

where $I_1 = G_1 \cdot (\omega_{,x}\omega^*_{,y} - \omega_{,y}\omega^*_{,x})$, $I_2 = D_x G_2$, $I_3 = D_y G_3$, $D_x \equiv \frac{d}{dx}$, $D_y \equiv \frac{d}{dy}$ and $G_k = G_k(\omega, \omega^*) \in \mathcal{C}^2$, $k = (1,2,3)$ are some functions, which are to be determined. After applying the CSNC to (8), we obtain the dual equations [17]:

$$\tilde{\mathcal{H}}_{,\omega} : 16\beta \frac{(\omega_{,x}\omega^*_{,y} - \omega_{,y}\omega^*_{,x})^2\omega^*}{(1 + \omega\omega^*)^5} + V_{,\omega} + G_{1,w} \cdot (\omega_{,x}\omega^*_{,y} - \omega_{,y}\omega^*_{,x})$$
$$+ D_x G_{2,\omega}(\omega, \omega^*) + D_y G_{3,\omega}(\omega, \omega^*) = 0, \tag{9}$$

$$\tilde{\mathcal{H}}_{,\omega^*} : 16\beta \frac{(\omega_{,x}\omega^*_{,y} - \omega_{,y}\omega^*_{,x})^2\omega}{(1 + \omega\omega^*)^5} + V_{,\omega^*} + G_{1,\omega^*} \cdot (\omega_{,x}\omega^*_{,y} - \omega_{,y}\omega^*_{,x})$$
$$+ D_x G_{2,\omega^*}(\omega, \omega^*) + D_y G_{3,\omega^*}(\omega, \omega^*) = 0, \tag{10}$$

$$\tilde{\mathcal{H}}_{,\omega_{,x}} : N_0\omega^*_{,y} + G_{2,\omega} = 0, \quad \tilde{\mathcal{H}}_{,\omega_{,y}} : -N_0\omega^*_{,x} + G_{3,\omega} = 0, \tag{11}$$

$$\tilde{\mathcal{H}}_{,\omega^*_{,x}} : -N_0\omega_{,y} + G_{2,\omega^*} = 0, \quad \tilde{\mathcal{H}}_{,\omega^*_{,y}} : N_0\omega_{,x} + G_{3,\omega^*} = 0, \tag{12}$$

where $N_0 = -8\beta \frac{(\omega_{,x}\omega^*_{,y} - \omega_{,y}\omega^*_{,x})}{(1+\omega\omega^*)^4} + G_1(\omega, \omega^*)$. Now, we must make the equations (9)–(12) self-consistent. To this end, we must reduce the number of independent equations by an appropriate choice of the functions G_k, $k = (1,2,3)$. Usually, such ansatzes exist only for some special $V(\omega, \omega^*)$, and in most cases of $V(\omega, \omega^*)$ for many nonlinear field models, it is impossible to reduce the system of corresponding dual equations to the Bogomolny equations. However, even then, such system can be used to derive at least some particular set of solutions of the Euler–Lagrange equations. We consider ω, ω^*, G_k, $k = (1, 2, 3)$, as equivalent dependent variables, governed by the system of equations (9)–(12). We make two operations (they were applied for the first time in [18] for the cases of hyperbolic and elliptic systems of nonlinear PDE's). At first, we integrate equations (9), (10), with respect to ω and to ω^*, respectively. We get [17]:

$$- 4\beta \frac{(\omega_{,x}\omega^*_{,y} - \omega_{,y}\omega^*_{,x})^2}{(1 + \omega\omega^*)^4} + V(\omega, \omega^*) + G_1 \cdot (\omega_{,x}\omega^*_{,y} - \omega_{,y}\omega^*_{,x})$$
$$+ D_x G_2(\omega, \omega^*) + D_y G_3(\omega, \omega^*) = F(\omega_{,x}, \omega_{,y}, \omega^*_{,x}, \omega^*_{,y}), \quad (13)$$

where F is some function, which will be determined later.

The second step is to make equations (11)–(12) self-consistent. After properly multiplying equations (11)–(12) by $\omega_{,x}$, $\omega_{,y}$, $\omega^*_{,x}$, $\omega^*_{,y}$, respectively and adding the sides of so obtained equations, we get the relations, including the divergences $D_x G_2(w, w^*)$ and $D_y G_3(w, w^*)$:

$$N_0 \cdot (\omega_{,x}\omega^*_{,y} - \omega_{,y}\omega^*_{,x}) + D_k G_{k+1} = 0, \ (k = 1, 2, \ D_1 \equiv D_x, D_2 \equiv D_y). \quad (14)$$

Hence,

$$D_x G_2(\omega, \omega^*) = D_y G_3(\omega, \omega^*). \quad (15)$$

Moreover, if we multiply again equations (11), (12), by $\omega_{,x}$, $\omega_{,y}$, $\omega^*_{,x}$, $\omega^*_{,y}$ and add the sides, but in such a way as to get the relations, including: $D_y G_2(\omega, \omega^*)$ and $D_x G_3(\omega, \omega^*)$, we get $D_y G_2(\omega, \omega^*) = 0$, $D_x G_3(\omega, \omega^*) = 0$. We call the two last relations and (13)–(14), the *divergence representation* of (9)–(12) (this technique was applied for the first time in [18]). Hence, and from (15)

$$G_2(\omega, \omega^*) = \text{const.}, \quad G_3(\omega, \omega^*) = \text{const.} \quad (16)$$

Then, from relations (14) we have [17]:

$$\omega_{,x}\omega^*_{,y} - \omega_{,y}\omega^*_{,x} = \frac{1}{8\beta} G_1(\omega, \omega^*)(1 + \omega\omega^*)^4. \quad (17)$$

Obviously, all solutions of (17) satisfy equations (11)–(12), when (16) hold. Now, we must investigate, when equation (13) is satisfied by the solutions of (17). Then, we insert (16) and (17), into equation (13) [17]:

$$V(\omega, \omega^*) + \frac{1}{16\beta} G_1^2(\omega, \omega^*)(1 + \omega\omega^*)^4 = F(\omega_{,x}, \omega_{,y}, \omega^*_{,x}, \omega^*_{,y}). \quad (18)$$

Now, in order to determine the function F, we compare (18) with Hamilton–Jacobi equation, which has the form (Ref. [18] and the references therein): $\tilde{\mathcal{H}} = 0$, where, of course $\tilde{\mathcal{H}}$ in general, for $\omega = \omega(x^\mu)$, $\omega^* = \omega^*(x^\mu)$, $\mu = (0, 1, 2, 3)$ and $x^0 = t$, is defined, as follows: $\tilde{\mathcal{H}} = \Pi_\omega \omega_{,t} + \Pi_{\omega^*}\omega^*_{,t} - \tilde{\mathcal{L}}$. Here $\Pi_\omega = \tilde{\mathcal{L}}_{\omega,t}$, $\Pi_{\omega^*} = \tilde{\mathcal{L}}_{\omega^*_{,t}}$ are canonical momenta and $\tilde{\mathcal{L}}$ is Lagrange density gauge-transformed on the invariants I_k, $k = (1, 2, 3)$. In our case $\tilde{\mathcal{H}} = -\tilde{\mathcal{L}}$. By inserting relations (16) and (17) into this equation and taking into account $\tilde{\mathcal{H}} = 0$, we get that: $F = 0$ and: $V(\omega, \omega^*) = -\frac{1}{16\beta} G_1^2(\omega, \omega^*)(1 + \omega\omega^*)^4$. Then, of course, $G_1 = \frac{4i\sqrt{\beta}}{(1+\omega\omega^*)^2} \sqrt{V(\omega, \omega^*)}$. We insert this last equation in (17) and we obtain Bogomolny decomposition for restricted baby Skyrme model in $(2+0)$ dimensions, for the given potential $V(w, w^*)$ [17]:

$$\omega_{,x}\omega^*_{,y} - \omega_{,y}\omega^*_{,x} = \frac{i}{2\sqrt{\beta}}\sqrt{V(\omega, \omega^*)}(1+\omega\omega^*)^2. \tag{19}$$

3.2. The case of full baby Skyrme model

We obtain gauge transformed functional $\tilde{\mathcal{H}}$, by making gauge transformation (analogical to (8)) of the functional (4), but now the invariants are: $G_1(u,v)(u_{,x}v_{,y} - u_{,y}v_{,x})$, $R_1(u,v)(u_{,x}v_{,y} - u_{,y}v_{,x})$, $D_x G_2(u,v)$, $D_y G_3(u,v)$, $D_x R_2(u,v)$, $D_y R_3(u,v)$, where $G_k(u,v)$, $R_k(u,v) \in \mathcal{C}^2$, $k = (1,2,3)$, are some functions, which are to be determined later. After applying the CSNC to such gauged functional, we obtain the dual equations (here: $k = 1,2$ and: $D_1 \equiv D_x$, $D_2 \equiv D_y$) [17]:

$$\tilde{\mathcal{H}}_{,u} : -2\lambda_1 \frac{(u_{,x}^2 + u_{,y}^2 + v_{,x}^2 + v_{,y}^2)u}{(1+u^2+v^2)^3} - 8\lambda_2 \frac{(u_{,x}v_{,y} - u_{,y}v_{,x})^2 u}{(1+u^2+v^2)^5} + V_{,u} \tag{20}$$
$$+ G_{1,u} \cdot (u_{,x}v_{,y} - u_{,y}v_{,x}) + R_{1,u} \cdot (u_{,x}v_{,y} - u_{,y}v_{,x}) + D_k G_{k+1,u} + D_k R_{k+1,u} = 0,$$

$$\tilde{\mathcal{H}}_{,v} : -2\lambda_1 \frac{(u_{,x}^2 + u_{,y}^2 + v_{,x}^2 + v_{,y}^2)v}{(1+u^2+v^2)^3} - 8\lambda_2 \frac{(u_{,x}v_{,y} - u_{,y}v_{,x})^2 v}{(1+u^2+v^2)^5} + V_{,v} \tag{21}$$
$$+ G_{1,v} \cdot (u_{,x}v_{,y} - u_{,y}v_{,x}) + R_{1,v} \cdot (u_{,x}v_{,y} - u_{,y}v_{,x}) + D_k G_{k+1,v} + D_k R_{k+1,v} = 0,$$

$$\tilde{\mathcal{H}}_{,u_{,x}} : \lambda_1 \frac{u_{,x}}{(1+u^2+v^2)^2} + 2\lambda_2 \frac{(u_{,x}v_{,y} - u_{,y}v_{,x})v_{,y}}{(1+u^2+v^2)^4}$$
$$+ G_1 v_{,y} + R_1 v_{,y} + G_{2,u} + R_{2,u} = 0, \tag{22}$$

$$\tilde{\mathcal{H}}_{,u_{,y}} : \lambda_1 \frac{u_{,y}}{(1+u^2+v^2)^2} - 2\lambda_2 \frac{(u_{,x}v_{,y} - u_{,y}v_{,x})v_{,x}}{(1+u^2+v^2)^4}$$
$$- G_1 v_{,x} - R_1 v_{,x} + G_{3,u} + R_{3,u} = 0, \tag{23}$$

$$\tilde{\mathcal{H}}_{,v_{,x}} : \lambda_1 \frac{v_{,x}}{(1+u^2+v^2)^2} - 2\lambda_2 \frac{(u_{,x}v_{,y} - u_{,y}v_{,x})u_{,y}}{(1+u^2+v^2)^4}$$
$$- G_1 u_{,y} - R_1 u_{,y} + G_{2,v} + R_{2,v} = 0, \tag{24}$$

$$\tilde{\mathcal{H}}_{,v_{,y}} : \lambda_1 \frac{v_{,y}}{(1+u^2+v^2)^2} + 2\lambda_2 \frac{(u_{,x}v_{,y} - u_{,y}v_{,x})u_{,x}}{(1+u^2+v^2)^4}$$
$$+ G_1 u_{,x} + R_1 u_{,x} + G_{3,v} + R_{3,v} = 0. \tag{25}$$

First, we integrate (20)–(21) with respect to u and v, respectively:

$$\tilde{\mathcal{H}} = F(u_{,x}, u_{,y}, v_{,x}, v_{,y}), \tag{26}$$

where F is some function, which is to be determined. Now, the first step of making equations (22)–(25) self-consistent, is by putting [17]:

$$\frac{u_{,x}v_{,y} - u_{,y}v_{,x}}{(1+u^2+v^2)^4} = -\frac{1}{2\lambda_2}G_1(u,v), \quad G_n(u,v) = \text{const.}, \quad n = 2,3. \tag{27}$$

However, now we need to make the second step, by proper choice of the functions R_1, R_2, R_3, when conditions (27) are satisfied. The details, how equations (22)–(25) are made self-consistent (when (27) hold) are described in ref. [19]. Namely, if (27) are satisfied, then the system (22)–(25) is some special case of the corresponding system of equations, for which, in [19], the divergence representation has been derived. Then, we apply the results from [19] and we get $D_y R_2 = D_x R_3$. We also obtain relations, including $D_x R_2$, $D_y R_3$. Next, if we put [17]:

$$R_1 = \frac{\lambda_1}{(1 + u^2 + v^2)^2}, \quad R_{2,u} = R_{3,v}, \quad R_{2,v} = -R_{3,u}, \tag{28}$$

and take into account (27), then equations (22)–(25) are reduced to [17]:

$$\lambda_1 \frac{u_{,x} + v_{,y}}{(1 + u^2 + v^2)^2} = -R_{2,u}, \quad \lambda_1 \frac{u_{,y} - v_{,x}}{(1 + u^2 + v^2)^2} = R_{2,v}. \tag{29}$$

From the two last relations in (28) it follows that $R_2(u, v)$ must satisfy the Laplace equation [17]:

$$R_{2,uu} + R_{2,vv} = 0. \tag{30}$$

Now, we determine F in (26), by using again the Hamilton–Jacobi equation: $\tilde{\mathcal{H}} = 0$ and again in our case: $\tilde{\mathcal{H}} = -\tilde{\mathcal{L}}$. Thus, we get that: $F = 0$ and hence, after taking into account the divergence representation and relations (26), (27), (28) and (29), we obtain the condition for the potential $V(u, v)$ [17]:

$$V(u, v) = \frac{(1 + u^2 + v^2)^4}{4\lambda_2} G_1^2(u, v) + \frac{(1 + u^2 + v^2)^2}{2\lambda_1} \left[R_{2,u}^2(u, v) + R_{2,v}^2(u, v) \right], \tag{31}$$

where $G_1(u, v) \in \mathcal{C}^2$ and $R_2(u, v)$ is some solution of the Laplace equation (30). Thus we have obtained the following system of three equations of the first-order (Bogomolny decomposition) [17]:

$$u_{,x} v_{,y} - u_{,y} v_{,x} = -\frac{1}{2\lambda_2} (1 + u^2 + v^2)^4 G_1(u, v),$$

$$\lambda_1 \frac{u_{,x} + v_{,y}}{(1 + u^2 + v^2)^2} = -R_{2,u}, \quad \lambda_1 \frac{u_{,y} - v_{,x}}{(1 + u^2 + v^2)^2} = R_{2,v}, \tag{32}$$

which constitute, with conditions (30) and (31), so-called Bogomolny relationship (this notion was used for the first time in [18]), for the full baby Skyrme model in $(2 + 0)$ dimensions. The same results can be obtained, by applying the results from ref. [20].

4. Summary

We have derived Bogomolny decompositions for the baby Skyrme models: restricted and full one, in $(2 + 0)$ dimensions by using the concept of strong necessary conditions (CSNC). One can check, by comparison, that the Bogomolny decomposition (19) for the first model mentioned above, is a generalization of the Bogomolny equation, obtained in [5]. Some physical features of solutions of the derived Bogomolny equations will be presented in a separate paper [21].

In [6] the Bogomolny equations for ungauged restricted baby Skyrme model in $(2 + 0)$ dimensions for the potential of the form $V = \frac{1}{2} U^2$, (where U is some non-negative function of the class $\mathcal{C}^1$ on any compact Riemann surface with isolated zeroes), were obtained in the language of differential forms and not by using the CSNC. We stress here that the Bogomolny decomposition (19) has been obtained *without any assumption* on the form of the potential. In case of the full baby Skyrme model, as we see, the set of solutions of the Bogomolny decomposition (32) is a subset of the set of solutions of the Bogomolny decomposition for the restricted baby Skyrme model. So, by adding the $O(3)$ term in the Lagrangian of the restricted baby Skyrme model, in order to get the full baby Skyrme model, we cannot get a wider set of solutions of the Bogomolny decomposition. The conclusion that the Bogomolny bound for this model cannot be saturated by non-trivial solutions [5], is based on the derivation of the Bogomolny bound for the full baby Skyrme model by using traditional technique of derivation of the Bogomolny equations, which in contrary to the CSNC, does not give any information for which potentials Bogomolny equations can be derived.

The traditional technique of derivation of the Bogomolny equations, gave many important results and belongs to classical methods in nonlinear field theory. However, the method of derivation of the Bogomolny equations, based on the CSNC, can be considered as a systematic procedure, what makes it easier to find Bogomolny equations (Bogomolny decomposition) for a given model in nonlinear field theory, of course, if it is possible.

Acknowledgment

The author thanks Dr. hab. A. Wereszczyński for interesting discussions about the restricted baby Skyrme models, carried out in 2010. The author also thanks Dr. Z. Lisowski for some interesting remarks.

The participation of the author, in "XXXI Workshop on Geometric Methods in Physics" (24–30 June 2012, Białowieża, Poland), was possible owing to the financial support, provided by The Pedagogical University of Cracow, within the research project (the leader of the project: Dr. K. Rajchel).

The computations were carried out by using Waterloo MAPLE 12 Software on the computer *mars* in ACK-CYFRONET AGH in Kraków (No. of grant: MNiI/IBM_BC_HS21/AP/057/2008) and also in the Interdisciplinary Centre for Mathematical and Computer Modelling (ICM), within the grant No. G31-6. This research was supported also by PL-Grid Infrastructure.

References

[1] T.H.R. Skyrme, *Proc. R. Soc. A* **260** (1961), 127; *Nucl. Phys.* **31** (1962), 556; *J. Math. Phys.* **12** (1971), 1735.

[2] W.G. Makhankov, Yu.P. Rybakov and W.I. Sanyuk, *Skyrme models and solitons in physics of hadrons*, Dubna 1989, (in Russian).

[3] R.A. Leese, M. Peyrard, and W.J. Zakrzewski, *Nonlinearity* **3** (1990), 773; B.M.A.G. Piette, B.J. Schroers, and W.J. Zakrzewski, *Z. Phys. C* **65** (1995), 165; *Nucl. Phys.*

B439 (1995), 205; *Chaos, Solitons and Fractals* **5** (1995), 2495; P.M. Sutcliffe, *Nonlinearity* **4** (1991), 1109; T. Weidig, *Nonlinearity* **12** (1999), 1489; P. Eslami, M. Sarbishaei, and W.J. Zakrzewski, *Nonlinearity* **13** (2000), 1867; M. Karliner and I. Hen, *Nonlinearity* **21** (2008), 399.

[4] C. Adam, P. Klimas, J. Sánchez-Guillén, and A. Wereszczyński, *Phys. Rev. D* **80** (2009), 105013.

[5] C. Adam, T. Romańczukiewicz, J. Sánchez-Guillén, and A. Wereszczyński, *Phys. Rev. D* **81** (2010), 085007.

[6] J.M. Speight, *J. Phys. A* **43** (2010), 405201.

[7] J. Jäykkä and M. Speight, *Phys. Rev. D* **82** (2010), 125030; J. Jäykkä, M. Speight, and P. Sutcliffe, *Proc. R. Soc. A* **468** (2012), 1085.

[8] A.A. Belavin and A.M. Polyakov, *JETP Lett.* **22** (1975), 245; S.L. Sondhi, A. Karlhede, S.A. Kivelson, and E.H. Rezayi, *Phys. Rev. B* **47** (1993), 16419; N.R. Walet and T. Weidig, *Europhys. Lett.* **55** (2001), 633.

[9] T. Ioannidou and O. Lechtenfeld, *Phys. Lett. B* **678** (2009), 508.

[10] T. Gisiger and M.B. Paranjape, in *Solitons: properties, dynamics, interactions, applications R. MacKenzie, M.B. Paranjape, W.J. Zakrzewski, eds.*, 183, Springer-Verlag, 2000.

[11] T. Gisiger and M.B. Paranjape, *Phys. Rev. D* **55** (1997), 7731.

[12] E.B. Bogomolny, *Sov. J. Nucl. Phys.* **24** (1976), 861.

[13] A.A. Belavin, A.M. Polyakov, A.S. Schwartz and Yu.S. Tyupkin, *Phys. Lett. B* **59** (1975), 85 (placed in: *Instantons in Gauge Theories*, edited by M. Shifman, World Scientific 1994).

[14] M. de Innocentis and R.S. Ward. *Nonlinearity* **14**, (2001), 663.

[15] K. Sokalski, *Acta Phys. Pol. A* **56** (1979), 571.

[16] K. Sokalski, T. Wietecha, and Z. Lisowski, *Acta Phys. Pol. B* **32** (2001), 17; *Acta Phys. Pol. B* **32** (2001), 2771.

[17] Ł.T. Stępień, *arXiv:1204.6194*, 27 April 2012.

[18] K. Sokalski, Ł. Stępień, and D. Sokalska, *J. Phys. A* **35** (2002) 6157.

[19] Ł. Stępień, *Bogomolny decomposition in the context of the concept of strong necessary conditions*, PhD thesis, Jagiellonian University, Kraków, Poland, 2003. (in Polish).

[20] Ł. Stępień, D. Sokalska, and K. Sokalski, *J. Nonl. Math. Phys.* **16** (2009), 25.

[21] Ł.T. Stępień, *paper in preparation.*

Ł.T. Stępień
The Pedagogical University of Cracow
Chair of Computer Science and Computer Methods
ul. Podchorazych 2
PL-30-084 Krakow, Poland
e-mail: sfstepie@cyf-kr.edu.pl
        stepien50@poczta.onet.pl
URL: http://www.ltstepien.up.krakow.pl/

Printed in the United States
By Bookmasters